Software Quality Assurance

About IEEE Computer Society

IEEE Computer Society is the world's leading computing membership organization and the trusted information and career-development source for a global workforce of technology leaders including: professors, researchers, software engineers, IT professionals, employers, and students. The unmatched source for technology information, inspiration, and collaboration, the IEEE Computer Society is the source that computing professionals trust to provide high-quality, state-of-the-art information on an on-demand basis. The Computer Society provides a wide range of forums for top minds to come together, including technical conferences, publications, and a comprehensive digital library, unique training webinars, professional training, and the TechLeader Training Partner Program to help organizations increase their staff's technical knowledge and expertise, as well as the personalized information tool myComputer. To find out more about the community for technology leaders, visit http://www.computer.org.

IEEE/Wiley Partnership

The IEEE Computer Society and Wiley partnership allows the CS Press authored book program to produce a number of exciting new titles in areas of computer science, computing, and networking with a special focus on software engineering. IEEE Computer Society members continue to receive a 15% discount on these titles when purchased through Wiley or at wiley.com/ieeecs.

To submit questions about the program or send proposals, please contact Mary Hatcher, Editor, Wiley-IEEE Press: Email: mhatcher@wiley.com, Telephone: 201-748-6903, John Wiley & Sons, Inc., 111 River Street, Hoboken, NJ 07030-5774.

Software Quality Assurance

Claude Y. Laporte
Alain April

WILEY

The rights of Claude Y. Laporte and Alain April to be identified as the authors of this work have been asserted in accordance with law.

Translated by Rosalia Falco.

Registered Office
John Wiley & Sons, Inc., 111 River Street, Hoboken, NJ 07030, USA

Editorial Office
111 River Street, Hoboken, NJ 07030, USA

For details of our global editorial offices, customer services, and more information about Wiley products visit us at www.wiley.com.

Wiley also publishes its books in a variety of electronic formats and by print-on-demand. Some content that appears in standard print versions of this book may not be available in other formats.

Library of Congress Cataloging-in-Publication Data

Names: Laporte, Claude Y., author. | April, Alain, author.
Title: Software quality assurance / by Claude Y. Laporte, Alain April.
Description: 1 | Hoboken, NJ : Wiley-IEEE Computer Society, Inc., 2018. |
  Includes bibliographical references and index. |
Identifiers: LCCN 2017036440 (print) | LCCN 2017041869 (ebook) | ISBN
  9781119312413 (pdf) | ISBN 9781119312420 (epub) | ISBN 9781118501825 (hardback)
Subjects: LCSH: Computer software–Quality control. | Computer software–Quality
  control–Standards. | BISAC: TECHNOLOGY & ENGINEERING / Quality Control.
Classification: LCC QA76.76.Q35 (ebook) | LCC QA76.76.Q35 L42 2018 (print) |
  DDC 005.3028/7–dc23
LC record available at https://lccn.loc.gov/2017036440

Cover image: ©naqiewei/Gettyimages
Cover design by Wiley

Set in 10/12pt TimesLTStd by Aptara Inc., New Delhi, India

10 9 8 7 6 5 4 3 2 1

Contents

5. Reviews   167

6. Software Audits   210

9. Policies, Processes, and Procedures   335

10. Measurement   397

Preface

This book addresses the global challenge of the improvement of software quality. It seeks to provide an overview of software quality assurance (SQA) practices for customers, managers, auditors, suppliers, and personnel responsible for software projects, development, maintenance, and software services.

In a globally competitive environment, clients and competitors exert a great deal of pressure on organizations. Clients are increasingly demanding and require, among other things, software that is of high quality, low cost, delivered quickly, and with impeccable after-sales support. To meet the demand, quality, and deadlines, the organization must use efficient quality assurance practices for their software activities.

Ensuring software quality is not an easy task. Standards define ways to maximize performance but managers and employees are largely left to themselves to decide how to practically improve the situation. They face several problems:

- increasing pressure to deliver quality products quickly;
- increasing size and complexity of software and of systems;
- increasing requirements to meet national, international, and professional standards;
- subcontracting and outsourcing;
- distributed work teams; and
- ever changing platforms and technologies.

We will focus on the issue of SQA in industry and in public organizations. Industry and public organizations do not have access to a complete and integrated reference (i.e., one book) that can help them with assessing and improving activities specific to SQA. The SQA department must meet service standards for its customers, the technical criteria of the field, and maximize strategic and economic impacts.

The purpose of this book is to enable managers, clients, suppliers, developers, auditors, software maintainers, and SQA personnel to use this information to assess the effectiveness and completeness of their approach to SQA. Some of the issues raised here include:

- What are the processes, practices, and activities of SQA and software improvement?
- Can the current standards and models serve as a reference?
- How do we ensure that managers and their staff understand the value of SQA activities and their implementation?

To answer these questions, we drew upon over 30 years of practical experience in software engineering and SQA in different organizations such as telecom, banking, defense, and transportation. This industry experience has convinced us of the importance of supporting the presentation of concepts and theory with references and practical examples. We have illustrated the correct and effective implementation of numerous quality assurance practices with real case studies throughout the book.

In many organizations, SQA is a synonym for testing. SQA, as presented in this book, covers a large spectrum of proven practices to provide a level of confidence that quality in software development and maintenance activities is independent of the life cycle selected by an organization or a project.

In this book, we will extensively use the term "software quality assurance" and the acronym SQA. As defined in the IEEE Standard for Software Quality Assurance Processes, IEEE 730-2014, a function is a set of resources and activities that achieve a particular purpose [IEE 14]. The SQA function can be executed by a software project team member. It could also be executed by an independent party (e.g., within a quality assurance (QA) department responsible for hardware, software, and supplier quality).

STRUCTURE AND ORGANIZATION OF THIS BOOK

The book is divided into 13 chapters that cover the basic knowledge of SQA as identified, among others, by the IEEE 730 Standard for SQA Processes of the Institute of Electrical and Electronics Engineers (IEEE), the ISO/IEC/IEEE 12207 software life cycle processes standard, the Capability Maturity Model® Integration for Development (CMMI®-DEV) developed by the Software Engineering Institute as well as the ISO Guide to the Software Engineering Body of Knowledge (SWEBOK®). Numerous practical examples are used to illustrate the application of SQA practices.

CHAPTER 1: SOFTWARE QUALITY FUNDAMENTALS

This chapter presents an overview of the knowledge required by SQA practitioners. From this overview, the book develops every aspect of the field and cites the important references that deepen each specific topic. We use the concept of business models to explain the significant differences in the selection of SQA practices. In this chapter, we also establish terms and their definitions as well as useful concepts that are used throughout the book.

CHAPTER 2: QUALITY CULTURE

This chapter introduces the concept of cost of quality, followed by practical examples. It also introduces the concept of quality culture and its influence on the SQA practices used. We also present five dimensions of a software project and how these dimensions can be used to identify the degrees of freedom a project manager has to ensure its success. In this chapter, we present an overview of software engineering ethics and the techniques to manage the expectations of managers and customers with respect to software quality.

CHAPTER 3: SOFTWARE QUALITY REQUIREMENTS

This chapter adds to the concepts and terminology already presented. It deals with software quality models as well as ISO standards on software quality models. These models propose classifications of software quality requirements and steps to define them. Practical examples describe how to use these models to define the quality requirements of a software project. Finally, we introduce the concept of requirements traceability and the importance of quality requirements for the SQA plan.

CHAPTER 4: SOFTWARE ENGINEERING STANDARDS AND MODELS

This chapter presents the most important international standards of ISO and models about software quality, such as the CMMI® developed by the Software Engineering Institute. A new ISO standard for very small organizations is also presented. The SQA practitioner and specialist will find proven practices from standards and models. This chapter provides the framework that can be useful for the following major software activities: (1) development, (2) maintenance, and (3) IT services. Finally, a short discussion on the standards specific to certain domains of application is presented, followed by recommendations for a SQA plan.

CHAPTER 5: REVIEWS

This chapter presents different types of software reviews: personal review, the "desk check," the walk-through, and the inspection. We describe the theory about reviews and then provide practical examples. It introduces reviews in an agile context. Subsequently, we describe other reviews specific to a project: the project launch review and lessons learned review. The chapter concludes with a discussion on the selection of one type of review depending on your business domain and how these techniques fit into the SQA plan.

CHAPTER 6: SOFTWARE AUDITS

This chapter describes the audit process and the software problem resolution process. Sooner or later in the career of a software practitioner, audits will be conducted in a software project. Standards and models describing audits are presented followed by a practical case. The chapter concludes with a discussion of the role of audits in the SQA plan.

CHAPTER 7: VERIFICATION AND VALIDATION

This chapter describes the concept of software verification and validation (V&V). It describes its benefits as well as the costs of using V&V practices. Then, the standards and models that impose or describe V&V practices for a project are described. Finally, the description of the contents of a V&V plan is presented.

CHAPTER 8: SOFTWARE CONFIGURATION MANAGEMENT

This chapter describes an important component of software quality: software configuration management (SCM). The chapter begins by presenting the usefulness of SCM and typical SCM activities. It presents repositories and branching techniques involved in source code management, as well as the concepts of software control, software status, and software audits. Finally, this chapter concludes with a proposal for the implementation of SCM in a small organization and ends with a discussion of the role of SCM in the SQA plan.

CHAPTER 9: POLICIES, PROCESSES, AND PROCEDURES

This chapter explains how to develop, document, and improve policies, processes, and procedures to ensure the effectiveness and efficiency of the software organization. It explains the importance of documentation presenting a few notations, as examples, to document processes and procedures. The chapter ends by presenting the Personal Software Process (PSP) developed by the Software Engineering Institute to ensure individuals have a disciplined and structured approach to software development that enables them to significantly increase the quality of their software products.

CHAPTER 10: MEASUREMENT

This chapter explains the importance of measurement, standards, and models, and presents a methodology to describe the requirements for a measurement process. It presents how measurement can be used by small organizations and small projects. Then, an approach to implement a measurement program, to detect the potential pitfalls, and the potential impact of human factors, when measuring, is discussed. The chapter concludes with a discussion of the role of measurement in a SQA plan.

CHAPTER 11: RISK MANAGEMENT

This chapter presents the main models and standards that include requirements for the management of risks. It discusses the risks that may affect the quality of software and techniques to identify, prioritize, document, and mitigate them. It also presents the roles of stakeholders in the risk management process and discusses the human factors to consider in the management of software risks. The chapter concludes with a discussion on the critical role of risk in the development of a SQA plan.

CHAPTER 12: SUPPLIER MANAGEMENT AND AGREEMENTS

This chapter deals with the important topic of supplier management and agreements. It discusses the major reviews and recommendations of the CMMI®. Subsequently, it lists the different types of software agreements and the benefits of the risk sharing agreement are illustrated using a practical example. This chapter concludes with recommendations for the content of the SQA plan when suppliers are involved.

CHAPTER 13: SOFTWARE QUALITY ASSURANCE PLAN

This chapter summarizes the topics presented in the whole book by using the concepts presented in each chapter to assemble a comprehensive SQA plan that conforms to the IEEE 730 recommendation. It ends by presenting additional recommendations and practical examples.

APPENDICES

Appendix 1 – Software Engineering Code of Ethics and Professional Practice (Version 5.2)

Appendix 2 – Incidents and Horror Stories involving Software

ICONS USED IN THE BOOK

Different icons are used throughout this book to illustrate a concept with a practical example; to focus on a definition; to present an anecdote, a tool, or checklist; or simply to provide a quote or a website. Consult the table below for the meaning of each icon.

Icon	Meaning
Ⓟ	Practical example: An example of the practical application of a theoretical concept
❝❝	Quote: A quote from an expert
❓	Definition: A definition of an important term
www	Reference on the Web: An internet site to learn more about a specific topic
⚙	Tools: Examples of tools that support the techniques presented
A	Anecdote: A short story of a little known fact, or a curious point on the subject discussed
✔	Checklist: A list of items to check, or not to be forgotten, during the execution of a presented technique
🔍	Tip: A tip from the authors or from another professional

WEBSITE

Supplementary material for teaching as well as for use in organizations (e.g., presentation material, solutions, project descriptions, templates, tools, articles, and links) is available on the website: www.sqabook.org.

Given that international standards are updated on a regular basis, the website will also highlight the latest developments that contribute to SQA practices.

EXERCISES

Each chapter contains exercises.

NOTES

Many software engineering standards from ISO and IEEE have been cited in this book. These standards are updated on a regular basis, typically every five years, to reflect evolving software engineering practices. The accompanying website, www.sqabook.org, contains complementary information as well as the latest developments that impact or contribute to SQA practices described in each chapter and will evolve over time.

Since software engineering standards can be cited in an agreement between a customer and a supplier and add additional legal requirements to the agreement, we have not paraphrased the text of standards in our book, we have directly quoted the text from the standards.

Acknowledgments

We would like to thank Professor Normand Séguin of the University of Quebec in Montreal (UQAM), Mr. Jean-Marc Desharnais for allowing us to use an excerpt that describes the implementation process of a measurement program, and many graduate students of the Masters in Software Engineering from the École de technologie supérieure (ÉTS) who reviewed the chapters of this book and contributed through their vast industry experience, analogies, and case studies to enrich the content.

We are also very grateful to Kathy Iberle for letting us use her description of business models and their application in different business domains [IBE 02, IBE 03]. The business models are very helpful in understanding the risks facing a specific business domain as well as the breadth and depth of software engineering practices used to mitigate the risks. Finally, we would like to thank Karl Wiegers and Daniel Galin for allowing us to use figures from their books.

Chapter 1

Software Quality Fundamentals

After completing this chapter, you will be able to:

- – use the correct terminology to discuss software quality issues;
- – identify the major categories of software errors;
- – understand the different viewpoints regarding software quality;
- – provide a definition of software quality assurance;
- – understand software business models as well as their respective risks.

1.1 INTRODUCTION

Software is developed, maintained, and used by people in a wide variety of situations. Students create software in their classes, enthusiasts become members of open-source development teams, and professionals develop software for diverse business fields from finance to aerospace. All these individual groups will have to address quality problems that arise in the software they are working with. This chapter will provide definitions for terminology and discuss the source of software errors and the choice of different software engineering practices depending on an organization's sector of business.

Every profession has a body of knowledge made up of generally accepted principles. In order to obtain more specific knowledge about a profession, one must either: (a) have completed a recognized curriculum or (b) have experience in the domain. For most software engineers, software quality knowledge and expertise is acquired in a hands-on fashion in various organizations. The Guide to the Software Engineering Body of Knowledge (SWEBOK) [SWE 14] constitutes the first international consensus developed on the fundamental knowledge required by all software engineers.

Software Quality Assurance, First Edition. Claude Y. Laporte and Alain April.
© 2018 the IEEE Computer Society, Inc. Published 2018 by John Wiley & Sons, Inc.

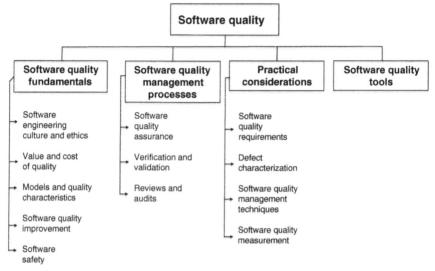

Figure 1.1   Software Quality in the SWEBOK® Guide [SWE 14].

Chapter 10 of SWEBOK is dedicated to software quality (see Figure 1.1) and its first topic is labeled "fundamentals" and introduces the concepts and terminology that form the underlying basis for understanding the role and scope of software quality activities. The second topic refers to the management processes and highlights the importance of software quality across the life cycle of a software project. The third topic presents practical considerations where various factors that influence planning, management, and selection of software quality activities and techniques are discussed. Last, software quality related tools are presented.

1.2 DEFINING SOFTWARE QUALITY

Before explaining the components of software quality assurance (SQA), it is important to consider the basic concepts of software quality. Once you have completed this section, you will be able to:

- define the terms "software," "software quality," and "software quality assurance";
- differentiate between a software "error," a software "defect," and a software "failure."

Intuitively, we see software simply as a set of instructions that make up a program. These instructions are also called the software's source code. A set of programs forms an application or a software component of a system with hardware components. An information system is the interaction between the software application and the information technology (IT) infrastructure of the organization. It is the information system or the system (e.g., digital camera) that clients use.

Is ensuring the quality of the source code sufficient for the client to be able to obtain a quality system? Of course not; a system is far more complex than a single program. Therefore, we must identify all components and their interactions to ensure that the information system is one of quality. An initial response to the challenge regarding software quality can be found in the following definition of the term "software."

Software

1) All or part of the programs, procedures, rules, and associated documentation of an information processing system.
2) Computer programs, procedures, and possibly associated documentation and data pertaining to the operation of a computer system.

ISO 24765 [ISO 17a]

When we consider this definition, it is clear that the programs are only one part of a set of other products (also called intermediary products or software deliverables) and activities that are part of the software life cycle.

Let us look at each part of this definition of the term "software" in more detail:

– Programs: the instructions that have been translated into source code, which have been specified, designed, reviewed, unit tested, and accepted by the clients;

– Procedures: the user procedures and other processes that have been described (before and after automation), studied, and optimized;

– Rules: the rules, such as business rules or chemical process rules, that had to be understood, described, validated, implemented, and tested;

– Associated documentation: all types of documentation that is useful to customers, software users, developers, auditors, and maintainers. Documentation enables different members of a team to better communicate, review, test, and maintain software. Documentation is defined and produced throughout the key stages of the software life cycle;

– Data: information that is inventoried, modeled, standardized, and created in order to operate the computer system.

Software found in embedded systems is sometimes called microcode or firmware. Firmware is present in commercial mass-market products and controls machines and devices used in our daily lives.

Firmware

Combination of a hardware device and computer instructions or computer data that reside as read-only software on the hardware device.

ISO 24765 [ISO 17a]

1.3 SOFTWARE ERRORS, DEFECTS, AND FAILURES

If you listen closely during various meetings with your colleagues, you will notice that there are many terms that are used to describe problems with a software-driven system. For example:

- The system crashed during production.
- The designer made an error.
- After a review, we found a defect in the test plan.
- I found a bug in a program today.
- The system broke down.
- The client complained about a problem with a calculation in the payment report.
- A failure was reported in the monitoring subsystem.

Do all of these terms refer to the same concept or to different concepts? It is important to use clear and precise terminology if we want to provide a specific meaning to each of these terms. Figure 1.2 describes how to use these terms correctly.

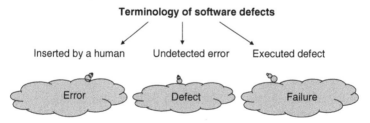

Figure 1.2 Terminology recommended for describing software problems.

Bug

Since the time of Thomas Edison, engineers have used the word "bug" to refer to failures in the systems that they have developed. This word can describe a multitude of possible problems. The first documented case of a "computer bug" involved a moth trapped in a relay of the Mark II computer at Harvard University in 1947. Grace Hopper, the computer operator, pasted the insect into the laboratory log, specifying it as the "First actual case of a bug being found" (see the page of this log in the photograph below).

In the early 1950s, the terms "bug," "debug," and "debugging," as applied to computers and computer programs, started to appear in the popular press [KID 98].

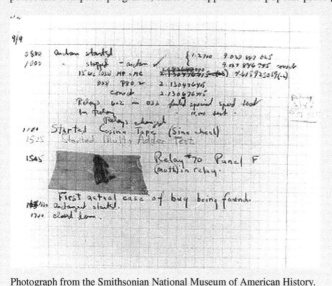

Photograph from the Smithsonian National Museum of American History.

A failure (synonymous with a crash or breakdown) is the execution (or manifestation) of a fault in the operating environment. A failure is defined as the termination of the ability of a component to fully or partially perform a function that it was designed to carry out. The origin of a failure lies with a defect hidden, that is, not detected by tests or reviews, in the system currently in operation. As long as the system in production does not execute a faulty instruction or process faulty data, it will run normally. Therefore, it is possible that a system contains defects that have not yet been executed. Defects (synonym of faults) are human errors that were not detected during software development, quality assurance (QA), or testing. An error can be found in the documentation, the software source code instructions, the logical execution of the code, or anywhere else in the life cycle of the system.

Error, Defect, and Failure

Error

A human action that produces an incorrect result (ISO 24765) [ISO 17a].

Defect

1) A problem (synonym of fault) which, if not corrected, could cause an application to either fail or to produce incorrect results. (ISO 24765) [ISO 17a].

2) An imperfection or deficiency in a software or system component that can result in the component not performing its function, e.g. an incorrect data definition or source code instruction. A defect, if executed, can cause the failure of a software or system component (ISTQB 2011 [IST 11]).

Failure

The termination of the ability of a product to perform a required function or its inability to perform within previously specified limits (ISO 25010 [ISO 11i]).

Figure 1.3 shows the relationship between errors, defects, and failures in the software life cycle. Errors may appear during the initial feasibility and planning stages of new software. These errors become defects when documents have been approved and the errors have gone unnoticed. Defects can be found in both intermediary products (such as requirements specifications and design) and the source code itself. Failures occur when an intermediary product or faulty software is used.

Case of Errors, Defects, and Failures

Case 1: A local pharmacy added a software requirement to its cash register to prevent sales of more than $75 to customers owing more than $200 on their pharmacy credit card. The programmer did not fully understand the specification and created a sales limit of $500 within the program. This defect never caused a failure since no client could purchase more than $500 worth of items given that the pharmacy credit card had a limit of $400.

Case 2: In 2009, a loyalty program was introduced to the clients of American Signature, a large furniture supplier. The specifications described the following business rules: a customer who makes a monthly purchase that is higher than the average amount of monthly purchases for all customers will be considered a Preferred Customer. The Preferred Customer will be identified when making a purchase, and will be immediately given a gift or major discount once a month. The defect introduced into the system (due to a poor understanding of the algorithm to set up for

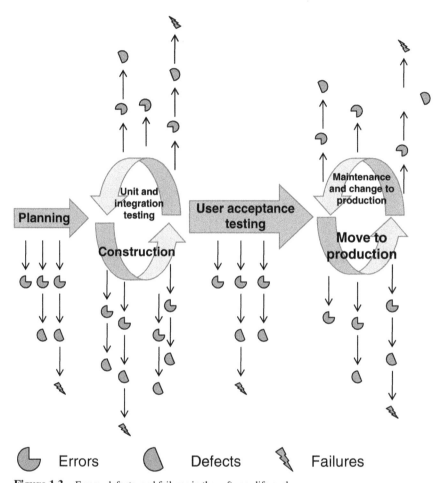

Figure 1.3   Errors, defects, and failures in the software life cycle.
Source: Galin (2017). [GAL 17]. Adapted with permission of Wiley-IEEE Computer Society Press.

this requirement) involved only taking into account the average amount of current purchases and not the customer's monthly history. At the time of the software failure, the cash register was identifying far too many Preferred Clients, resulting in a loss for the company.

Case 3: Peter tested Patrick's program when Patrick was away. He found a defect in the calculation for a retirement savings plan designed to apply the new tax-exemption law for this type of investment. He traced the error back to the project specification and informed the analyst. In this case, the test activity correctly identified the defect and the source of the error.

The three cases above correctly use the terms to describe software quality problems. They also identify issues that are investigated by researchers in the field of software quality in order to discover means to help eliminate these problems:

– Errors can occur in any of the software development phases throughout the life cycle.
– Defects must be identified and fixed before they become failures.
– The cause of failures, defects, and errors must be identified.

Life Cycle
Evolution of a system, product, service, project, or other human-made entity from conception through retirement.
Development Life Cycle
Software life cycle process that contains the activities of requirements analysis, design, coding, integration, testing, installation, and support for acceptance of software products.
ISO 12207 [ISO 17]

During software development, defects are constantly being involuntarily introduced and must be located and corrected as soon as possible. Therefore, it is useful to collect and analyze data on the defects found as well as the estimated number of defects left in the software. By doing so, we can improve the software engineering processes and in turn, minimize the number of defects introduced in new versions of software products in the future.

Methods for classifying defects have been created for this purpose, one of which is explained in the chapter on verification and validation.

 Undetected Hole in the Ozone Layer

The hole in the ozone layer over Antarctica went unnoticed for a long period of time because the TOMS data analysis software used by NASA as part of its project to map the ozone layer had been designed to ignore values that deviate significantly from the anticipated measurements.

The project was launched in 1978, but it was only in 1985 that the hole was discovered, and not by NASA. Following data analysis, NASA confirmed this design error.

http://earthobservatory.nasa.gov/Features/RemoteSensingAtmosphere/remote_sensing5.php

Depending on the business model of your organization, you will have to allow for varying degrees of effort in identifying and correcting defects. Unfortunately, there exists today a certain culture of tolerance for software defects. However, there is no question that we all want Airbus, Boeing, Bombardier, and Embraer to have identified and corrected all the defects in the software for their airplanes before we board them!

Many researchers have studied the source of software errors and have published studies classifying software errors by type in order to evaluate the frequency of each type of error. Beizer (1990) [BEI 90] provides a study that has combined the result of several other studies to provide us with an indication of the origin of errors. The following is a summarized list of this study's results [BEI 90].

- 25% structural;
- 22% data;
- 16% functionalities implemented;
- 10% construction/coding;
- 9% integration;
- 8% requirements/functional specifications;
- 3% definition/running tests;
- 2% architecture/design;
- 5% unspecified.

Researchers also try to determine how many errors can be expected in a typical software. McConnell (2004) [MCC 04] suggested that this number varied based on the quality and maturity of the software engineering processes as well as the training and competency of the developers. The more mature the processes are, the fewer errors are introduced into the development life cycle of the software. Humphrey (2008) [HUM 08] also collected data from many developers. He found that a developer involuntarily creates about 100 defects for each 1000 lines of source code written. In addition, he noted large variations for a group of 800 experienced developers, that is, from less than 50 defects to more than 250 defects injected per 1000 lines of code. At Rolls-Royce, the manufacturer of airplane engines, the variation published is from 0.5 to 18 defects per 1000 lines of source code [NOL 15]. The use of proven processes, competent and well-trained developers, and the reuse of already proven software components can considerably reduce the number of errors of a software.

McConnell also referenced other studies that have come to the following conclusions:

- The scope of most defects is very limited and easy to correct.
- Many defects occur outside of the coding activity (e.g., requirement, architecture activities).

– Poor understanding of the design is a recurrent problem in programming error studies.

– It is a good idea to measure the number and origin of defects in your organization to set targets for improvement.

Therefore, errors are the main cause of poor software quality. It is important to look for the cause of error and identify ways in which to prevent these errors in the future. As we have shown in Figure 1.3, errors can be introduced at each step of the software life cycle: errors in the requirements, code, documentation, data, tests, etc. The causes are almost always human mistakes made by clients, analysts, designers, software engineers, testers, or users. SQA will need to develop a classification of the causes of software error by category which can be used by everyone involved in the software engineering process.

For example, here are eight popular error-cause categories:

1) problems with defining requirements;

2) maintaining effective communication between client and developer;

3) deviations from specifications;

4) architecture and design errors;

5) coding errors (including test code);

6) non-compliance with current processes/procedures;

7) inadequate reviews and tests;

8) documentation errors.

Each of the eight categories of error causes listed above is described in more detail in the following sections.

1.3.1 Problems with Defining Requirements

Defining software requirements is now considered a specialty, which means a business analyst or a software engineer specialized in requirements. Requirements definition is the topic of interest groups as well as the subject of professional certification programs (see http://www.iiba.org).

There are a certain number of problems related to the clear, correct, and concise writing of requirements so that they can be converted into specifications that can be directly used by colleagues, such as architects, designers, programmers, and testers.

It must also be understood that there are a certain number of activities that must be mastered when eliciting requirements:

– identifying the stakeholders (i.e., key players) who must participate in the requirements elicitation;

– managing meetings;

– interview techniques that can identify differences between wishes, expectations, and actual needs;

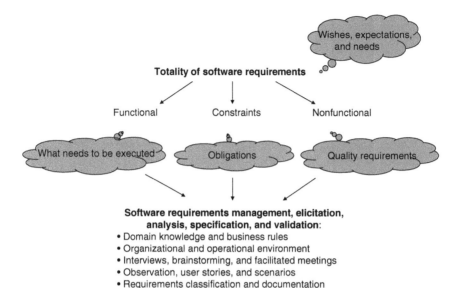

Figure 1.4   Context of software requirements elicitation.

– clear and concise documentation of functional requirements, performance requirements, obligations, and properties of future systems;

– applying systematic techniques for requirement elicitation;

– managing priorities and changes (e.g., changes to requirements).

It is clear that errors can arise when eliciting requirements. It can be difficult to cater to the wishes, expectations, and needs of many different user groups at the same time (see Figure 1.4). Therefore, it is important to pay particular attention to erroneous requirement definitions, the lack of definitions for critical obligations and software characteristics, the addition of unnecessary requirements (e.g., those not requested by the customer), the lack of attention to business priorities, and fuzzy requirements descriptions.

Ordering Soup

Let us say you order soup at a restaurant. Your expressed requirement is "I would like the soup of the day." But in fact, unexpressed wishes, expectations, and needs include: not too hot, not too cold; soup that is not too salty; utensils, salt, and pepper available on the table; clean washrooms; a well situated table; a quiet environment.

And that was a simple requirement, imagine a complex software!

A requirement is said to be of good quality when it meets the following characteristics:

- correct;
- complete;
- clear for each stakeholder group (e.g., the client, the system architect, testers, and those who will maintain the system);
- unambiguous, that is, same interpretation of the requirement from all stakeholders;
- concise (simple, precise);
- consistent;
- feasible (realistic, possible);
- necessary (responds to a client's need);
- independent of the design;
- independent of the implementation technique;
- verifiable and testable;
- can be traced back to a business need;
- unique.

We will present techniques to help detect defects in requirements documentation in a later chapter concerning reviews.

We must also ensure that we are not looking for the Holy Grail of the perfect specification, since we do not always have the time or means, or the budget, to achieve this level of perfection.

The article by Ambler [AMB 04] entitled "Examining the Big Requirements Up Front Approach" suggests that it is sometimes ineffective to write detailed requirements early in the life cycle of a software project. He claims that this traditional approach increases the risk of a project's failure. He stipulates that a large percentage of these specifications are not integrated in the final version of the software and that the corresponding documentation is rarely updated during the project. He thus asserts that this way of working is outdated. In his article, he recommends using more recent agile techniques, such as Test-Driven Development, in order to produce a minimum amount of paper documentation.

We have observed that software analysts and designers also often use prototyping, which helps to partially eliminate the traditional requirements document and replace it with a set of user interfaces and test cases that describe the requirements, architecture, and software design to be developed. Prototypes prove useful for pinpointing what the client is envisioning and getting valuable feedback early in the project. In the next section, the development practices adopted by different business sectors will be discussed.

In a system with hardware and software components, requirements are developed at the system level and then allocated to hardware, software, and sometimes to an operator. The following figure illustrates the interactions between system, hardware, and software life cycle processes.

Systems engineering must work in close collaboration with hardware and software engineering during the allocation of system requirements (for example, functionalities and quality requirements such as safety and performance) to hardware and software.

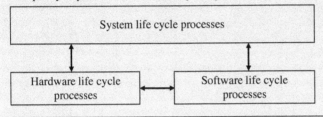

1.3.2 Maintaining Effective Communications Between Client and Developer

Errors can also occur in intermediary products due to involuntary misunderstandings between software personnel and clients and users from the outset of the software project. Software developers and software engineers must use simple, non-technical language and try to take into account the user's reality. They must be aware of all signs of lack of communication, on both sides. Examples of these situations are:

– poor understanding of the client's instructions;
– the client wants immediate results;
– the client or the user does not take the time to read the documentation sent to him;
– poor understanding of the changes requested from the developers during design;
– the analyst stops accepting changes during the requirements definition and design phase, given that for certain projects 25% of specifications will have changed before the end of the project.

To minimize errors:

– take notes at each meeting and distribute the minutes to the entire project team;
– review the documents produced;

– be consistent with your use of terms and develop a glossary of terms to be shared with all stakeholders;

– inform clients of the cost of changing specifications;

– choose a development approach that allows you to accept changes along the way;

– number each requirement and implement a change management process (as it will be presented in another chapter).

This book includes a glossary that could be used to develop the glossary for a specific project.

HVCCR is a company dedicated to the online sale of specialized ventilation, refrigeration, and air-conditioning tools. A client contacted the company that maintained its web catalog to update a few images and add some recent products. The task was estimated to take 10–20 minutes of work.

The person in charge of maintaining the web catalog contacted the client and informed him that to complete the update of the changes requested, the server would have to be restarted, which could cancel any current sessions. This work should preferably be done at night. The client, who did not fully grasp the impact of his request, insisted on having this update performed immediately. At the time of the update, several buyers were online processing payments. Shutting down the system interrupted bank transactions, causing customer dissatisfaction as well as data corruption. HVCCR took several days to address buyers' complaints and fix the problems.

1.3.3 Deviations from Specifications

This situation occurs when the developer incorrectly interprets a requirement and develops the software based on his own understanding. This situation creates errors that unfortunately may only be caught later in the development cycle or during the use of the software.

Other types of deviations are:

– reusing existing code without making adequate adjustments to meet new requirements;

– deciding to drop part of the requirements due to budget or time pressures;

– initiatives and improvements introduced by developers without verifying with clients.

1.3.4 Architecture and Design Errors

Errors can be inserted in the software when designers (system and data architects) translate user requirements into technical specifications. The typical design errors are:

– an incomplete overview of the software to be developed;

– unclear role for each software architecture component (responsibility, communication);

– unspecified primary data and data processing classes;

– a design that does not use the correct algorithms to meet requirements;

– incorrect business or technical process sequence;

– poor design of business or process rule criteria;

– a design that does not trace back to requirements;

– omission of transaction statuses that correctly represent the client's process;

– failure to process errors and illegal operations, which enables the software to process cases that would not exist in the client's sector of business—up to 80% of program code is estimated to process exceptions or errors.

1.3.5 Coding Errors

Many errors can occur in the construction of software. McConnell (2004) [MCC 04] devotes a substantial part of his book *"Code Complete"* to describing effective techniques for creating quality source code. He describes common programming errors and inefficiencies. According to McConnell, the typical programming errors are:

– inappropriate choice of programming language and conventions;

– not addressing how to manage complexity from the onset;

– poor understanding/interpretation of design documents;

– incoherent abstractions;

– loop and condition errors;

– data processing errors;

– processing sequence errors;

– lack of or poor validation of data upon input;

- poor design of business rule criteria;
- omission of transaction statuses that are required to truly represent the client process;
- failure to process errors and illegal operations, which enables the software to process cases that would not exist in the client's sector of business;
- poor assignment or processing of the data type;
- error in loop or interfering with the loop index;
- lack of skills in dealing with extremely complex nestings;
- integer division problem;
- poor initialization of a variable or pointer;
- source code that does not trace back to design;
- confusion regarding an alias for global data (global variable passed on to a subprogram).

1.3.6 Non-Compliance with Current Processes/Procedures

Some organizations have their own internal methodology and internal standards for developing/acquiring software. This internal methodology describes processes, procedures, steps, deliverables, templates, and standards (e.g., coding standard) that must be considered for software acquisition, development, maintenance, and operations. Of course, in a less mature organization, these processes/procedures will not be clearly defined.

We can therefore ask ourselves the following question: How can not fulfilling the requirements related to an internal methodology lead to defects in software? We must think in terms of the total life cycle (e.g., over many decades for subways and commercial airplanes) of the software, and not just of its initial development. It is clear that someone who only programs code appears to be far more productive than someone who develops intermediary products, such as requirements, test plans, and user documentation, as prescribed by the internal methodology of an organization. However, the immediate productivity would be disadvantageous in the long run.

Undocumented software will give rise to the following problems sooner or later:

- When members of the software team need to coordinate their work, they will have difficulty understanding and testing poorly documented or undocumented software.
- The person who replaces or maintains the software will only have the source code as a reference.

– SQA will find a large number of non-conformities (with respect to the internal methodology) regarding this software.

– The test team will have problems developing test plans and scenarios, primarily because the specifications are not available.

1.3.7 Inadequate Reviews and Tests

The purpose of software reviews and tests is to identify and check that errors and defects have been eliminated from the software. If these activities are not effective, the software delivered to the client will likely be prone to failure.

All kinds of issues can crop up when reviewing and testing software:

– reviews only cover a very small part of the software's intermediate deliverables;

– reviews do not identify all errors found in the documentation and software code;

– the list of recommendations stemming from reviews is not implemented or followed up on adequately;

– incomplete test plans do not adequately cover the entire set of functions of the software, leaving parts untested;

– the project plan has not left much time to perform reviews or tests. In some cases, this step is shortened because it is wedged between coding and the final delivery. Delays in the early steps of the project do not always mean the delivery date will be extended, to the detriment of proper testing;

– the testing process does not correctly report the errors or defects found;

– the defects found are corrected, but are not subject to adequate regression testing (i.e., retesting the complete corrected software) thereafter.

1.3.8 Documentation Errors

It has been recognized that obsolete or incomplete documentation for software being used in an organization is a common problem. Few development teams enjoy spending time preparing and reviewing documentation.

We would be inclined to say no to the question "does software wear out?" Indeed, the 0s and 1s found in the memories do not wear out from use as with hardware. In addition to classifying types of errors, it is important to understand the typical reliability curve for software. Figure 1.5 describes the reliability curve for computer hardware as a function of time. This curve is called a U-shaped or bathtub curve. It represents the reliability of a piece of equipment, such as a car, throughout its life cycle.

With regard to software, the reliability curve resembles more of what is shown in Figure 1.6. This means that software deterioration occurs over time due to, among other things, numerous changes in requirements.

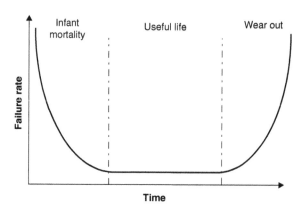

Figure 1.5 Reliability curve for hardware as a function of time.
Source: Adapted from Pressman 2014. [PRE 14].

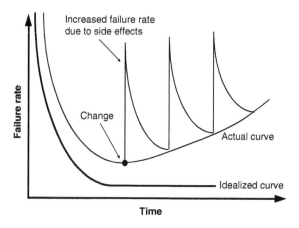

Figure 1.6 Reliability curve of software.
Source: Adapted from Pressman 2014. [PRE 14].

Professor April worked in the Middle East between 1998 and 2003 in a large telecommunications company. When he arrived, he noticed that the original documentation for critical application software for the telephone company had not been updated in over 10 years. There was no software quality assurance function in the information technology division of that company.

In conclusion, we see that there are many sources of potential errors, and that without SQA, these defects may result in failures if not discovered.

1.4 SOFTWARE QUALITY

The previous section, which presented the issues with identifying defects, has laid the ground work for our next discussion, namely software quality. How do we define software quality? The standards groups suggest the following definition.

Software Quality
Conformance to established software requirements; the capability of a software product to satisfy stated and implied needs when used under specified conditions (ISO 25010 [ISO 11i]).
   The degree to which a software product meets established requirements; however, quality depends upon the degree to which those established requirements accurately represent stakeholder needs, wants, and expectations [Institute of Electrical and Electronics Engineers (IEEE 730)] [IEE 14].

The second definition in the text box is very different, despite appearances. The first part of the definition comes from the perspective of Crosby who reassures the software engineer with its strictness. This perspective is: "If I deliver all that is specified in the requirements document, then I will have delivered quality software." However, the second part of this definition is from the quality perspective of Juran, which specifies that one must satisfy the client's needs, wants, and expectations that are not necessarily described in the requirements documentation!

These two points of view force the software engineer to establish the kind of agreement that must describe client's requirements and attempt to faithfully reflect his needs, wants, and expectations. Of course, there is a practical element to the functional characteristics that need to be described, but also implicit characteristics, which are expected of any professionally developed piece of software.

In this context, the software engineer can be inspired by the standards in his field, just as his colleagues in construction engineering or other engineering specialties, in order to identify his obligations. Process conformance can be achieved and measured. As an example, Professor April published an example of the measurement, in Ouanouki and April (2007) [OUA 07], where the software testing process had to be assessed for Sarbanes-Oxley conformance for the largest Canadian hardware retailer.

Software quality is recognized differently depending on each perspective, including that of the clients, maintainers, and users. Sometimes, it is necessary to differentiate between the client, who is responsible for acquiring the software, and the users, who will ultimately use it.

Users seek, among other things, functionalities, performance, efficiency, accurate results, reliability, and usability. Clients typically focus more on costs and deadlines, with a view to the best solution at the best price. This can be considered an external point of view with regard to quality. To draw a parallel with the automobile industry, the user (driver) will go to the garage that provides him with fast service, quality, and a good price. He has a non-technical point of view.

As for software specialists, they focus more on meeting obligations based on the allocated budget. Therefore, they see their obligations from the point of view of meeting requirements and the terms and conditions of the agreement. The choice of the right tools and modern techniques are often at the heart of concerns, and is therefore an internal point of view like that of a mechanic who is interested in the engine technology and knows it in detail. To him or her, quality is equally important with regard to the choice and assembly of components. We will consider these two points of view (external versus internal) when discussing the software product quality models.

Therefore, quality software is software that meets the true needs of the stakeholders while respecting any predefined cost and time constraints.

The client's need for software (or more generally any kind of system) may be defined at four levels:

- True needs

- Expressed needs

- Specified needs

- Achieved needs

The ability of software to meet (or not meet) the needs of the client can be described in the differences between these four levels. Throughout the development of a project, there will be factors that will affect the final quality.

For each level, Table 1.1 describes the typical factors that can affect the satisfaction of the client requirements.

1.5 SOFTWARE QUALITY ASSURANCE

This section presents a definition of SQA. This section also aims to describe the objectives of SQA. In order to put these definitions into perspective, here is a reminder of the general definition of software engineering:

Software Engineering
The systematic application of scientific and technological knowledge, methods, and experience for the design, implementation, testing, and documentation of software.
                                                               ISO 24765 [ISO 17a]

To be a recognized profession, software engineering must have its own body of knowledge for which there is consensus. As with most other engineering fields, recognized knowledge, methods, and standards must be used for the development, maintenance/evolution, and infrastructure/operation of software. The body of

Table 1.1   Factors that can Affect Meeting the True Requirements of the Client [CEG 90] (© 1990 - ALSTOM Transport SA)

Type of requirement	Origin of the expression	Main causes of difference
True	Mind of the stakeholders	– Unfamiliarity with true requirements
		– Instability of requirements
		– Different viewpoints of ordering party and users
Expressed	User requirements	– Incomplete specification
		– Lack of standards
		– Inadequate or difficult communication with the ordering party
		– Insufficient quality control
Specified	Software Specification Document	– Inappropriate use of management and production methods, techniques, and tools
Achieved	Documents and Product Code	– Insufficient tests
		– Insufficient quality control techniques

knowledge for software engineering is published in the SWEBOK guide (www.swebok.org). An entire chapter is dedicated to SQA.

Quality Assurance

1) a planned and systematic pattern of all actions necessary to provide adequate confidence that an item or product conforms to established technical requirements;

2) a set of activities designed to evaluate the process by which products are developed or manufactured;

3) the planned and systematic activities implemented within the quality system, and demonstrated as needed, to provide adequate confidence that an entity will fulfill requirements for quality.

ISO 24765 [ISO 17a]

Software Quality Assurance

A set of activities that define and assess the adequacy of software processes to provide evidence that establishes confidence that the software processes are appropriate for and produce software products of suitable quality for their intended purposes. A key attribute of SQA is the objectivity of the SQA function with respect to the project. The SQA function may also be organizationally independent of the project; that is, free from technical, managerial, and financial pressures from the project.

IEEE 730 [IEE 14]

The term "software quality assurance" could be a bit misleading. The implementation of software engineering practices can only "assure" the quality of a project, since the term "assurance" refers to "grounds for justified confidence that a claim has been or will be achieved." In fact, QA is implemented to reduce the risks of developing a software that does not meet the wants, needs, and expectations of stakeholders within budget and schedule.

This perspective of QA, in terms of software development, involves the following elements:

- the need to plan the quality aspects of a product or service;
- systematic activities that tell us, throughout the software life cycle, that certain corrections are required;
- the quality system is a complete system that must, in the context of quality management, allow for the setting up of a quality policy and continuous improvement;
- QA techniques that demonstrate the level of quality reached so as to instill confidence in users; and lastly,
- demonstrate that the quality requirements defined for the project, for the change or by the software department have been met.

In addition to software development, SQA can also focus on the maintenance/evolution and infrastructure/operations of software. A typical quality system should include all software processes from the most general (such as governance) to the most technical (e.g., data replication). QA is described in standards such as ISO 12207 [ISO 17], IEEE 730 [IEE 14], ISO 9001 [ISO 15], and exemplary practices models, such as CobiT [COB 12] and the Capability Maturity Model Integration (CMMI) models that will be presented in a later chapter.

1.6 BUSINESS MODELS AND THE CHOICE OF SOFTWARE ENGINEERING PRACTICES

In this section, Iberle (2002) [IBE 02], a senior test engineer at Hewlett-Packard, describes her experience in two business sectors of the same company: cardiology products and printers. Different business models are then described to help us understand the risks and the respective needs of each business sector with regards to software practices. These business models will be used in the following chapters to help choose or adapt software practices according to the context of a specific project or application domain.

Business Model

A business model describes the rationale of how an organization creates, delivers, and captures value (economic, social, or other forms of value). The essence of a business model is that it defines the manner by which the business enterprise delivers value to customers, entices customers to pay for value, and converts those payments to profit.

Adapted from Wikipedia

Knowledge of the business models and organizational culture will help the reader to [IBE 02]:

- evaluate the effectiveness of new practices for an organization or specific project;
- learn software practices from other fields or cultures;
- understand the context that promotes collaboration with members of other cultures;
- more easily integrate into a new job within another culture.

This section concludes with a brief discussion of exemplary software practices.

1.6.1 Description of the Context

Medical products belong to a field known for its very high quality standards. During a mandate in the cardiology products sector, Iberle (2002) [IBE 02] used a large number of traditional practices described in software engineering manuals, for example: detailed written specifications, intensive use of inspections and reviews throughout the life cycle, and exhaustive tests for requirements. Exit criteria were created at the beginning of the project and a product could not be shipped as long as the exit criteria were not met.

In this field, a project end date can be missed by weeks and even months. These delays are acceptable in order to fix any last-minute problems using a long checklist. It was far from painless. Iberle (2002) [IBE 02] explains that she worked many extra hours to try to be on schedule (and not exceed the deadline too much). There were heated debates as to whether a specific defect should be qualified as severe (level 1 severity) or average (level 2–5 severity). However, in the end, quality always won out over the schedule.

After 8 years of working on medical products, Iberle (2002) [IBE 02] was assigned to the business sector that produced printers and served small businesses and consumers. Practices in this business sector of the company were very different.

For example, specifications were far shorter, project exit criteria significantly less formal, but making the delivery date was very important. While Iberle was working in testing, she noticed differences in test practices. The main test effort was not focused on tests related to specifications. They were not trying to test all possible entry combinations. There was far less test documentation. In fact, some testers had no test procedures. This was a huge culture shock. At first, Iberle would walk around shaking her head, and grumbling "These people don't care about quality!" After a while, she started to see that her definition of quality was different and was based on her experience in a different field. It was time for her to revisit her beliefs about software quality.

1.6.2 Anxiety and Fear

When Iberle (2003) [IBE 03] worked on defibrillators and cardiographs, missing a delivery date was not the worst thing that could happen. What really scared the team was what could kill a patient or technician due to an electrical shock, cause a person to come to the wrong diagnosis, or that the device could not be used in an emergency situation. If the team raised the possibility of a failure, the delivery date was automatically pushed forward, without any discussion whatsoever. Lengthy and costly efforts to find and definitely eliminate the cause of the defect were systematically approved. It was obvious that, for an organization in this business sector, shirking one's legal responsibility or being blamed by the American Food and Drug Administration definitely contributed to these decisions. Delivery dates could be changed and production completed with overtime.

In the consumer products division, the reality was quite different. The potential for injury was very low, even in the worst conditions imaginable. The real concern was not respecting schedules or exceeding costs. When software has to be packaged in hundreds of thousands of boxes and these boxes must be sent to resellers on time for the day of a major sale, there is not much room to "play catch up." Another fear was having thousands of users unable to install their new printer and calling customer support lines the day after Christmas. Incompatibility between the most popular software and hardware was another source of concern.

So these two business divisions had different definitions of "quality." Clients valued different things: clients from the medical sector favored accuracy and reliability above all, whereas printer customers looked for user-friendliness and compatibility far more than reliability. Of course, everyone wants reliability. However, whether they are aware of it or not, people value reliability as a function of the pain that certain problems may cause them. People are not happy when they have to restart their computer from time to time, but their misfortune is nothing in comparison with the anguish of a patient faced with a functional problem with a heart defibrillator. When someone goes into fibrillation, there is a 5–6 minute window for saving the patient. So there is no time to lose with equipment problems.

The definition of "reliability" is therefore also very different in these two business sectors. When it was understood that no one would die from a printer software error, the team examined the software practices in the medical products division to determine whether they were also useful in the printer sector [IBE 03]. It would take Iberle several months to realize that what seemed shoddy in the printer sector was a way of dealing with different priorities that did not carry the same weight as for medical products.

1.6.3 Choice of Software Practices

As expected, people from both business sectors chose software engineering practices that would lower the probability of their worst fears. Since their apprehensions are different, their practices are also different. In fact, in light of their fears, the choice of practices starts to make sense. The fear of a false diagnosis leads to many detailed reviews and various types of tests. However, the fear of confusing printer users results in more usability tests.

It is not surprising to see that people who work in the same business sectors have similar concerns and use similar practices. Certain concerns can also be found in other organizations. For example, the aerospace sector and medical sector are very closely related. It is also possible for the same organization to have different fears and values in different business sectors, as Iberle (2003) [IBE 03] described above of her employment at Hewlett-Packard.

Software organizations or software specialists are divided into groups that appreciate similar things or share the same concerns, based on similarities in client and business community expectations. These cultures are called "practice groups," that is, software development groups, which share common definitions of quality and tend to use similar practices.

1.6.4 Business Model Descriptions

The following models were developed by Iberle to better understand the need for QA in different business sectors, given that the way in which money flows through an organization (e.g., contract income, cost of products delivered, and losses) and how profits are generated affect the choice of the software practices used to develop products for an organization. The five main business models in the software industry are [IBE 03]:

- – Custom systems written on contract: The organization makes profits by selling tailored software development services for clients (e.g., Accenture, TATA, and Infosys).
- – Custom software written in-house: The organization develops software to improve organizational efficiency (e.g., your current internal IT organization).

- Commercial software: The company makes profits by developing and selling software to other organizations (e.g., Oracle and SAP).
- Mass-market software: The company makes profits by developing and selling software to consumers (e.g., Microsoft and Adobe).
- Commercial and mass-market firmware: The company makes profits by selling software in embedded hardware and systems (e.g., digital cameras, automobile braking system, and airplane engines).

1.6.5 Description of Generic Situational Factors

Each business model has a set of attributes or factors that are specific to it. Here is a list of situational factors that seem to influence the choice of software engineering practices in general [IBE 03]:

- Criticality: The potential to cause harm to the user or prejudice the interests of the purchaser varies depending on the type of product. Some software may kill a person if it shuts down; other software programs may result in major money losses for many people; others will make a user waste time.
- Uncertainty of users' wants and needs: The requirements for software that implements a familiar process in an organization are better known than the requirements for a consumer product that is so new that the end-users do not even know what they want.
- Range of environments: Software written for use in a specific organization only has to be compatible with its own computer environment, whereas software sold to a mass market must work in a wide range of environments.
- Cost of fixing errors: Distributing corrections for certain software applications (e.g., embedded software of an automobile) is usually far more costly than fixing a website.
- Regulations: Regulatory bodies and contractual clauses may require the use of software practices other than those that would normally be adopted. Certain situations require process audits to check whether a process was followed at the time of producing the software.
- Project size: Projects that take several years and require hundreds of developers are common in certain organizations, whereas in other organizations, shorter projects developed by a single team are more typical.
- Communication: There are a certain number of factors, in addition to project scope, that can increase the quantity of person-to-person communication or make communications more difficult. Certain factors seem to occur more often within certain cultures, whereas others happen at random:
  o Concurrent developer–developer communication: Communication with other people on the same project is affected by the way in which the work is

distributed. In certain organizations, senior engineers design the software and junior staff carries out the coding and unit tests (instead of having the same person carrying out the design, coding, and unit tests for a given component). This practice increases the quantity of communications between developers.
  o Developer–maintainer communication: Maintenance and enhancements require communication with the developers. Communication with developers is greatly facilitated when they work in the same area.
  o Communication between managers and developers: Progress reports must be sent to upper management. However, the quantity of information and form of communication that managers believe they need may vary substantially.
- Organization's culture: The organization has a culture that defines how people work. There are four types of organizational cultures:
  o Control culture: control cultures, such as IBM and GE, are motivated by the need for power and security.
  o Skill culture: A culture of skill is defined by the need to make full use of one's skills: Microsoft is a good example.
  o Collaborative culture: A collaborative culture, as illustrated by Hewlett-Packard, is motivated by a need to belong.
  o Thriving culture: A thriving culture is motivated by self-actualization, and can be seen in start-up organizations.

1.6.6 Detailed Description of Each Business Model

This section goes into more detail about each of the five main business models. A single business model, contract-based development for made-to-measure systems, is described as an in-depth case study. For this business model, we describe the following four perspectives:

- context;
- situational factors;
- concerns; and
- software practices predominately used in this business model.

For the other four business models, we will only consider the context and concerns.

1.6.6.1 Custom Systems Written on Contract

In a fixed-price contract, Iberle (2003) [IBE 03] indicated that the client specifies exactly what he wants and promises the supplier a given sum of money. The profits made by the supplier depend on his ability to remain within budget and to deliver on schedule, as defined in the contract, a product that performs as intended. Large-scale applications and military software are often written under contract. The software produced in this business culture is often critical software. The cost of distributing fixes after delivery is manageable because the corrections are provided to an environment that is known and accessible, and to a reasonable number of sites.

Critical Software

Software having the potential for serious impact on the users or environment due to factors including safety, performance, and security.

Adapted from ISO 29110 [ISO 16f]

Critical System

System having the potential for serious impact on the users or environment due to factors including safety, performance, and security.

ISO 29110 [ISO 16f]

Following is the list of dominating factors in this business model [IBE 03]:

– Criticality: Software failures in financial systems can seriously compromise the client's business interests. Software defects in planes and military systems may endanger lives, even if many software programs purchased by the Defense Department are business software applications whose failure would have the same impact as that in financial systems.

– Uncertainty of user needs and requirements: Since buyers and users are an identifiable group, they can be contacted to find out what they are looking for. In general, they have a relatively clear idea of what they want. However, the process to put this into place is not always well documented, and users may not agree on the steps in the process, their demands may require technology that does not exist, business needs may change during the project, and sometimes people completely change their minds.

– Range of environments: In general, the purchasing organization has identified a small set of target environments in order to avoid cost increases. The result is a range of environments that are clearly defined and relatively small, compared with other cultures.

– Cost of fixing errors: In general, there are inexpensive ways of distributing fixes—a large portion of the software will be on servers in a given building and the client's software location is generally known.

– Regulations: Defense software (e.g., for a fighter or commercial plane) must comply with a huge list of regulations, most of which concern the software development process. Financial software is not subject to regulations in the same way. It is common that the contract will stipulate process audits to prove that the organization followed its development process. The client expects to receive regular progress reports on the project.

– Project size: Often large or even immense. Several dozen people work for more than 2 years on the average-size project, but hundreds of people over several years are required for large projects. There is also some data that indicates that small projects are far more common than large projects.

– Communication: The practice involving dividing architecture and coding between senior and junior professionals is occasionally observed in this culture. Given that the systems and projects are large, often different people and even separate departments are used for analysis, design, etc. Moreover, maintenance contracts can go to people other than the original developers. This may produce competition and make communications more complex. Organizations that develop software are often large, whether their projects are small or large, which means additional hierarchical levels.

– Organizational culture: Organizations that write software on contract often have a control culture. This seems logical since most of them have ties with the military.

The concerns of the developers of these systems are often:

– incorrect results;
– exceeding budget;
– penalties for late delivery, and
– not delivering what the client asked for (which may lead to legal proceedings).

These situational factors lead to certain assumptions regarding this business model:

– delivery on schedule and within budget is imperative;
– reliable, correct software is imperative;
– requirements must be known and detailed from the project onset;
– projects are typically large scale with many communication channels;
– it is necessary to show that what was promised has indeed been delivered;
– plans must be developed, and regular progress reports prepared (which are sent to project management and the client).

In the text above, we presented three perspectives: context; situational factors; and concerns about the first business model.

In the next few paragraphs, we present the prevailing practices used with the business model of this case study.

These practices are taken from [IBE 03]:

– A lot of documentation

Documentation is a valuable way of communicating when the project size is large and when external suppliers are involved. Written documentation is often far more effective than discussions around the cooler when the communication channels are complex, which occurs when people are geographically remote and in different organizations. In addition, certain documents are often necessary to prove that we are doing what was set out in the contract. Lastly, in order for the requirements to be known in detail at the start of the project, documentation and many reviews of the requirements are necessary before responding to the call for tenders.

– Lists of exemplary practices

Lists of exemplary practices, such as the CobiT [COB 12] and CMMI models, developed by the Software Engineering Institute, are used to develop contractual clauses. For example, in this business model, the focus is on project estimating and management in order to be on schedule and within budget as stipulated in the contract, and regular progress reports are necessary.

– Waterfall development cycle

The waterfall development life cycle was invented in the 1950s to provide enough structure for large IT projects to be able to plan and strategize on-time delivery. The new iterative and agile development cycles plan out development in smaller increments, which allow for planning while offering more flexibility as to delivery. However, as it has been observed, in this business model, cascade development is often the preferred method.

– Project audits

Audits are often specified in the contract for this business model. The audit is used to prove to the client, or during legal proceedings, that the contractual clauses, such as respecting schedules, quality, and functions, have been fulfilled.

We have now described the four perspectives for this business model: the context; situational factors; concerns; and predominant practices. In the next section, we present, as described by Iberle (2003) [IBE 03], only the context and concerns regarding the other four business models.

1.6.6.2 Custom Systems Written In-House

When using one's own employees to develop software, economic aspects are different than for those who have their software developed on a contract basis. The value of the work depends on improving efficacy or efficiency of operations within the organization. Less focus is put on scheduling meetings since projects are often pending or postponed depending on the budget. The systems can be critical for the organization or of an experimental nature. Fixes are distributed to a limited number of sites.

Developers of these systems often are concerned with the following:

– producing incorrect results;

– limiting the ability of other employees to do their own work;

– their project being cancelled.

1.6.6.3 Commercial Software

Commercial software is software sold to other organizations rather than to an individual consumer. Profits depend on the familiar economic model, which involves selling many copies of the same piece of software for more than the cost of developing and making the copies. Instead of meeting the specific needs of a single client, the developer aims to satisfy many clients. The software is often critical for the organization or at least very important for the client's organizational operations. Since the software is in the hands of many clients in many places, the distribution of corrections can be very costly. These clients also tend to instigate legal proceedings if the software is deficient, which increases the cost of errors.

Business system vendors are generally fearful of:

– court cases;

– recalls;

– tarnishing their reputation.

1.6.6.4 Mass-Market Software

This software is sold to individual consumers often at a very high volume. Profits are made by selling products at higher than development cost, often in a niche market or at certain times of the year, such as at Christmas. The potential effects of software failures for the client are generally less serious than those in the previous models and clients are less likely to demand reparation for any damage incurred. The failure of certain software may considerably affect the user's well-being, such as in the case of tax preparation software. However, for most, a failure is simply a source of frustration.

The typical concerns in this culture are:

– missing the marketing opportunity;

– a high level of support calls;

– bad reviews in the press.

The cost of fixing errors, for mass-market product manufacturers, could be significantly reduced when the owner can update their products. Unfortunately, the customer will be left to search for and perform these upgrades.

1.6.6.5 Commercial and Mass-Market Firmware

Given that profits depend on the sale of the product for more than the manufacturing cost, the cost of distributing fixes is extremely high, since electronic circuits must often be changed on site. Corrections cannot simply be sent to the client. The impact of down time with mass-market embedded software is potentially more serious than the impact of software failures, since the software is controlling a device. Although the destructive potential of small objects, such as digital watches, is low, in certain cases, software failures could have fatal consequences.

The typical concerns of this culture are:

- incorrect behavior of the software in certain situations;
- recalls;
- court cases.

Business Model—Open-Source Software

Open-source software is software that is distributed with its source code and the authorization to modify and distribute it freely under the condition that it is also provided as open-source software once modified.

This business model is becoming an influential economic model in the software industry. It permits its users to collaborate, essentially over the Internet, by adding improvements to a software and distributing it once modified. This approach allows others to benefit from these innovations. However, open-source software does not permit its developers to be paid for these improvements.

The concerns associated with this model are:

- undemonstrated quality;
- lack of support;
- delays in providing fixes.

Adapted from Wikipedia

1.7 SUCCESS FACTORS

Implementing practices to improve software quality can be facilitated or slowed down based on factors inherent to the organization. The following text boxes list some of these factors.

Factors that Foster Software Quality

1) SQA techniques adapted to the environment.
2) Clear terminology with regards to software problems.
3) An understanding and specific attention to each major category of software error sources.
4) An awareness of the SQA body of knowledge of the SWEBOK as a guide for SQA.

Factors that may Adversely Affect Software Quality

1) A lack of cohesion between SQA techniques and environmental factors in your organization.
2) Confusing terminology used to describe software problems.
3) A lack of understanding or interest for collecting information on software error sources.
4) Poor understanding of software quality fundamentals.
5) Ignorance or non-adherence with published SQA techniques.

1.8 FURTHER READING

ARTHUR L. J. *Improving Software Quality: An Insider's Guide to TQM*. John Wiley & Sons, New York, 1992, 320 p.

CROSBY P. B. *Quality Is Free*. McGraw-Hill, New York, 1979, 309 p.

DEMING W. E. *Out of the Crisis*. MIT Press, Cambridge, MA, 2000, 524 p.

HUMPHREY W. S. *Managing the Software Process*. Addison-Wesley, Reading, MA, 1989, Chapters 8, 10, and 16.

JURAN J. M. *Juran on Leadership for Quality*. The Free Press, New York, 1989.

SURYN W., ABRAN A., and APRIL A. ISO/IEC SQuaRE. The Second Generation of Standards for Software Product Quality. In: Proceedings of the 7th IASTED international conference on Software Engineering and Applications (ICSEA'03), Montreal, Canada, 2003, pp. 1–9.

VINCENTI W. G. *What Engineers Know and How They Know It—Analytical Studies from Aeronautical History*. John Hopkins University Press, Baltimore, MD, 1993, 336 p.

1.9 EXERCISES

1.1 Describe the difference between a defect, an error, and a failure.

1.2 According to the studies of Boris Beizer, when do the greatest number of software errors occur in the software development life cycle?

1.3 Describe the difference between the software and hardware reliability curves.

1.4 Eight categories for causes of errors describe the development and maintenance environment, as experienced in organizations:

a) Identify and describe these situations.

b) What situations more specifically influence software engineers who develop and maintain the software?

c) What situations more specifically influence the effort of the software engineering managers who develop and maintain the software?

1.5 Describe the different perspectives of software quality from the point of view of the client, the user, and the software engineer.

1.6 Describe the types of needs, their origin, and the causes for differences that may be due to a discrepancy between the needs expressed by the client and those carried out by the software engineer.

1.7 Describe the concept of business models and how it creates different perspectives for SQA requirements.

1.8 Describe the main differences between QA and quality control.

Chapter 2

Quality Culture

After completing this chapter, you will be able to:

– understand the cost of quality;
– recognize the signs of a quality culture;
– identify the compromises of the five dimensions of a software project;
– know and follow the software engineer's code of ethics.

2.1 INTRODUCTION

In this chapter, we consider the concepts of the cost of quality, quality culture, and the code of ethics for software engineers. The issues related to quality will be applied to the context of software development. The principles set out in this book apply to those who develop, maintain, and work on information technology infrastructure, be it computer technicians, management computer specialists, or software and IT engineers. Typically, software engineers graduate from a school of engineering and are members of a professional order. We know that the term "software engineer" is currently used freely in both universities and the business world. Very few software engineers have obtained certified university degrees. For the engineer who is a member of a professional association, quality is part of the civil liabilities for the profession and must be managed prudently.

Non-quality in software has led to several catastrophes. In this chapter, we present examples where poor quality culture and the lack of a code of ethics have led to disastrous situations involving poor quality software and irreversible damage to people and the environment.

The following description is of the case of a Canadian medical device for treating cancer (Therac-25), which caused the death of several patients from massive

Software Quality Assurance, First Edition. Claude Y. Laporte and Alain April.
© 2018 the IEEE Computer Society, Inc. Published 2018 by John Wiley & Sons, Inc.

irradiation. Similar incidents have occurred with devices developed by other companies throughout the world. We describe the Therac-25 case because it has been extensively studied and documented.

A large Japanese electronic product manufacturing company uses a considerable number of suppliers. The supplier pyramid is made up of a first level with some sixty suppliers, a second level with a few hundred suppliers, and a third level with a few thousand very small suppliers.

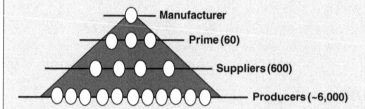

A software defect for a component produced by a third-level supplier caused a loss of over $200 million for this manufacturer.

Adapted from Shintani (2006) [SHI 06]

We believe that setting up a quality culture and software quality assurance (SQA) principles, as stipulated in the standards, could help solve these problems.

It is clear that quality, which is influenced by your organization's senior management and the organization's culture, has a cost, has a positive effect on profits, and must be governed by a code of ethics.

"In today's software market, the main focus is on costs, schedule and functions, and quality is lost in the noise. This is regrettable since mediocre quality performance is at the heart of most software costs and scheduling problems."

Watts S. Humphrey

In the software industry, there are still too many quality problems that are becoming increasingly severe and with more consequences (consult the Incidents and Horror Stories Involving Software in the appendix at the end of this book). In the following text box, we describe the case of a defective medical device, the Therac-25, which injured and killed patients. Its software had many defects.

The Therac-25 Medical Device

Computers are increasingly being introduced into safety-critical systems and, as a consequence, have been involved in accidents. Some of the most widely cited software-related accidents in safety-critical systems involved a computerized radiation therapy machine called the Therac-25. Between June 1985 and January 1987, six known accidents involved massive overdoses, by the Therac-25, resulting in deaths and serious injuries.

In 1982, the Therac-25 replaced older models of the Therac-6 and Therac-20 radiation therapy machines at Atomic Energy Canada Limited (AECL). The new model was computer/software controlled, whereas the older models were controlled by electromechanical components. In the new model, the software was supposed to immediately stop the machine/treatment in the event of a malfunction. This is very complex and expensive equipment, costing around $1 million. Some software of Therac-25 had been recently created, whereas other parts had been reused from the previous models. However, reused software from older models was not essential to the safety of the treatment since it was the electrotechnical components that ensured the additional safeguard.

During the incidents, the patients complained of severe burns, and radiation overdose was not initially identified as a possible cause of these accidents. In one case, the patient complained of having a burning sensation. The technician told the patient that there was nothing indicating that a problem had occurred during the treatment. This same patient had to have a mastectomy and subsequently lost the use of one arm and her shoulder.

Another incident occurred in Ontario. During the treatment, the machine shut down after a few seconds and indicated that no dose was delivered to the patient. The operator then pressed the Proceed command key. The machine shut down again. The operator repeated the process a few times and the machine displayed "no dose" delivered. The patient complained of a burning sensation. The patient died a few months later of an extremely virulent cancer. An AECL technician later estimated the patient had received between 13,000 and 17,000 rads. Typical single therapeutic doses are in the 200-rad range. Doses of 1000 rads can be fatal if delivered to the whole body.

Later that year, another patient was burned while the unit indicated it was in pause mode. This patient died of his injury 5 months later. Failing to reproduce the problem, AECL concluded that the patient was probably burned by an electrical shock unrelated to the radiation treatment. An independent company was asked to check the device and found no electrical problems.

After this incident, the US Food and Drug Administration reported to AECL that this device did not operate in accordance with the law and required AECL to inform all of its users about these problems and possible consequences to the patients. Users of the Therac-25 formed a user group and questioned AECL about the lack of transparency concerning these accidents. Sometime later that year, the newspapers started publishing stories about two of the accidents that had occurred.

It seemed that AECL did not consider that the software could have been the cause of these incidents since AECL had reused software that worked well in previous models. It was thought that by reusing software that was functioning and verified, this would automatically ensure its reliability. They had forgotten that certain parameters of the new model were different than the older models. Electromechanical devices in the older models were overriding software defects and preventing equipment malfunctions. With the Therac-25 model, the company had, for cost reasons, removed the electromechanical devices and thus exposed the faulty software instructions and harmed patients by inadvertently removing the additional safety feature.

Basic software-engineering principles that apparently were violated with the Therac-25 include:

– documentation should not be an afterthought;

– software quality assurance practices and standards should be established;

– designs should be kept simple;

– ways to get information about errors, for example, software audit trails, should be designed into the software from the beginning;

– the software should be subjected to extensive testing and formal analysis at the module and software level; system testing alone is not adequate.

Furthermore, these problems are not limited to the medical industry. It is still a common belief that any good engineer can build software, regardless of whether he or she is trained in state-of-the-art software-engineering procedures. Many companies building safety-critical software are not using proper procedures from a software engineering and safety-engineering perspective.

Most accidents are system accidents; that is, they stem from complex interactions between various components and activities. To attribute a single cause to an accident is usually a serious mistake. In this article, we hope to demonstrate the complex nature of accidents and the need to investigate all aspects of system development and operation to understand what has happened and to prevent future accidents.

The Therac-25 accidents are the most serious computer-related accidents to date (at least non-military and admitted) and have even drawn the attention of the popular press.

Adapted from Leveson and Turner (1993) [LEV 93]

Readers are invited to read the 24-page article, available on the internet, written by Leveson and Turner (1993) [LEV 93], describing the six accidents involving massive overdoses to patients in more detail as well as a description of the investigations conducted to discover the multiple causes of the Therac-25 accidents.

2.2 COST OF QUALITY

One of the major factors that explains the resistance to implementing quality assurance is the perception of its high cost. In organizations that develop software, it is rare to find data on the cost of non-quality.

   In this part of the book, we define the components of the cost of quality. Next, we discuss the cost of quality used in a major American company working in the military industry, and the data collected by Professor Claude Y. Laporte while working with different organizations. Lastly, we present a case study using data collected from a major transportation equipment manufacturer in Canada, Bombardier Transport.

"We never have time to do work correctly the first time, but we always manage to find the time to redo the work one or two times."

Anonymous

   Thinking in terms of costs helps attract the attention of administrators. This approach means better positioning of the software quality file and confronting the sometimes negative perceptions of company administrators, project managers, and often engineers from other disciplines (e.g., mechanical engineer). This approach is preferred to a technological discussion since the terminology of costs is common to almost all administrators, and a constant concern for management. Thus, SQA must strive to present the software quality improvement file in a way that is homogenous and consistent with the company's other investments, that is, as a business case for the organization. It is also a professional way to present an investment file.

   As software engineers, you are responsible for informing administrators of the risks that a company takes when not fully committed to the quality of its software. A practical manner of beginning the exercise is to identify the costs of non-quality. It is easier to identify potential savings by studying the problems caused by software.

To convince administrators of the necessity of quality software, they must be shown the impacts of not having quality software. This can be done by collecting data on the experiences of a company that invested in quality programs versus one that did not. They must be presented with the positive results and benefits obtained from all aspects of applying quality, as well as the negative effects and failures due to a lack of quality or a concern for quality.

The costs of a project can be grouped into four categories: (1) implementation costs, (2) prevention costs, (3) appraisal costs, and (4) the costs associated with failures or anomalies.

If we are certain that all development activities are error free, then 100% of costs could be implementation costs. Given that we make mistakes, we need to be able to identify them. The costs of detecting errors are appraisal costs (e.g., testing).

Costs due to errors are anomaly costs. When we want to reduce the cost of anomalies, we invest in training, tools, and methodology. These are prevention costs.

"Quality is not only an important concept, it is also free. And it is not only free, it is the most profitable product that we produce! The real question is not how much a quality management system costs, but what is the cost of not having one."

Harold Geneen
CEO ITT Corporation

The "cost of quality" is not calculated in the same way in all organizations. There is a certain amount of ambiguity between the notion of the cost of quality, the cost of non-quality, and the cost of obtaining quality. In fact, this is a non-issue, regardless of what it is called, it is important to clearly identify the different costs taken into account in our calculations. To do so, we can see how other companies have dealt with this.

In the most commonly used model at this time, costs of quality take into account the following five perspectives: (1) prevention costs, (2) appraisal costs, (3–4) failure costs (internal during development, external on the client's premises), and (5) costs associated with warranty claims and loss of reputation caused by non-quality.

The only difficulty with this model is clearly identifying the activities associated with each perspective and then tracking the actual costs for the company. We must not underestimate the complexity of this second step! The calculation of the cost of quality in this model is as follows:

Quality costs = Prevention costs
+ Appraisal or evaluation costs
+ Internal and external failure costs
+ Warranty claims and loss of reputation costs

– Prevention costs: This is defined as the cost incurred by an organization to prevent the occurrence of errors in the various steps of the development or maintenance process. For example, the cost of training employees, the cost of maintenance to ensure a stable manufacturing process, and the cost of making improvements.

– Appraisal costs: The cost of verifying or evaluating a product or service during the different steps in the development process. Monitoring system costs (their maintenance and management costs).

– Internal failure costs: The cost resulting from anomalies before the product or service has been delivered to the client. Loss of earnings due to non-compliance (cost of making changes, waste, additional human activities, and the use of extra products).

– External failure costs: The cost incurred by the company when the client discovers defects. Cost of late deliveries, cost of managing disputes with the client, logistical costs for storing the replacement product or for delivery of the product to the client.

– Warranty claims and loss of reputation costs: Cost of warranty execution and damage caused to a corporate image as well as the cost of losing clients.

In the late 1990s, a hospital in the Montreal region developed a clinical research management platform. Before a manager who was specialized in medicine and was an advocate of software quality joined the hospital, between 40% and 50% of defects were identified by the doctor-users of the software for each version. After the manager set up software quality processes, defects dropped by 20% within 1 year. The development team was delighted with the results, and especially with the idea of not having to rework half the functions with each version. However, management found these processes to be too long and decided to lay off the manager. After his departure, the process was abolished, and several of the best employees left the hospital because they were demotivated when old habits set in again.

In addition to the costs listed above, an organization may suffer other repercussions following the development of a defective software such as lawsuits, penalties imposed by the court, a deterioration of the market value of their shares, the withdrawal of financial partners, and the departure of key employees.

Krasner [KRA 98] published a table to better understand the activities and costs for three quality perspectives (see Table 2.1).

Cost of quality models started appearing in the 1950s based on the work of Feigenbaum and colleagues. They proposed a method for classifying the costs associated with the quality assurance carried out on a product. They proposed an economic point of view and empirically studied the relationship of each perspective on total cost.

Table 2.1   Categories of Software Quality Costs

Major categories	Subcategories	Definition	Typical costs
Prevention cost	Quality basis definition	Effort to define quality, and to set quality goals, standards, and thresholds. Quality trade-off analysis	Definition of release criteria for acceptance testing and related quality standards
	Project and process-oriented interventions	Effort to prevent poor product quality or improve process quality	Process improvement, updating of procedures and work instructions; metric collection and analysis; internal and external quality audits; training and certification of employees
Appraisal or evaluation cost	Discovery of the condition of the product	Discovery of the level of non-conformance	Test, walk-through, inspection, desk-check, quality assurance
	Ensuring the achievement of quality	Quality control gating	Contract/proposal review, product quality audits, "go" or "no go" decisions to release or proceed, quality assurance of subcontractor
Cost of anomalies or non-conformance	Internal anomalies or non-conformance	Problem detected before delivery to the customer	Rework (e.g., recode, retest, re-review, re-document, etc.)
	External anomalies or non-conformance	Problem detected after delivery to the customer	Warranty support, resolution of complaints, reimbursement damage paid to customer, domino effect (e.g., other projects are delayed), reduction of sales, damage to reputation of enterprise, increased marketing

Source: Adapted from Krasner 1998 [KRA 98].

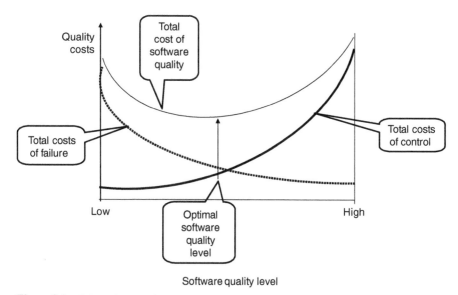

Figure 2.1 Balance between the software quality level and the cost of quality [GAL 17].

As illustrated in Figure 2.1, increased detection and prevention costs lead to reduced costs related to failures (internal or external) and vice versa: a drop in the costs of detection and prevention leads to an increase in the costs related to failures. As well, there is a relationship between the level of software quality and the total cost of detection and prevention. The discussion in the previous section on the different business fields enables us to see that software with high criticality will need a higher quality level than software with lower criticality, which will mean increased detection and prevention costs.

Acme Communications Inc. received a new project. The project involved automating a survey system for a company that provides consulting services to contractors. The company already had experience in developing survey software.

After meeting with the client, the project manager analyzed the information and prepared a two-page document that specified his understanding of the requirements. The estimate was 40 hours of work for the first delivery. The project was considered to be so simple that it was assigned to an intern. The project manager, without having consulted with the previous developers, asked an intern to use the source code of an existing application that he believed to be similar. Thirty hours after the development began, the intern had not yet understood the source code. The manager had already promised the first delivery within the time estimated, so there remained only 10 hours and the entire weekend for the version to be ready for the following Monday. The manager felt it necessary to ask a former developer to do the work.

The developer's analysis showed that this application was new and that there was no similar one existing in the company. In other words, the project had to be started from scratch. The manager, anxious after hearing this information, asked the developer whether he could work the entire weekend to deliver something to the client. The developer saw this as a tremendous opportunity and asked for a 25% salary raise.

The first version was delivered, but with only 70% of the features required. The final version was completed after requiring an extra 20-hour workload. The project initially estimated at 40 hours instead took a total of 84 hours.

The first objective of the SQA is to convince management that there are proven benefits to SQA activities. To do this, he must be convinced of this himself. We all have been taught the following: "identifying an error early in the process can save a lot of time, money and effort."

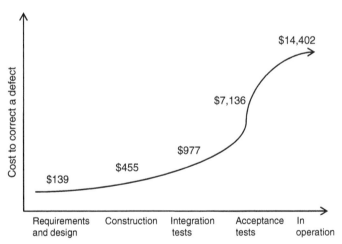

Figure 2.2   Costs of propagating an error[1] [JON 00].

After years spent studying and measuring software practices, Capers Jones [JON 00] estimated the average costs of fixing defects in the software industry (see Figure 2.2). The simulations show that it is profitable to identify and fix a defect as soon as possible. A defect that arises during the assembly phase will cost three times more to fix than one that is corrected during the previous phase (during which we should had been able to find it). It will cost seven times more to fix the defect in the

---

[1]These costs are based on the value of the American dollar in 1981.

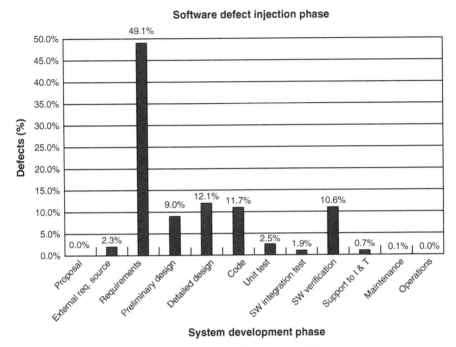

Figure 2.3   Defect injection distribution during the life cycle [SEL 07].

next phase (test and integration), 50 times more in the trial phase, 130 times more in the integration phase, and 100 times more when it is a failure for the client, and has to be repaired during the operational phase of the product.

In fact, Boehm [BOE 01] also published that it would only cost $25, in 2001, to correct a problem identified in a requirements and specifications document. This cost increases to $139 if the problem is discovered during programming. The origin of software defects ended up being the subject of many studies. For example, Figure 2.3 presents the origin of the defects throughout the development life cycle of software in a case study published by Selby and Selby (2007) [SEL 07].

In this case study, we see that approximately 50% of defects occurred during the requirements phase. If these defects are not corrected early on, they will be very expensive to fix during the test and operational phases. As well, it is quite possible that correcting these defects will involve extending the delivery schedule, since all too often the effort involved in correcting the defects was not factored into the initial plan. If, however, the software must be delivered by a set date, it is quite possible that a large number of defects will not be corrected in order to meet the delivery date. Therefore, the client will have to use software containing a large number of defects while awaiting a new version.

In the defense industry, the American company Raytheon carried out a study on the cost of quality [DIO 92, HAL 96]. This study measured the cost of quality over a

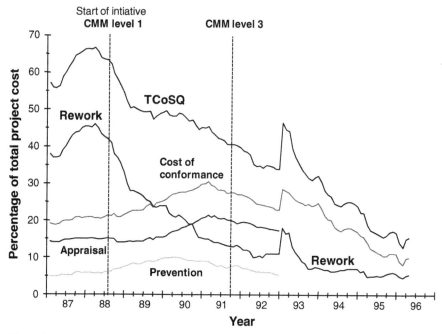

Figure 2.4   Data on software quality costs [HAL 96].

9-year period. Figure 2.4 shows that, at the beginning of the study, the rework cost for this company was evaluated at approximately 41% of the total cost of the projects, at 18% when process maturity had been improved and 11% when the process maturity was at level 3, and then 6% when it was at level 4 maturity. The figure also describes that with prevention costs of less than 10% of the total cost of the projects, the rework costs were reduced from 45% to less than 10% of the total cost of the projects.

The study by Dion concluded that the cost of quality seems to be correlated with the implementation of increasingly mature processes. Another study on quality costs published by Krasner proposes that the rework costs vary between 15% and 25% of the development costs for a level 3 maturity organization (see Table 2.2).

The Case of Bombardier Transport

A project to measure software quality costs was carried out within the software development group at Bombardier Transport located in Quebec. A team, composed of 15 specialized software engineers in this group, was commissioned to develop control software for the subway system in a large American city. They decided to set up data collection to measure the quality costs for this project. The measuring activity took place in four steps: drawing up a list of typical activities related to software quality costs; categorizing these activities (prevention, evaluation, and correcting anomalies); developing and applying weighting rules; and finally, measuring software quality costs. In all, 27 weighting rules were developed, and one weighting rule was assigned to each project task. More than 1121 software activities were analyzed for a project amounting to 88,000 hours of work. The results obtained showed that, for this project, the software quality cost represented 33% of the total project cost. The rework, or anomaly costs were 10%, prevention costs were 2% and evaluation costs were 21% of the total development costs. The following figure illustrates these results.

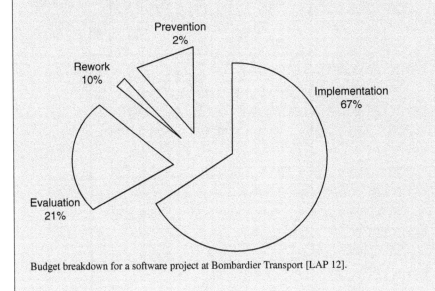

Budget breakdown for a software project at Bombardier Transport [LAP 12].

Table 2.2   Relationship Between the Process Maturity Characteristic and Rework [KRA 98]

Process maturity	Rework (percent of total development effort)
Immature	$\geq 50\%$
Project controlled	$25\% - 50\%$
Defined organizational process	$15\% - 25\%$
Management by fact	$5\% - 15\%$
Continuous learning and improvement	$\leq 5\%$

The Cost of Quality

Over many years, Professor Laporte collected data on the costs of quality. Engineers and managers of multinational organizations and Master of software engineering students working for organizations in the Montreal area provided these data. The following table shows that the average rework cost is approximately 30%.

	Site A American engineers (19)	Site A American managers (5)	Site B European engineers (13)	Site C European engineers (14)	Site D European engineers (9)	Course A 2008 (8)	Course B 2008 (14)	Course C 2009 (11)	Course C 2010 (8)	Course E 2011 (15)	Course F 2012 (10)	Course G 2013 (14)	Course H 2014 (11)	Course I 2015 (13)	Course J 2016 (14)
Performance costs	41%	44%	34%	31%	34%	29%	43%	45%	45%	34%	40%	44%	36%	37%	40%
Rework costs	30%	26%	23%	41%	34%	28%	29%	30%	25%	32%	31%	25%	29%	27%	27%
Appraisal costs	18%	14%	32%	21%	26%	24%	18%	14%	20%	27%	20%	19%	20%	22%	20%
Prevention costs	11%	16%	11%	8%	7%	14%	10%	11%	10%	8%	9%	12%	15%	14%	13%
Quality*	71	8	23	35	17	43	19	48	35	60	55	72	44	23	

* Number of defects per 1000 lines of source code.

Adapted from Laporte et al. (2014) [LAP 14]

We hope that this section, which discussed the concepts of the cost of quality, has convinced you of the importance of implementing improvements in software processes and quality assurance so as to better understand the cost categories to include in your estimates. It is indeed possible to produce quality software while minimizing rework costs (since rework costs are, if the organization has made the right choices, either losses or profits). The use of the cost of quality concept could affect the level of a company's competitiveness. We have shown that a company that invests in defect prevention can offer a product at a lower cost, with less risk of failures and can gradually make strides ahead of their competitors.

Defect prevention is an activity that has not always been popular with software developers. This activity is wrongly perceived as a waste of time and energy. The following section presents a model that evaluates the efficacy of the defect elimination activity and its associated costs. The objective is to convince you that adding a quality activity is worthwhile, and once you are convinced, you will then be able to convince your bosses.

2.3 QUALITY CULTURE

Tylor [TYL 10] defined human culture as "that complex whole which includes knowledge, belief, art, morals, law, custom, and any other capabilities and habits acquired by man as a member of society." It is culture that guides the behaviors, activities, priorities, and decisions of an individual as well as of an organization.

Wiegers (1996) [WIE 96], in his book "*Creating a Software Engineering Culture*," illustrates the interaction between the software engineering culture of an organization, its software engineers, and its projects (see Figure 2.5).

According to Wiegers, a healthy culture is made up of the following elements:

– The personal commitment of each developer to create quality products by systematically applying effective software engineering practices.

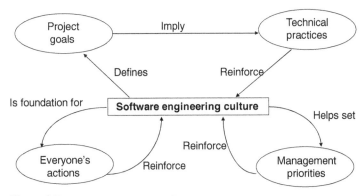

Figure 2.5   Software engineering culture.
Source: Adapted from Wiegers 1996 [WIE 96].

– The commitment to the organization by managers at all levels to provide an environment in which software quality is a fundamental factor of success and allows each developer to carry out this goal.

– The commitment of all team members to constantly improve the processes they use and to always work on improving the products they create.

"The challenge of designing the Boeing 777 was 20% technical and 80% cultural."

John Warner
President (retired), Boeing Computer Systems

As software engineers, why should we be concerned with this aspect? First of all, quality culture cannot be bought. The organization's founders must develop it from the creation of the company. Then, when employees are selected and hired, the founders' corporate culture will slowly begin to change, as illustrated in Figure 2.5. Quality culture cannot be an afterthought; it must be firmly integrated from the outset, and constantly consolidated. The objective of management is to instill a culture that will promote the development of high quality software products and offer them at competitive prices in order to produce income and dividends within an organization where employees are engaged and happy.

The second reason why a software engineer should be interested in the cultural aspects of quality is that effecting change within an organization does not boil down to placing an order with employees. The organization must work to change its culture with the help of change agents. We now know that one of the main stumbling blocks to change within an organization is the organization's culture.

Employees must feel involved and be able to see the benefits of any change. For example, if the change provides no benefit to a worker and has only been made to satisfy a manager's whim, the worker will not be interested in this change. The actions of the employee in our example will not foster the reinforcement of corporate culture and its degree of maturity will not be affected. It is imperative that cultural change is managed in order to obtain the desired results [LAP 98].

You may be asked to "cut corners" regarding the quality of your work or the way in which things are done (comic strip in Figure 2.6). Managers and clients may increase pressure and ask you to skip steps. In the most desperate situations, you may be asked to start programming even before having identified the needs. This old way of thinking about IT still prevails in some companies.

It is not always easy to resist pressure from people who are paying the bills or your salary. In certain cases, you will feel that you have no choice: do what you are being asked to do or leave (see the comic strips in Figures 2.7 and 2.8).

It may be difficult to imagine spending your career in this type of environment. When a quality culture is well established, employees will get involved even in times

Figure 2.6 Start coding ... I'll go and see what the client wants!
Source: Reproduced with permission of CartoonStock ltd.

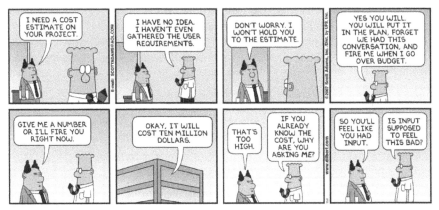

Figure 2.7 Dilbert is threatened and must provide an estimate on the fly. DILBERT © 2007 Scott Adams. Used By permission of UNIVERSAL UCLICK. All rights reserved.

of crisis. It is for this reason that software developers and managers must adopt principles that motivate them to stick to their processes even during these difficult periods. Of course, we are always more flexible during a crisis, but not to the point of dropping all quality assurance activities and our good judgment.

In difficult situations, you must never let your boss, a colleague, or your client convince you do to bad work; proceed with integrity and intelligence with your boss and clients.

The client is not always right. Yes, you heard right! However, he probably has a valid point of view and you have a duty to listen to the client's concerns. Having said that, you are the person solely responsible for interpreting his needs and converting them into specifications and reasonable estimates of effort and duration. The

Figure 2.8 Dilbert tries to negotiate a change in his project. DILBERT © 2009 Scott Adams. Used By permission of UNIVERSAL UCLICK. All rights reserved.

highest added value of a business analyst in a project is to mitigate expectations and find a solution that will meet requirements and is realistic, feasible, and practical. Be aware of overly aggressive salespersons. One of my friends used to say "Nothing is impossible for those who do not have to do it."

Be honest with your clients and make sure that you clearly communicate the limitations and scope of the work that will be done. Wiegers lists fourteen principles to follow to develop a culture that fosters quality (see Table 2.3).

The culture of an organization is a defining factor of successful efforts to improve processes. "Culture" includes a set of shared values and principles that guide

Table 2.3 Cultural Principles in Software Engineering [WIE 96, p. 17]

1. Never let your boss or client cause you to do poor work.
2. People must feel that their work is appreciated.
3. Continuing education is the responsibility of each team member.
4. Participation of the client is the most critical factor of software quality.
5. Your greatest challenge is to share the vision of the final product with the client.
6. Continuous improvement in your software development process is possible and essential.
7. Software development procedures can help establish a common culture of best practices.
8. Quality is the number one priority; long-term productivity is a natural consequence of quality.
9. Ensure that it is a peer, not a client, who finds the defect.
10. A key to software quality is to repeatedly go through all development steps except coding; coding should only be done once.
11. Controlling error reports and change requests is essential to quality and maintenance.
12. If you measure what you do, you can learn to do it better.
13. Do what seems reasonable; do not base yourself on dogma.
14. You cannot change everything at the same time. Identify changes that will reap the most benefits, and start to apply them as of next Monday.

behaviors, activities, priorities, and decisions of a group of people working in the same field. When colleagues share beliefs, it is easier to bring about changes to increase group efficiency as well as the chances of survival.

A development team was set up. A few of the team members were experts in the field, while others were new. The development tools and processes that guided the project were new and therefore not yet mature. The project manager emphasized team unity to create a learning environment. This allowed the team to practice these new processes. Mistakes were allowed, and thus improvement was continuous. The manager's style combined humor and enthusiasm, which also fostered team spirit. Team activities were carried out to promote individual success as well as the teams' success. The manager also set up a culture of expertise, which meant that each team member had his or her area of expertise and focused all efforts on improving this expertise.

Quality involves an important social aspect in which the level of involvement and collaboration of each member in the company becomes essential. It is imperative to promote, and continuously support, beliefs and practices in terms of quality in order to preserve and enrich this critical aspect of corporate culture.

It is important to understand the context in which the organization is developing software in order to be able to understand why it is using a specific practice and not another [IBE 03]. Indeed, a company that develops and distributes software in a highly regulated sector is quite different from a company that develops its own applications for in-house operations. Moreover, other factors within these organizations may also be influencing practices, such as risk and project scope, mastering business rules, and laws for the sector. By analyzing the situation, the software engineer can better evaluate the changes that should be made to promote the development of a quality culture.

Quality culture must, above all, be expressed by the willingness to cultivate it emanating from the company's upper management and not only from the engineering team who, in a way, plays the role of quality police. In the most extreme cases where accidents occurred, corporate management very often seemed involved in choosing non-quality in order to maximize shareholder profits without having a long-term vision. The next section will provide tools for managing these situations.

2.4 THE FIVE DIMENSIONS OF A SOFTWARE PROJECT

Wiegers developed an approach at Eastman Kodak in New York, in order to better frame project start-up discussions. He believed there are five dimensions that must be managed as part of a software project (see Figure 2.10). These dimensions are not independent. For example, if you add staff members to the project team, this may (but not always) shorten the *deadline*; however, this will increase *costs*. A common compromise is to shorten the deadlines by reducing *features* or reducing *quality*. Compromises that need to be made among these five dimensions are not easily made, but can be illustrated so as to have more realistic discussions and better document these decisions. For each project, you must have an idea of which dimensions have more constraints, and which are able to offset these constraints.

Three roles can be associated with each dimension: (1) a driver, (2) a constraint, and (3) a degree of freedom.

A driver is the specific and major reason for which the project must be carried out. For a product that has competition in the market and must offer new features that are expected by clients, the deadline is the main driver. For a project to update software, the driver could be a specific feature. Defining the project driver and associating it with one of the five dimensions allows us to focus on the status of each dimension.

A constraint is an external factor that is not under the control of the project manager. If you have a set number of resources, the "staff" dimension will be a constraint. Often cost will be a constraint for a project under contract. Quality is also a software constraint for medical devices or helicopters, for example.

Certain dimensions, such as features, may have roles as both drivers and constraints when a feature is non-negotiable. Any dimension that is not a driver or a constraint provides a certain degree of freedom. These dimensions can be adjusted to contribute to achieving an objective under a given constraint. The five dimensions cannot simultaneously be either all drivers or all constraints. Therefore, the priority of each dimension must be negotiated with the client right from the start of the project.

Here are two examples of how to use this negotiation model proposed by Wiegers. A Kiviat diagram with five dimensions allows us to illustrate the model using a graduated 10-point scale where 0 means total constraint and 10 means total freedom.

Figure 2.9 describes a project developed in-house with a set team size and a schedule with little leeway. The team is free to determine which features to implement and the quality level for the first version. Since the cost is linked for the most part to staff expenses, it will differ depending on the amount of resources used. Given the schedule allows for some leeway, deadlines may vary slightly.

Figure 2.10 describes the flexibility diagram for a critical software project in which quality is a constraint and the schedule has a high degree of freedom. The patterns shown in these diagrams provide an idea of the type of project you are dealing with. A project cannot focus all its dimensions at 0. There must be some latitude with certain levels to ensure some project success.

All too often in short discussions we tend to speak only of budget and schedules. It is this attitude that generally leads to overruns and chronic dissatisfaction in our industry. We therefore have to get our administrators and clients accustomed to the reality of not always having the features requested, without defects, delivered quickly by a small team at a low cost.

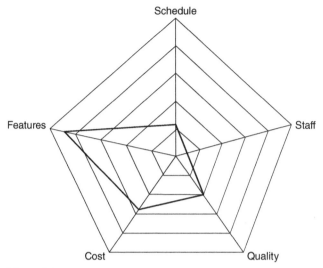

Figure 2.9 Diagram of the flexibility of an in-house project.
Source: Wiegers 1996. [WIE 96, p. 30]. Reproduced with permission of Karl E. Wiegers.

Table 2.4 can be drawn up to summarize the results of negotiation at the start of a project. For each dimension, the table describes the driver, the constraints, and the degree of freedom. The software engineer will attempt to describe the values given to each dimension. The example in Table 2.4 is related to the case in Figure 2.9.

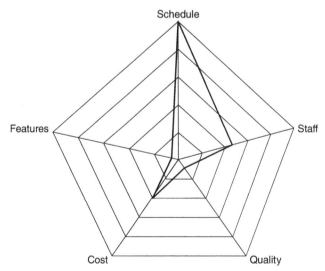

Figure 2.10 Diagram of the flexibility of a critical software project.
Source: Wiegers 1996. [WIE 96]. Reproduced with permission of Karl E. Wiegers.

Table 2.4   Summary Table of Dimensions Negotiated for an In-House Project [WIE 96, p. 32]

Dimension	Driver	Constraint	Degree of freedom
Cost			20% overrun acceptable
Features			60–90% of priority 1 features must be in release 1.0
Quality			Release 1.0 can contain up to five known major defects
Schedule	Release 1.0 must be delivered within 4 months		
Staff		Four people	

A major characteristic of a quality culture and software engineering principles is that expectations and promises are established professionally and in a negotiated manner. Of course, this type of step may at first meet with resistance, but this approach will help you to avoid accepting projects that are not realistic or impossible to carry out in the conditions laid down. Therefore, you should get into the habit of taking a pass on these unrealistic projects and not committing to unavoidable disaster! You can also use other tools that are available to you to help make the right decisions. The next section presents the code of ethics that can be used to convince your clients and superiors of the importance of having a quality culture. Referring to a code of ethics can help the engineer better understand and communicate a quality culture.

2.5   THE SOFTWARE ENGINEERING CODE OF ETHICS

The first draft of the software engineering code of ethics was developed in cooperation with the Institute of Electrical and Electronics Engineers (IEEE) Computer Society and the Association for Computing Machinery (ACM) in 1996. Following this, the draft was widely circulated to elicit comments and suggestions for improvement. The IEEE and the ACM approved the current version in 1998 [GOT 99a; GOT 99b; IEE 99]).

Different perspectives clashed as the code of ethics was being developed. One such confrontation took place in regards to how to approach ethics. The first approach was based on the inherent virtuous nature of humans. This implies that we simply have to indicate the proper direction and people will follow it. Proponents of this approach wanted a code that inspires good behavior with few details. The other approach, based on rights and obligations, requires an outline of all rights and all obligations. The supporters of the latter approach wanted a very detailed code.

Another source of tension stemmed from conflicts in terms of what priority to give to the code's principles. For example, should we favor the employer or the public? This was resolved by indicating that the public good comes before loyalty to the employer or to the profession.

Some people would have preferred to list the standards to be used. It was decided not to list any and instead specify within the code that the currently accepted standards should be used.

On the other hand, everyone agreed that software engineers must be proactive when they become aware of potential problems within the system. Therefore, clauses were added to require engineers to communicate potentially dangerous situations, and to allow software engineers to denounce these types of situations.

To satisfy those who wanted a high-level code and those who wanted a more detailed code, two versions of the code were developed: an abridged version and a complete version.

The code has already been adopted by many organizations. For example, the code is a document that is part of an employee contract. The employee must sign it upon hiring. Over the years, the code became a de facto standard. In some cases, companies may decide not to do business with a company that does not adhere to this code.

"When the devil comes to visit us, he will not have big horns. He will not do any harm, he will not hurt a living being. He will just encourage us to lower our standards and ethics, just a little bit, and the rest will follow ... "

Albert Brooks in the film "Broadcast News"

The code describes eight commitments with which peers, the public, and legal organizations can measure the moral behavior of a software engineer (see Table 2.5).

Each commitment, called a principle here, is described in one sentence and supported by a certain number of clauses that include examples and details that help in its interpretation. Software engineers and software developers who adopt the code agree to respect the eight principles of quality and morality.

Following are some examples of the code of ethics: Principle 3 (product) declares that software engineers will ensure that their products and any related changes meet the highest possible professional standards. This principle is supported by 15 clauses. Clause 3.10 states that software engineers will carry out the tests, debugging and reviews of software and the related documents on which they are working.

The code has been translated into nine languages: German, Chinese, Croatian, English, French, Hebrew, Italian, Japanese, and Spanish. Many organizations have publically adopted the code of ethics and a few universities have included it as part of their software engineering study program. The complete version is found in Appendix 1.

Table 2.5 The Eight Principles of the IEEE's Software Engineering Code of Ethics [IEE 99]

Principle	Description
1. The public	Software engineers shall act consistently with the public interest.
2. Client and Employer	Software engineers shall act in a manner that is in the best interests of their client and employer, consistent with the public interest.
3. Product	Software engineers shall ensure that their products and related modifications meet the highest professional standards possible.
4. Judgment	Software engineers shall maintain integrity and independence in their professional judgment.
5. Management	Software engineering managers and leaders shall subscribe to and promote an ethical approach to the management of software development and maintenance.
6. Profession	Software engineers shall advance the integrity and reputation of the profession consistent with the public interest.
7. Colleagues	Software engineers shall be fair to and supportive of their colleagues.
8. Self	Software engineers shall participate in lifelong learning regarding the practice of their profession and shall promote an ethical approach to the practice of the profession.

The software engineer's code of ethics, translated in 9 languages, is available on this site: http://seeri.etsu.edu/Codes/default.shtm

2.5.1 Abridged Version: Preamble

The abridged version of the code states the main aspirations, whereas the articles in the complete version of the code provide examples and more information as to the manner in which these aspirations must be reflected in the behavior of the software engineer. Together, the abridged and complete versions of the code form a consistent whole. Indeed, without the statement of aspirations, the code risks seeming boring and full of legalese, and without any detailed development included, aspirations could seem abstract and devoid of meaning.

Software engineers who are professionally involved in carrying out analysis, specifications, design, development, tests, and maintenance must respect the principles outlined in the code of ethics.

In accordance with their commitment to the health, safety, and the public good, software engineers must adhere to eight principles presented in Table 2.5.

To use the code professionally, a simple usage procedure is proposed. It is important to understand that the practices are presented in a specific order: from most important to least important. Therefore, you must examine any conflict and go through the code one article at a time to see which article relates to the situation.

As soon as you identify an article in the code of ethics that represents the situation described, you should note that the article has been infringed. Then you should briefly explain why this situation violates the code. Continue in this way, going through all the articles one by one, since the situation may violate more than one article in the code.

Case Study—Confirm [OZ 94]

In 1988, a consortium made up of the Hilton and Marriott hotel chains and of the Budget Rent-A-Car car rental company decided to develop a centralized reservation system to make reservations for airplane tickets, hotel rooms, and car rentals. This project was assigned to AMR Information Services (AMRIS), a company belonging to the American Airlines Corporation. This company had already successfully developed an airplane ticket reservation system used by large companies, that is, the SABRE system.

The partners agreed on a budget not exceeding $55.7 million and a schedule of no more than 45 months. This system had to perform transactions at a cost not exceeding $1.05 per reservation.

At the end of the design phase in September 1989, AMRIS put forth a development plan, which could cost $72.6 million. The cost per reservation was $1.39 instead of $1.05.

In the summer of 1990, two partners were concerned about the system delivery date. Employees working on the CONFIRM project estimated that the project was not respecting deadlines. They had been asked by their supervisors to modify their dates for updates to reflect the original dates of the project.

In February 1991, AMRIS presented a new plan for $92 million. The president stepped down, and, in 1992, more than 20 employees also left. The employees were not happy with project management. They also thought that the managers imposed unrealistic delivery dates and lied about the project status.

In the summer of 1991, a consultant, hired by the developer to evaluate the project, submitted a report. One vice-president was not happy with the consultant's observations. He "buried" the report and fired the consultant. During this period, the Marriott hotel chain was being billed $1 million a month.

In April 1992, AMRIS admitted that it was 2–6 months behind schedule. Hilton was having serious problems with the beta version. Again in April, AMRIS wrote to the partner companies about the managers deliberately concealing a number of major technical and performance problems. It also announced that system development was behind by 15–18 months. As well, eight executives were fired, and 15 employees transferred.

In May 1992, it was announced to the partners that the CONFIRM system did not meet the performance and reliability requirements.

In July 1992, after more than 3.5 years of development and having spent $125 million, the project was abandoned. It was also determined that should the system crash, the database would not be recoverable.

AMRIS settled out of court with the partner companies. This company was said to have been sued for more than $500 million, and that it settled the entire dispute for around $160 million.

2.5.2   The Example of the Code of Ethics of the Ordre des ingénieurs du Québec

Since the code of ethics of the Ordre des ingénieurs du Québec, an association of professional engineers, is quite similar to the code of ethics of the software engineer presented in this chapter, we will illustrate only one of the consequences that an engineer who does not respect this code of ethics may be subject to.

The following text box contains an example of a sanction incurred by an engineer.

Example of a Notice of Permanent Removal from a Society of Professional Engineers

Pursuant to section 180 of the Professional Code, notice is hereby given that on June 4, 2015, the Disciplinary Council of the Society of Software Engineers has declared that Mr. Paul Roberts, at business address 12345 Near Here Street, is guilty of various offenses, including:

In Denver, on or about October 14, 2014, as part of an inspection mandate located at 12345, Client Avenue in Denver, the engineer Paul Roberts issued opinions that were not based on sufficient factual knowledge about software design in his report thus violating Article 4.02 of the Code of ethics of software engineers.

Under this decision, the Commission has ordered Mr. Paul Roberts temporarily removed from the roll of the Society for 6 months. This decision is enforceable from its delivery to the respondent, that is, as of June 11, 2015.

Denver, June 4, 2015
Secretary of the Disciplinary Board
Society of Software Engineers

Another possible consequence is the limitation of the right to practice for a given period, the obligation to undergo training, fines, and the obligation to redo the professional engineering certification.

The following form can be used for the software staff in an organization to demonstrate their commitment to the code of ethics.

 Employee Commitment to the Code of Ethics

(Date) … ….

I have read the Software Engineer's Code of Ethics developed by the IEEE and ACM.

Last Name	First Name	Signature

You can also change the text of the form above to modify the level of commitment for the personnel using these statements:

– I agree to adhere to the Software Engineer's Code of Ethics developed by the IEEE and ACM.

– I agree to respect the Software Engineer's Code of Ethics developed by the IEEE and ACM.

A ceremony could be held where all software professionals would swear, in front of their colleagues, to respect the code. The ceremony would end with the signing of the form above. This ceremony could be organized annually to remind everyone, especially new employees, of the importance of respecting this code.

The signed form could also be made available on the organization's intranet.

Remember that engineers from certain countries must first respect the code of ethics imposed by their professional order.

2.5.3 Whistle Blowers

Sometimes, in an organization, a person must make certain situations public. The whistle blower seriously thinks that the interest of the client or public is at stake, and denounces the situation either internally or externally. Internally, he may communicate with the ombudsman or security group, which represents upper management, whereas externally he can contact his professional order or a journalist. When the whistle blower denounces a situation internally, he may be subject to pressure or attacks from his superiors and colleagues. He might also be fired. This fear of reprisals is quite real. However, the engineer is protected; otherwise the practice of whistle blowing would not be encouraged. The professional tribunal ensures that the identity of the whistle blower is protected by law, if he so wishes. As well, there is an article in the code that stipulates the engineer being accused does not have to communicate with the complainant should he be told his name.

2.6 SUCCESS FACTORS

In the following text box, several factors in relation to the culture of an organization which are likely to foster the development of quality software are listed.

Factors that Foster Software Quality

1) Good team spirit.

2) The skills of the members in the organization (it is imperative that managers select competent people in their organization to carry out the different tasks; if not, even with good team spirit, good managers, good communication, and good processes that are correctly applied, if people are not competent, nothing will really be effective).

3) Managers who set a good example.

4) Effective communication between colleagues, managers, and the client.

5) Recognizing and valuing initiatives to improve quality.

6) Highlighting the notion of organizational culture by qualifying it as the key factor in guaranteeing quality.

7) Defining the culture of an organization as being a set of shared values and principles guiding the behaviours, activities, priorities, and decisions of a group of people who work in the same sector.

8) Including the notion of culture in an organization's strategy and ensuring that it is respected by all personnel.

9) In small and medium-sized businesses, there is often a discrepancy between the perception and ideas of quality that the managers have as opposed to the technical teams; therefore, it is up to the software engineer to educate managers and other members of the organization about the implications of quality, while proposing adapted solutions to enable the organization's objectives to be met.

10) According to Wiegers, the level of involvement of the client throughout the project is the factor that will have the most impact on software quality (indeed, for the client to consider having been provided with a quality product, the product must first and foremost accurately meet the client's requirements. If certain requirements were not well understood or were poorly interpreted at the beginning of the project, the costs related to correcting the situation will only increase as the project advances): having

a client representative who is involved throughout the project can only help to clear up any ambiguities as they arise.

11) Clearly defined roles and responsibilities.

12) Allocating the necessary budgets.

In the following text box, some factors related to organizational culture which could adversely affect the development of quality software are listed.

Factors that may Adversely Affect Software Quality

1) Give employees the responsibility but not the authority to take the actions necessary to ensure the project's success.

2) When we "shoot the messenger."

3) When managers hide their heads in the sand rather than solve problems.

4) Lack of knowledge in quality assurance.

5) Unrealistic time frames.

6) A lack of common working methodology between team members.

7) A manager who says yes to everyone.

2.7 FURTHER READING

GOTTERBARN D. How the new software engineering code of ethics affects you, *IEEE Software*, vol. 16, issue 6, November/December 1999, pp. 58–64.

LEVERSON N. G. and TURNER C. S. An investigation of the Therac-25 accidents, *IEEE Computer*, vol. 26, issue 7, July 1993, pp. 18–41.

WIEGERS K. E. Standing on principles, *The Journal of the Quality Assurance Institute*, vol. 11, issue 3, July 1997, pp. 1–8.

2.8 EXERCISES

2.1 Give arguments to convince management that it is necessary and profitable to invest in SQA.

2.2 In the most commonly used model today, the costs of quality consider five perspectives. Describe the quality cost formula and give some examples to illustrate each perspective.

2.3   What is the relationship between software quality and the total cost of detection and prevention according to observations made by researchers?

2.4   What are the benefits to identifying and eliminating defects early on in the software life cycle?

2.5   How can we use the results of the Raytheon study to convince management of the benefits of setting up SQA?

2.6   According to Weigers, what elements make up a healthy quality culture in an organization?

2.7   Name 5 of the 10 cultural principles of software engineering.

2.8   What are the five dimensions of a software project according to Wiegers?

2.9   Draw up the summary table of dimensions negotiated for the project described in Figure 2.10.

2.10   Apply the code of ethics and indicate the two main clauses that have been infringed in the following situations:

a) Peter, a software engineer, developed software for his company. His company develops and sells inventory software. After months of work, he finds he is stuck regarding several parts of the software. His manager, not understanding the complexity of the problem, wants the work finished this week. Peter remembers that a colleague, Francine, had showed him modules from a commercial software package that she developed at another company. After studying this package, Peter directly incorporates some of these modules into his software. However, he did not tell Francine or his boss, and did not mention this in the software documentation. What clauses in the code did Peter violate?

b) A company developed software to manage a nuclear power plant. The software was designed to manage the plant's reactor. While inspecting the code, Marie found major errors in the software. Frank, Marie's boss, said that the software must be delivered to the client this week. Marie knows that the errors will not be corrected in time. What clauses in the code require Marie to take action?

c) In a company with approximately 10 people, you were named by the president to apply the code of ethics. You are head of the development and maintenance team. Describe the steps in your action plan to apply the code of ethics.

2.11   Name factors within an organization that make it easier to apply the code of ethics.

2.12   What clauses in the code of ethics require a software engineer to denounce a potentially dangerous situation?

2.13   What clause in the code of ethics clearly states that a software engineer must not have pirated software in his possession?

2.14   Read the CONFIRM case [OZ 94], which describes the development of a reservation system that became a money pit, and:

a) Identify the clauses in the code that were infringed.

b) What should the AMRIS directors have done differently?

c) What could the AMRIS developers have done differently?

d) How can a consumer know whether software will cause him irreparable damage before it is actually installed on his workstation?

e) If the software causes major errors a few months after being installed, how could the consumer have protected himself?

2.15 Read the case of Therac-25 [LEV 93], a medical device that has caused the death of several patients, and:

a) Identify the clauses in the code that were infringed.

b) What could you have said to a representative of AECL who has made the usual excuses in the software industry in order to avoid responsibility (complexity, testability, and development process)? To back up your answer, explain in concrete terms what the company could have done to reduce the risks inherent in these three characteristics of the software product?

c) You were recently named director of SQA for the new Therac-30 project. This project will reuse Therac-25 technology to produce a more efficient version. What SQA precautions should be taken?

d) You talked with the Software Quality Director for the Therac-25. He shared the lessons learned from the incidents caused by Therac-25. List four of these lessons.

2.16 You have just purchased a new computer. The technician tells you that he installed demo software on your computer. When you start up your computer, you notice that commercial software was installed. What should you do?

Chapter 3

Software Quality Requirements

After completing this chapter, you will be able to:

- describe the history of the concepts conveyed by software quality models;
- understand the different characteristics and sub-characteristics of software quality as outlined in the ISO 25010 international standard;
- use the concepts to specify the software quality requirements of a software product;
- explain the positive and negative interaction between the quality characteristics of software as described in the ISO 25010 international standard;
- understand the concept of software traceability.

3.1 INTRODUCTION

All software is an element of a system, whether it be a computer system in which the software may be used on a personal computer, or in an electronic consumer product like a digital camera. The needs or requirements of these systems are typically documented either in a request for quote (RFQ) or request for proposal (RFP) document, a statement of work (SOW), a software requirements specification (SRS) document or in a system requirements document (SRD). Using these documents, the software developer must extract the information needed to define specifications for both the functional requirements and performance or non-functional requirements required by the client. The term "non-functional," as applied to requirements, is deprecated and is not used in the Institute of Electrical and Electronics Engineers (IEEE) 730 standard [IEE 14].

Software Quality Assurance, First Edition. Claude Y. Laporte and Alain April.
© 2018 the IEEE Computer Society, Inc. Published 2018 by John Wiley & Sons, Inc.

Functional Requirement

A requirement that specifies a function that a system or system component must be able to perform.

ISO 24765 [ISO 17a]

Non-Functional Requirement

A software requirement that describes not what the software will do but how the software will do it. Synonym: design constraint.

ISO 24765 [ISO 17a]

Performance Requirement

The measurable criterion that identifies a quality attribute of a function or how well a functional requirement must be accomplished (IEEE Std 1220™-2005). A performance requirement is always an attribute of a functional requirement.

IEEE 730 [IEE 14]

Software quality assurance (SQA) must be able to support the practical application of these definitions. To achieve this, many concepts proposed by software quality models must be mastered. This chapter is dedicated to presenting the models as well as the software engineering standards available for correctly defining performance or non-functional (i.e., quality) requirements of software. Using these SQA practices early in the software development life cycle will ensure that the client receives a quality software product that meets his needs and expectations.

Quality Model

A defined set of characteristics, and of relationships between them, which provides a framework for specifying quality requirements and evaluating quality.

ISO 25000 [ISO 14a]

The above definition of a quality model implies that the quality of software can be measured. In this chapter, we describe the research carried out over the years that culminated in a definition of a software quality model. Quality assurance, in certain industries and business models overseen by the software developer, requires a more formal management of software quality throughout the software life cycle. If it is not possible to evaluate the quality of the resulting software, how can the client be in a position to accept it? Or at least, how can we prove that the quality requirements were met?

In order to support this condition, a software quality model is used so that the client can:

- define software quality characteristics that can be evaluated;
- contrast the different perspectives of the quality model that come into play (i.e., internal and external perspectives);
- carefully choose a limited number of quality characteristics that will serve as the non-functional requirements for the software (i.e., quality requirements);
- set a measure and its objectives for each of the quality requirements.

It is therefore necessary for the model to demonstrate its ability to support the quality requirements definition, and subsequently, their measurement and evaluation. We have seen in the previous chapter that quality is a complex concept. Quality is often measured by considering a specific perception of quality. Humankind has mastered measuring physical objects for centuries, but even today, many questions remain about measuring a software product and the ability to measure it in an objective manner. The role and importance of each quality characteristic are still difficult to clearly define, identify, and isolate. What's more, software quality is often a subjective concept that is perceived differently depending on the point of view of the client, user, or software engineer.

Evaluation

A systematic examination of the extent to which an entity is capable of fulfilling specified requirements.

ISO 12207 [ISO 17]

This chapter will provide the SQA practitioner with the knowledge to use the ISO 25010 software quality model concepts. In this way, he can initiate processes and support software engineers in their development, maintenance, and software acquisition projects. We will begin by presenting the history of the different models and standards designed to characterize software quality. This will be followed by a discussion of the concept of software criticality and its value. We will then introduce the concept of quality requirements and present a process for defining quality requirements. Lastly, the technique of software traceability, which assures that a requirement has truly been integrated into the software, will be outlined. In a later chapter, we will cover traceability in greater detail.

3.2 SOFTWARE QUALITY MODELS

Unfortunately, in software organizations, software quality models are still rarely used. A number of stakeholders have put forth the hypothesis that these models do not clearly identify all concerns for all of the stakeholders involved and are difficult to use. In the next two sections, we will see that this is simply an excuse for not formally defining and evaluating the quality of software before it is delivered to the client.

Definition of Requirements: The Essential First Step

"Overall, the government has made limited progress since our last audit of IT projects in 1997. Although since 1998 the Treasury Board Secretariat has established a framework of best practices for managing IT projects, many of the problems we cited in past reports have persisted.

In our 1995 audit, it was noted that the government has begun to develop systems before clearly defining the system requirements for reasonable and realistic periods.

We estimate cost overruns of $250 million for every billion spent in two projects. Senior management's interest and intervention in large IT development projects is urgent."

Report of the Auditor General of Canada [AGC 06]

Let us begin by considering the five quality perspectives described by Garvin (1984) [GAR 84]. His study makes a link between the work of renowned software quality experts and the proposals of quality models of the era. He questioned whether these models take the different perspectives of quality into account:

- Transcendental approach to quality: The transcendental view of quality can be explained as follows: "Although I can't define quality, I know it when I see it." The main problem with this view is that quality is a personal and individual experience. You would have to use the software to get a general idea of its quality. Garvin explains that software quality models all offer a sufficient number of quality characteristics for an individual or an organization to identify and evaluate within their context. In other words, the typical model sets out the quality characteristics for this approach, and it only takes time for all users to see it.

- User-based approach: A second approach studied by Garvin is that quality software performs as expected from the user's perspective (i.e., fitness for purpose). This perspective implies that software quality is not absolute, but can change with the expectations of each user.

– Manufacturing-based approach: This view of software quality, in which quality is defined as complying with specifications, is illustrated by many documents on the quality of the development process. Garvin stipulates that models allow for defining quality requirements at an appropriate level of specificity when defining the requirements and throughout the life cycle. Therefore, this is a "process-based" view, which assumes that compliance with the process leads to quality software.

– Product-based approach: The product-based quality perspective involves an internal view of the product. The software engineer focuses on the internal properties of the software components, for example, the quality of its architecture. These internal properties correspond to source code characteristics and require advanced testing techniques. Garvin explains that if the client is willing to pay, then this perspective is possible with the current models. He describes the case of NASA, who was willing to pay an extra thousand dollars per line of code to ensure that the software aboard the space shuttle met high quality standards.

– Value-based approach: this perspective focuses on the elimination of all activities that do not add value, for example the drafting of certain documents as described by Crosby (1979) [CRO 79]. In the software domain, the concept of "value" is synonymous with productivity, increased profitability, and competitiveness. It results in the need to model the development process and to measure all kinds of quality factors. These quality models can be used to measure these concepts, but really only for insiders and mature organizations.

We will now describe these quality models. Over the past 40 years, researchers have attempted to define "THE" model of software quality. Of course, it has taken a while to get there. The following section describes the prior initiatives that influenced the software quality standard ISO 25000 [ISO 14a] that we use today: McCall, Richards, and Walter, and the IEEE 1061 [IEE 98b] standard.

"An important concept conveyed in all models of the quality of the software is that the software does not directly manifest its quality attributes. Most of the proposed software quality models fail to make the link between quality attributes and corresponding product specifications."

McCall et al. (1977) [MCC 77]

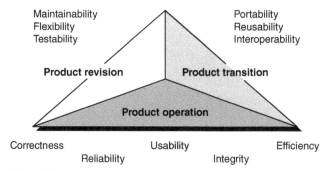

Figure 3.1   The three perspectives and 11 quality factors of McCall et al. (1977) [MCC 77].

3.2.1 Initial Model Proposed by McCall

McCall and his colleagues have been attributed with the original idea for a software quality model [MCC 77]. This model was developed in the 1970s for the United States Air Force and was designed for use by software designers and engineers. It proposes three perspectives for the user (see Figure 3.1) and primarily promotes a product-based view of the software product:

- Operation: during its use;
- Revision: during changes made to it over the years;
- Transition: for its conversion to other environments when the time comes to migrate to a new technology.

Each perspective is broken down into a number of quality factors. The model proposed by McCall and his colleagues lists 11 quality factors.

Each quality factor can be broken down into several quality criteria (see Figure 3.2). In general, the quality factors considered by McCall are internal attributes that fall under the responsibility and control of software engineers during software development. Each quality factor (on the left side of Figure 3.2) is associated with two or more quality criteria (which are not directly measurable in the software). Each quality criterion is defined by a set of measures. For example, the quality factor "reliability" is divided into two criteria: accuracy and error tolerance. The right side of Figure 3.2 presents the measurable properties (called "quality criteria"), which can be evaluated (through observation of the software) to assess quality. McCall proposes a subjective evaluation scale of 0 (minimum quality) to 10 (maximum quality).

The McCall quality model was primarily aimed at software product quality (i.e., the internal perspective) and did not easily tie in with the perspective of the user who is not concerned with technical details. Take, for example, a car owner who is not concerned with the metals or alloys used to make the engine. He expects the car to be well designed so as to minimize frequent and expensive maintenance costs. Another

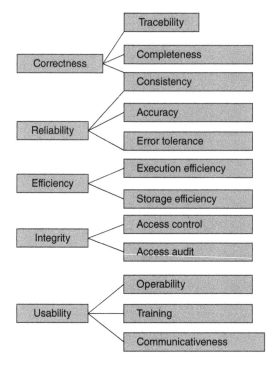

Figure 3.2   Quality factors and criteria from McCall et al. (1977) [MCC 77].

criticism regarding this model was that it involved far too many measurable properties (approximately 300).

What Constitutes the External Quality and Internal Quality of Software?

We should clarify the notions of external and internal quality. These two perspectives of software quality are often presented in quality models. From the external point of view, the focus is on presenting characteristics that are important to people who do not know the technical details. For example, a user is interested in the speed in which the maintainer can make a requested change to the software because the effort dedicated to this task has an impact on cost and waiting time. The user neither knows nor cares about technical details of the software, so his perspective is external.

From an internal point of view, the focus is on measuring the attributes of maintainability, which will influence: (1) the effort to identify where to make the change, (2) changes to the current structures (attempting to reduce the impact of a change), (3) testing the change, and (4) its release into production. If software is not well documented and is poorly structured, it will have low maintainability (internal quality). This low internal quality will affect the time required for making the change, which is the external characteristic that interests the user. Therefore, we can see here that external quality is not a directly observable measure, but it is derived from internal quality. Internal quality however, can be directly measured in the software.

3.2.2 The First Standardized Model: IEEE 1061

The IEEE 1061 standard, that is, the *Standard for a Software Quality Metrics Methodology* [IEE 98b], provides a framework for measuring software quality that allows for the establishment and identification of software quality measures based on quality requirements in order to implement, analyze, and validate software processes and products. This standard claims to adapt to all business models, types of software, and all of the stages of the software life cycle. The IEEE 1061 standard presents examples of measures without formally prescribing specific measures.

Figure 3.3 illustrates the structure of the major concepts proposed in this quality model. At the top tier, we can see that software quality requires prior specification of a certain number of quality attributes, which serve to describe the final quality desired in the software. The attributes desired by clients and users allow for the definition of

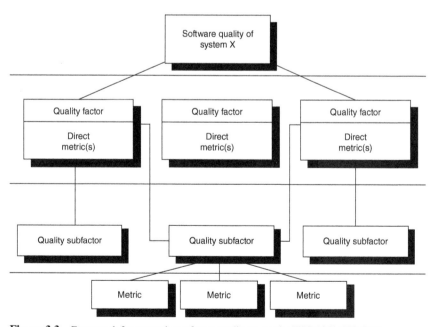

Figure 3.3 Framework for measuring software quality as per the IEEE 1061 [IEE 98b].

the software quality requirements. There must be a consensus with respect to these requirements within the project team, and their definitions should be clearly written in both the project and technical specifications.

Quality factors suggested by this standard are assigned attributes at the next tier. At the tier below that, and only if necessary, subfactors can then be assigned to each quality factor. Lastly, measures are associated with each quality factor, allowing for a quantitative evaluation of the quality factor (or subfactor).

 Quality Measures within the IEEE 1061 Model

As an example, users choose availability as a quality attribute. It is defined in the requirements specifications as being the ability of a software product to maintain a specified level of service when it is used under specific conditions. The team establishes a quality factor, such as mean time between failures (or MTBF).

Users wish to specify that the software should not crash too often, since it needs to perform important activities for the organization. The measurement formula established for Factor A = hours available/(hours available + hours unavailable). It is necessary to identify target values for each directly measured factor. It is also recommended to provide an example of the calculation to clearly define the measure. For example, the work team, when preparing the system specifications, indicates that the MTBF should be 95% to be acceptable (during service hours). If the software must be made available during work hours, that is, 37.5 hours per week, it should therefore not be down for more than 2 hours a week: 37.5/(37.5 + 2) = 0.949%.

Note that if you do not set a target measure (i.e., an objective), there will be no way of determining whether the quality level for the factor was reached when implementing or accepting the software.

This model is interesting since it provides defined steps to use quality measures in the following situations:

– Software program acquisition: In order to establish a contractual commitment regarding quality objectives for client-users and verify whether they were met by allowing their measurement when adapting and releasing the software.

– Software development: In order to clarify and document quality characteristics on which designers and developers must work in order to respect the customer's quality requirements.

– Quality assurance/quality control/audit: In order to enable those outside the development team to evaluate the software quality.

– Maintenance: Allow the maintainer to understand the level of quality and service to maintain when making changes or upgrades to the software.

– Client/user: Allow users to state quality characteristics and evaluate their presence during acceptance tests (i.e., if the software does not meet the specifications agreed upon by the developer, the client can negotiate to his advantage conditions to have them met; for example, the client could negotiate free software maintenance for a specific period of time or the addition of functions at no additional cost).

The following steps are proposed under the IEEE 1061 [IEE 98b] standard:

– Start by identifying the list of non-functional (quality) requirements for the software as of the beginning of the specifications elicitation. To define these requirements, contractual stipulations, standards, and company history must be taken into account. Set a priority for these requirements and try not to resolve any conflicting quality requirements at this time. Make sure that all participants can share their opinion when collecting information.

– Make sure that you meet everybody involved to discuss the factors that should be considered.

– Make a list and make sure to resolve any conflicting points of view.

– Quantify each quality factor. Identify the measures to evaluate this factor and the desired objective in order to meet the threshold and level of quality expected.

– Have measures and thresholds approved. This step is important, since it will identify the possibility of carrying out these measurements within your organization. The standard suggests that you customize and document Figure 3.4 for your specific project.

– Perform a cost–benefit study to identify the costs of implementing the measures for the project. This may be necessary due to:
  o additional costs to enter information, automate calculations, interpret, and present the results;
  o costs to modify support software;
  o costs for software assessment specialists;
  o the purchase of specialized software to measure the software application;
  o training required to apply the measurement plan.

– Implement the measurement method: Define the data collection procedure, describing storage, responsibilities, training, etc. Prototype the measurement process. Choose which part of the software on which to apply the measures. Use the result to improve the cost-benefit analysis. Collect data and calculate values observed from quality factors.

– Analyze the results: Analyze the differences between the measurements obtained and the expected values. Analyze significant differences. Identify measures outside the expected limits for further analysis. Make decisions based

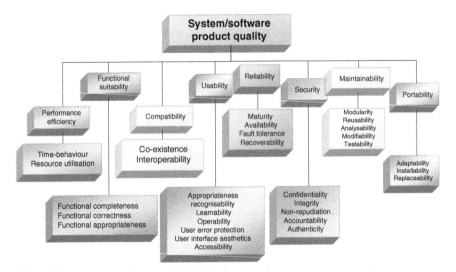

Figure 3.4   Quality model for an ISO 25010 software product.
Source: ISO/IEC 25010. Reproduced with permission of the Standards Council of Canada.

on quality (redo or continue). Use validated measures to make predictions during development. Use measures to validate the quality during tests.

- Validate the measures: It is necessary to identify measures that can predict the value of quality factors, which are numeric representations of quality requirements. Validation is not universal, but must be done for each project. To validate measures, the standard recommends using the following techniques:
  - *Linear correlation*: If there is a positive correlation, the measure can be used as a substitute for the factor.
  - *Identification of variations*: If a factor goes from F_1 (time t_1) to F_2 (time t_2), the measure must change in the same way. This criterion ensures that the measure chosen can detect changes in quality.
  - *Consistency*: If $F_1 > F_2 > F_3$, then $M_1 > M_2 > M_3$. This allows us to sort products based on quality.
  - *Foreseeability*: $(F_a - F_p)/F_p < A$ (F_a actual factor at time t, F_p anticipated factor at time t, A constant). This factor evaluates whether the measurement formula can predict a quality factor with the desired accuracy (A).
  - *Power of discrimination*: The measures must be able to distinguish between high-quality software and low-quality software.
  - *Reliability*: This measure must show that in $P\%$, the correlation, identification, consistency, and foreseeability are valid.

The IEEE 1061 [IEE 98b] standard has allowed us to put measurement into practice and to link product measures with client-user requirements. Since this standard was classified as a guide, it was not very popular outside of the military for several reasons:

– seen as being too expensive;

– some did not see its usefulness;

– the industry was not ready to use it;

– suppliers did not want to be measured in this way by their clients.

This American standard influenced international software quality standards. The ISO 25000 [ISO 14a] standard is presented next.

3.2.3 Current Standardized Model: ISO 25000 Set of Standards

It was during the eighth international conference in 1980, at The Hague in the Netherlands, that Japan proposed setting up an ISO committee to standardize an internationally recognized software quality model. A work group was created, workgroup 6, and it was assigned to Professor Motoei Azuma (an Emeritus Professor at Waseda University, Tokyo, Japan) who, in turn, asked for help from the international community to study the proposals and possible solutions.

The international standardization of a software quality model, the ISO 9126 standard [ISO 01], was published for the first time in 1991. As we can see by the terminology used by McCall et al. (1977) [MCC 77] and the IEEE 1061 [IEE 98b] standard, there are a number of definitions and terms that were reused. The ISO 9126 [ISO 01] standard has attempted to promote the systematic implementation of software quality measures since 1991. However, it is little known and not often used in industry and was replaced by the ISO/IEC 25000 standard [ISO 14a]. We continue to see that manufacturers, suppliers, and major consulting firms avoid the formal use of the standard in their services. The main reason is that it creates requirements for quality and for guarantees that they are trying to avoid. Gradually and inevitably, this standard will become essential for software professionals.

On February 11, 1993, the Treasury Board of Canada issued an internal directive, Directive NCTTI 26 entitled "*Software Evaluation—Quality Characteristics of Software and Usage Guidelines*" [CON 93]. This directive proposes the practical use of the International Organization for Standardization (ISO) standard. This internal standard for information technology at the Treasury Board supports the government's policy for improving the management of information technology, which requires the adoption of quality management practices, including control systems, in order to promote the acquisition, use and management of information technology resources in an innovative and cost effective manner. Since the software industry had gained a certain level of maturity and that software was now an essential component of a number of government products and services, it was necessary to be concerned with their quality. They went on to say that given the increasing demands for quality and safety, in the future, evaluation of software quality should be completed using a quality model proposed by ISO.

The Treasury Board concluded that there were basically two ways to determine the quality of a software product: (1) assess the quality of the development process, and (2) assess the quality of the final product. The ISO 25000 [ISO 14a] standard allows for the evaluation of the quality of the final software product.

In some cases, ministries and agencies may decide not to use this standard, particularly when it is much more advantageous in regards to performance or cost, it is for the general advantage of the Government of Canada, or if the ministry or agency has a contractual obligation or is a partner of an international treaty (such as NATO).

Since 2005, ISO 25000 [ISO 14a] has provided a series of standards for the evaluation of software quality. The purpose of this standard is to provide a framework and references for defining the quality requirements of software and the way in which these requirements will be assessed.

The ISO 25000's series of standards recommends the following four steps [ISO 14a]:

– set quality requirements;

– establish a quality model;

– define quality measures;

– conduct evaluations.

Note that the ISO 25000 [ISO 14a] standard was selected by the Software Engineering Institute (SEI) as a useful reference for the improvement of performance processes described in the Capability Maturity Model Integration (CMMI®) model. The CMMI model, which describes software engineering models and standards, will be presented in Chapter 4.

The ISO 25010 standard identifies eight quality attributes for software, as illustrated in Figure 3.4.

Software Quality Characteristics

The set of attributes of a software product by which the quality of this product is described, verified, and validated.

ISO 25000 [ISO 14a]

To illustrate how this standard is used, we will describe the characteristic of maintainability, which has five sub-characteristics: modularity, reusability, analyzability, modifiability, and testability (see Table 3.1). Maintainability is defined, under ISO 25010 [ISO 11i], as being the level of efficiency and efficacy with which software can be modified. Changes may include software corrections, improvements, or adaptation to changes in the environment, requirements, or functional specifications.

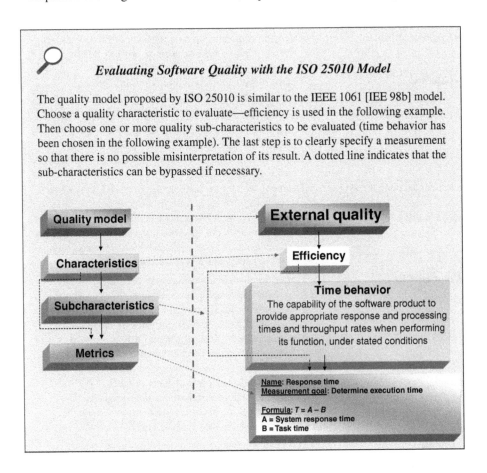

Evaluating Software Quality with the ISO 25010 Model

The quality model proposed by ISO 25010 is similar to the IEEE 1061 [IEE 98b] model. Choose a quality characteristic to evaluate—efficiency is used in the following example. Then choose one or more quality sub-characteristics to be evaluated (time behavior has been chosen in the following example). The last step is to clearly specify a measurement so that there is no possible misinterpretation of its result. A dotted line indicates that the sub-characteristics can be bypassed if necessary.

Quality model — — — — External quality

Characteristics — — — — Efficiency

Time behavior
The capability of the software product to provide appropriate response and processing times and throughput rates when performing its function, under stated conditions

Subcharacteristics

Metrics

Name: Response time
Measurement goal: Determine execution time

Formula: $T = A - B$
A = System response time
B = Task time

Two different perspectives, the internal and external points of view, of the maintainability of the software are often presented in software engineering publications [LAG 96]. Approached from an external point of view, maintainability attempts to measure the effort required to troubleshoot, analyze, and make changes to specific software. From an internal point of view, maintainability usually involves measuring the attributes of the software that influence this change effort. The internal measurement of maintainability is not a direct measurement, that is, a single measurement for

Table 3.1   Quality Factors of the ISO 25000 Standard [ISO 14a]

Factor	Description
Performance efficiency	Performance relative to the amount of resources used under stated conditions
• Time behavior	Degree to which the response and processing times and throughput rates of a product or system, when performing its functions, meet requirements (benchmark)
• Resource utilization	Degree to which the amounts and types of resources used by a product or system when performing its functions meet requirements
• Capacity	Degree to which the maximum limits of a product or system parameter meet requirements
Functional suitability	Degree to which a product or system provides functions that meet stated and implied needs when used under specified conditions
• Functional completeness	Degree to which the set of functions covers all the specified tasks and user objectives
• Functional correctness	Degree to which a product or system provides the correct results with the needed degree of precision
• Functional appropriateness	Degree to which the functions facilitate the accomplishment of specified tasks and objectives. As an example: a user is only presented with the necessary steps to complete a task, excluding any unnecessary steps
Compatibility	Degree to which a product, system or component can exchange information with other products, systems or components, and/or perform its required functions, while sharing the same hardware or software environment
• Coexistence	Degree to which a product can perform its required functions efficiently while sharing a common environment and resources with other products, without detrimental impact on any other product
• Interoperability	Degree to which two or more systems, products, or components can exchange information and use the information that has been exchanged
Usability	Degree to which a product or system can be used by specified users to achieve specified goals with effectiveness, efficiency, and satisfaction in a specified context of use
• Appropriateness recognizability	Degree to which users can recognize whether a product or system is appropriate for their needs. Appropriateness recognizability will depend on the ability to recognize the appropriateness of the product or system's functions from initial impressions of the product or system and/or any associated documentation.

Table 3.1 (*Continued*)

Factor	Description
• Learnability	Degree to which a product or system can be used by specified users to achieve specified goals of learning to use the product or system with effectiveness, efficiency, freedom from risk, and satisfaction in a specified context of use
• Operability	Degree to which a product or system has attributes that make it easy to operate and control
• User error protection	Degree to which a system protects users against making errors
• User interface aesthetics	Degree to which a user interface enables pleasing and satisfying interaction for the user
• Accessibility	Degree to which a product or system can be used by people with the widest range of characteristics and capabilities to achieve a specified goal in a specified context of use
Reliability	Degree to which a system, product or component performs specified functions under specified conditions for a specified period of time
• Maturity	Degree to which a system meets needs for reliability under normal operation
• Availability	Degree to which a system, product or component is operational and accessible when required for use
• Fault tolerance	Degree to which a system, product, or component operates as intended despite the presence of hardware or software faults
• Recoverability	Degree to which, in the event of an interruption or a failure, a product or system can recover the data directly affected and re-establish the desired state of the system
Security	Degree to which a product or system protects information and data so that persons or other products or systems have the degree of data access appropriate to their types and levels of authorization
• Confidentiality	Degree to which a product or system ensures that data are accessible only to those authorized to have access
• Integrity	Degree to which a system, product or component prevents unauthorized access to, or modification of, computer programs or data.
• Non-repudiation	Degree to which actions or events can be proven to have taken place, so that the events or actions cannot be repudiated later
• Accountability	Degree to which the actions of an entity can be traced uniquely to the entity

Table 3.1   (*Continued*)

Factor	Description
• Authenticity	Degree to which the identity of a subject or resource can be proved to be the one claimed
Maintainability	Degree of effectiveness and efficiency with which a product or system can be modified by the intended maintainers
• Modularity	Degree to which a system or computer program is composed of discrete components such that a change to one component has minimal impact on other components
• Reusability	Degree to which an asset can be used in more than one system, or in building other assets
• Analyzability	Degree of effectiveness and efficiency with which it is possible to assess the impact on a product or system of an intended change to one or more of its parts, or to diagnose a product for deficiencies or causes of failures, or to identify parts to be modified
• Modifiability	Degree to which a product or system can be effectively and efficiently modified without introducing defects or degrading existing product quality
• Testability	Degree of effectiveness and efficiency with which test criteria can be established for a system, product, or component and tests can be performed to determine whether those criteria have been met
Portability	Degree of effectiveness and efficiency with which a system, product, or component can be transferred from one hardware, software or other operational or usage environment to another
• Adaptability	Degree to which a product or system can effectively and efficiently be adapted for different or evolving hardware, software or other operational or usage environments
• Installability	Degree of effectiveness and efficiency with which a product or system can be successfully installed and/or uninstalled in a specified environment
• Replaceability	Degree to which a product can be replaced by another specified software product for the same purpose in the same environment

software cannot be used and multiple attributes must be measured to draw conclusions about its internal maintainability [PRE 14].

Note that the ISO 25010 standard proposes a wide range of measures for maintainability: size (e.g., number of lines of modified code), time (internal—in terms of software execution, and external—as perceived by the client), effort (individual or for a task), units (e.g., number of production failures, number of attempts to correct a production failure), rating (results of formulas, percentages, or ratios for several characteristics or types of measures, e.g., correlation between the complexity of the software and the testing effort when modifying the software).

From an internal point of view, in order for software to be maintainable, designers must pay special attention to its architecture and internal structure. Architecture and software structure measures are generally extracted from the source code through observations of the characteristics represented in the form of a graph[1] describing its classes, methods, programs, and functions of the source code of programs. Graph studies help to determine the level of complexity of the software. Even today, there are a large number of publications focused on the static and dynamic evaluation of source code. These studies are inspired by those performed in the 1970s by McCabe (1976) [MCC 76], Halstead (1978) [HAL 78], and Curtis (1979) [CUR 79].

Today we see a large number of commercial software programs and open-source, such as Lattix, Cobertura, and SonarQube, which can measure the internal characteristics of source code. These products contain ready-to-use measures and also enable the user to design new measures based on their specific requirements. Boloix [BOL 95] specified that it is the interpretation of these measures that is difficult, since they are very specialized, and there are not many mechanisms to summarize information for decision making. Software and SQA practitioners often end up with highly technical measures without a lot of added value that can be communicated directly to management or their clients.

Professor April [APR 00] describes the experience Cable & Wireless Inc. had using three commercial programs used to measure the internal quality of software (Insure++, Logiscope, and Codecheck) and how certain characteristics of maintainability can be quantified from source code. In this way, a well-designed software can be built with independent, modular, and concealed components with clear boundaries and standardized interfaces.

By using the principle of concealing information when designing the software, greater benefits can be obtained when changes are needed during tests or after release. This technique has the added advantage of protecting the other parts of the application when changes are needed. Any "side effects" can thus be reduced as a result of the change [BOO 94].

Routine coupling measures, for their part, help determine whether the software components are independent or not. It is specifically this code measure that helps to identify whether there will be "side effects" upon modifying the source code. A measure of the internal documentation of a given software may also be automatically carried out using these and similar tools. The internal documentation of software helps the maintenance programmer to understand, in more detail, the meaning of the variables and the logic behind a group of instructions. Some measures point to

---

[1] A program is associated with a graph having an input and an output, with each vertex corresponding to a set of sequential instructions.

the necessity of using a simple programming style. Certain measures will evaluate whether the programmers upheld programming standards over time.

A substantial part of maintenance costs goes toward functional adaptations, which become necessary given the changes required by users. "According to our observations and data, we also see that well-structured software makes it easier to make changes than with poorly designed software" [FOR 92].

We know that structured programming techniques are based on breaking down a complex problem into simpler parts and force each component to have a single input and a single output. The most common source code measures evaluate complexity and size, and help programmers form an opinion on the number of sets of test cases required as well as the complexity of the decisions in the source code.

"The ideal plan for unit tests is one that thoroughly executes all control flow paths of a program. In reality, it is not practical or often nearly impossible because the number of possible pathways is infinitely huge."

Humphrey (1989) [HUM 89]

Note that certain non-functional requirements can have a negative interaction with each other. For example, for usability and efficiency characteristics, the additional code and the time required to execute this code to increase usability will take up more storage space and entail greater processing times, potentially negatively affecting the efficiency of the code.

Table 3.2 illustrates, for other quality characteristics, the positive (+), negative (−), or neutral (0) interactions. It is important to explain to users that they have choices to make, and that each choice will have implications.

In conclusion, all software quality models have a similar structure with the same goals. The presence of more or fewer factors, however, is not indicative of a good or bad model. The value of a model of software quality is revealed in its practical use. It is important to completely understand how it works as well as the interaction between factors. Today, the model to use is the ISO model, since it represents an international consensus.

What is important for software developers and SQA is using a standardized model on which they can rely. It is important to use the definitions of the model proposed by ISO 25010 with suppliers. In this way, it is not the personal proposal of a specialist, but the use of a model and definitions that are published internationally and are indisputable.

Table 3.2 Examples of Interactions Between Quality Attributes [EGY 04]

Requirement attribute	Effect							
	Functionality	Efficiency	Usability	Reliability	Security	Recoverability	Accuracy	Maintainability
Functionality	+	−	+	−	−	0	0	−
Efficiency	0	+/−	+	−	−	0	−	−
Usability	+	+/−	+	+	0	+	+	0
Reliability	0	0	+	+	0	0	0	0
Security	0	−	−	+	+	0	0	0
Recoverability	0	−	+	+	0	+	0	0
Accuracy	0	−	+	0	0	0	+	0
Maintainability	0	0	0	+	+	0	0	+

3.3 DEFINITION OF SOFTWARE QUALITY REQUIREMENTS

We have provided an overview of the use of software quality models. In this section, we look at the process of defining quality requirements for software (i.e., a process that supports the use of a software quality model). However, before delving into quality requirements, we will first discuss all requirements expressed by stakeholders during software development projects.

In engineering, especially in public and private RFPs, requirements are the expression of a documented need of what a product or service should be or should provide. They are most often used formally (specified formally), especially in systems engineering and critical system software engineering.

A vision document or an operational concept document, or a specifications document (requirements definition) will be prepared in order to define the high-level problem areas and solutions for the software system to be created. Typically, this document will describe the context of the application with elements such as: the description and business objectives of the stakeholders and key users; the target market and the possible alternatives; the assumptions, dependencies, and constraints; the inventory of the product features to be developed as well as their priority; and requirements for infrastructure and documentation. This document should be clear and concise. It serves as a guide and a reference throughout the software requirements analysis.

Concept of Operations (ConOps) Document

A user-oriented document that describes a system's operational characteristics from the end-user's viewpoint.

IEEE 1362 [IEE 07]

In the classical engineering approach, requirements are considered to be prerequisites to the design and development stages of a product. The requirements development phase may have been preceded by a feasibility study, or a design analysis phase for the project.

Once the stakeholders have been identified, activities for software specifications can be broken down into:

- gather: collect all wishes, expectations, and needs of the stakeholders;
- prioritize: debate the relative importance of requirements based on, for example, two priorities (essential, desirable);
- analyze: check for consistency and completeness of requirements;

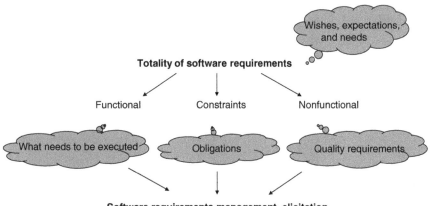

Figure 3.5 Context of software requirements elicitation.

– describe: write the requirements in a way that can be easily understood by users and developers;

– specify: transform the business requirements into software specifications (data sources, values and timing, business rules).

Requirements for software can be summarized as presented in Figure 3.5. Requirements are generally grouped into three categories:

1) Functional Requirements: These describe the characteristics of a system or processes that the system must execute. This category includes business requirements and functional requirements for the user.

2) Non-Functional (Quality) Requirements: These describe the properties that the system must have, for example, requirements translated into quality characteristics and sub-characteristics such as security, confidentiality, integrity, availability, performance, and accessibility.

3) Constraints: Limitations in development such as infrastructure on which the system must run or the programming language that must be used to implement the system.

Requirements are notoriously difficult to present to an ideal degree. Often, specialized business analysts are used to bridge the gap between software users and software specialists. These analysts have the experience, training and certification required to run meetings with users, express requirements in such a way that they can be easily conveyed as software specifications and also be understood by the end users.

For a few years now, an interest group, the International Institute of Business Analysts, offers certification for business analysts.

As part of its request for proposals for a system to sell and reload transit passes, a transportation company in a large city listed the non-functional requirements to be respected:

– maximum processing time of each machine information exchange must not exceed one second;
– time for reading or coding the pass must not exceed 250 milliseconds;
– success rate for reading and coding the passes must be 99.99%.
– ability to work without being connected to the central system for up to 4 days;
– reduce equipment down time to a minimum, option of working in failsafe mode.

www.stm.info

BABOK (Business Analyst Body Of Knowledge), available at www.theiiba.org, describes all the knowledge required to clearly define and manage business requirements during a software project.

BABOK [BAB 15]

Requirement elicitation techniques often take into account the needs, desires, and expectations of all parties involved that must be prioritized. All stakeholders must be identified before this exercise can begin. At the very beginning of the process, an initial analysis activity aims to identify the users' business requirements. Software requirements are generally documented using text, diagrams, and vocabulary that will be easily understood by users and clients. Business requirements describe certain events that take place during the carrying out of a business process and seek to identify business rules and activities that could be taken into account by the software.

Business requirements are then expressed in terms of functional requirements. A functional requirement describes the function that will be set up to meet a business requirement. They must be clearly expressed and be consistent. No requirement should be redundant or conflict with other requirements. Each requirement must be uniquely identified and be easily understood by the client. They should also be documented. The requirements that will not be taken into account shall be clearly identified as being excluded along with the reason for the exclusion. It is clear that the management of requirements and functional specifications, and of their quality, is an important factor in customer satisfaction.

 Agile Specifications and Requirements

Conventional methods of requirements elicitation may generate a large amount of documentation. Instead of generating written documents, agile specifications and requirements use prototypes, rapid iterations, images, and other multimedia elements to verify that functional requirements have been met. Agile methodologies are used in certain business sectors and are becoming increasingly popular.

Can the quality of a requirement be defined? Yes, good software requirements will have the following characteristics:

– Necessary: They must be based on necessary elements, that is, important elements in the system that other system components cannot provide.

– Unambiguous: They must be clear enough to be interpreted in only one way.

– Concise: They must be stated in a language that is precise, brief, and easy to read, which communicates the essence of what is required.

– Coherent: They must not contradict the requirements described upstream or downstream. Moreover, they must use consistent terminology throughout all requirements statements.

– Complete: They must all be stated fully in one location and in a manner that does not oblige the reader to refer to other texts to understand what the requirement means.

– Accessible: They must be realistic regarding their implementation in terms of available finances, available resources, and within the time available.

– Verifiable: They must allow for the determination of whether they are met or not based on four possible methods—inspection, analysis, demonstration, or tests.

The software developer should take a course in software requirements. One chapter of the SWEBOK® Guide is dedicated to this topic. The software engineer, for his part, has specific standards, such as ISO 29148 [ISO 11f] or IEEE 830 [IEE 98a] to which he may refer. These standards describe the recommended activities for listing all requirements in sufficient detail to allow for the designing of the product and implementing of its quality assurance (including tests). These standards present the activities that must be carried out by the software engineer: the description of each stimulus (input), of each response (output) and of all processing (functions) of the software.

ISO/IEC/IEEE 29148:2011—Systems and Software Engineering—Life Cycle Processes—Requirements Engineering

ISO/IEC/IEEE 29148:2011 contains provisions for the processes and products related to the engineering of requirements for systems and software products and services throughout the life cycle. It defines the construct of a good requirement, provides attributes and characteristics of requirements, and discusses the iterative and recursive application of requirements processes throughout the life cycle. ISO/IEC/IEEE 29148:2011 provides additional guidance in the application of requirements engineering and management processes for requirements-related activities in ISO/IEC 12207:2008 and ISO/IEC 15288:2008. Information items applicable to the engineering of requirements and their content are defined. The content of ISO/IEC/IEEE 29148:2011 can be added to the existing set of requirements-related life cycle processes defined by ISO/IEC 12207:2008 or ISO/IEC 15288:2008, or can be used independently.

ISO 29148 [ISO 11f]

All requirements must be identifiable and traceable. A different format is proposed for writing documentation based on software criticality. Therefore, the software engineer must have a thorough grasp of business analysis and make sure to carry out the following activities regarding software requirements:

- plan and manage the requirements phase;
- elicit the requirements;
- analyze the requirements and related documentation;
- communicate and ensure approval;
- evaluate the solution and validate requirements.

Many specialized studies exist in the field of software requirements. This book does not attempt to reiterate this knowledge, but rather to provide an overview. The next section shows how the software engineer must proceed in order to identify quality requirements.

3.3.1 Specifying Quality Requirements: The Process

Quality is often specified or described informally in RFPs, a requirements document, or in a systems requirement document. The software designer must interpret each functional and non-functional requirement to prepare quality requirements from these documents. To do this, he must follow a process. The quality requirements specifications process will allow for:

– correctly describing quality requirements;
– verifying whether the practices in place will allow for the development of software that will meet the client's needs and expectations;
– verifying or assessing that the software developed meets the quality requirements.

To identify quality requirements, the software engineer will have to carry out the steps described in Figure 3.6. These steps can be done at the same time as defining the functional requirements.

A non-functional requirement is something required but that does not add business rules in a business process. Some examples of a non-functional requirement include the number of users and transactions supported, transaction response time, disaster recovery time, and security. Therefore, all ISO 25010 quality characteristics and sub-characteristics fall under non-functional or performance requirements.

We will focus here on describing the quality aspects expected by clients that would not necessarily be discussed by business analysts during business requirements discussions.

The importance of non-functional requirements must not be underestimated. When developing the requirements of a system, greater importance is often given to business and functional requirements and describing the processing that the system should provide so that users can complete their tasks. For example, an unstable system or a system whose interface is greatly lacking but that fully meets functional requirements cannot be deemed a success.

"In real systems, meeting non-functional requirements is often more important than meeting functional requirements when defining the success or failure of a system."

Dr. Robert Charette

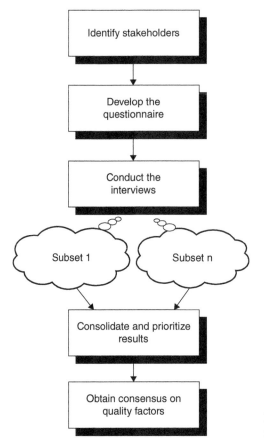

Figure 3.6   Steps suggested for defining non-functional requirements.

It is the responsibility of the software engineer to look at each quality character-istic in the ISO 25010 model and discuss whether it should be taken into account in the project. Quality requirements should also be verifiable and stated clearly in the same way as functional requirements. If the quality requirements are not verifiable, they could be interpreted, applied, and evaluated differently by different stakeholders. If the software engineer does not clarify non-functional requirements, they could be overlooked during software development.

The first activity in Figure 3.6 describes identifying stakeholders. In fact, stake-holders are any person or organization that has a legitimate interest in the quality of the software. A certain number of stakeholders will express different needs and hopes based on their own perspective. These needs and hopes may change during the system life cycle and must be checked when there is any change.

Stakeholders rarely specify their non-functional requirements since they only have a vague understanding of what quality really is in a software product. In practice, they perceive them more as general costs. It is important that you remember to include a representative from the infrastructure group, security group, and any other functional group within the company.

For subcontracted projects, quality requirements will often appear as part of the contractual agreement between the acquirer and the representative.

The acquirer may also require a specific evaluation of the product quality when a certain level of software criticality is required and human lives could be endangered.

Evaluation of the Functional Capacity of Software

Users often choose this characteristic. It is defined in the specifications as the ability of a software product to carry out all specified requirements. The *ability* sub-characteristic is chosen by the team and described as the percentage of requirements described in the specification document that must be delivered (%E). The measure established is

$$\%E = \text{(Number of functionalities requested/(Number of functionalities delivered))} \times 100.$$

It is necessary to identify target values as an objective for each measure. It is also recommended to provide an example of the calculations (i.e., measurement) to clearly illustrate the measure. For example, during the writing of the specification, the project team indicates that the %E should be 100% of the requirements described in the specifications document and that they are functional and delivered, without defects, before the final acceptance of the software for production.

Alternatively, with a lack of measurable objectives, the general policy is to accept the most stable version of the code having the necessary functionality.

ISO 25010 [ISO 11i]

The second activity in Figure 3.6 involves developing the questionnaire that presents the external quality characteristics in terms that are easy to understand for managers. This questionnaire introduces the quality characteristics and asks the respondent to choose a certain number of them and identify their importance (see Table 3.3).

The third activity consists of meeting with stakeholders to explain to them what is involved in identifying quality characteristics. The next activity consists of consolidating the different quality requirements put forth and incorporating the decisions and descriptions into a requirements document or a project quality plan. Quality characteristics may be presented in a summary table by specifying their importance as, for example, "Indispensable," "Desirable," or "Non-applicable."

Table 3.3   Example of Quality Criteria
Documentation

Quality characteristics	Importance
Reliability	Indispensable
User-friendliness	Desirable
Operational safety	Non-applicable

 Measuring Quality Equals User Satisfaction

Satisfaction indicates the extent to which users are free of discomfort and is their attitude to using the product. Satisfaction can be determined and measured using subjective indicators such as whether the product is liked, satisfaction during product use, acceptability of the workload when carrying out different tasks, or the extent to which usage objectives, such as productivity or ease of learning, were met. Other satisfaction measures could include the number of positive and negative comments recorded during use. Additional information can be obtained from long-term measures such as absenteeism rates, workload observations or reports on the software's "health" problems.

Subjective measurements of satisfaction are produced by trying to quantify reactions, attitudes, or opinions expressed by users. This quantification process may be done through different methods, for example, by asking the user to give a number corresponding to the strength of their feeling at any given time while using the product, asking users to classify products in order of preference, or by using an attitude scale in a questionnaire.

ISO 25010 [ISO 11i]

Next, for each characteristic, the quality measure must be described in detail and include the following information (see also the cases described in this chapter):

– quality characteristic;

– quality sub-characteristic;

– measure (i.e., formula);

– objectives (i.e., target);

– example.

The last step in defining quality requirements involves having these requirements authorized through consensus. After the quality requirements have been accepted, at different milestones of the project, typically when assessing a project step, these measures will be evaluated and compared with fixed and agreed upon objectives in the specifications.

Choosing Quality Measures

It is a good idea to consider measurements that are easy to set up before committing to measuring them. Take the time to check with your colleagues, the SQA group and the infrastructure group regarding whether the measures identified can be retrieved from the processes, management software and monitoring programs already installed.

ISO 25010 [ISO 11i]

Of course, in order to do this, the measurement must be implemented. Challenges to implementing a measurement program will be described in a later chapter.

3.4 REQUIREMENT TRACEABILITY DURING THE SOFTWARE LIFE CYCLE

Throughout the life cycle, client needs are documented and developed in different documents, such as specifications, architecture, code, and user manuals. Moreover, throughout the life cycle of a system, many changes regarding client needs should be expected. Every time a need changes, it must be ensured that all documents are updated. Traceability is a technique that helps us follow the development of needs as well as their changes.

Traceability is the association between two or more logical entities, such as requirements and the elements of a system. Bidirectional traceability allows us to follow how the requirements have changed and been refined over the life cycle. Traceability is the common thread that links the elements together: when an element is linked upstream to one element and this element is itself linked downstream to another, a chain of cause and effect is thereby formed. In the chapter on audits and validation, we will look at this topic in more detail.

3.5 SOFTWARE QUALITY REQUIREMENTS AND THE SOFTWARE QUALITY PLAN

The IEEE 730 standard specifies, in section 5.4 entitled *"Assessing the Product for its Conformity with the Requirements Established"* that, in the context of an acquisition of software, the function of SQA should be to ensure that the software products comply with established requirements. It is therefore necessary to produce, collect, and validate evidence to prove that the software product meets the required functional and non-functional requirements.

Moreover, section 5.4.6 of IEEE 730 is entitled *"Measuring Products"* and specifies that the selected measures of software quality and documentation accurately represent the quality of software products. The following tasks are required by the IEEE 730 [IEE 14]:

- Identify the necessary standards and procedures;
- Describe how the measures and attributes chosen adequately represent the quality of the product;
- Use these measures and identify gaps between objectives and results;
- Ensure the quality of the product measurement procedures and efficiency throughout the project.

The IEEE 730 recommends that the SQA plan define the concept of product quality for continuous improvement. This document should address the fundamental issues of functionality, external interfaces, performance, quality characteristics, and computational constraints imposed by a specific implementation. Each requirement must be identified and defined so that its implementation can be validated and verified objectively.

3.6 SUCCESS FACTORS

Factors that Foster Software Quality

1) Good understanding of non-functional quality.
2) Good process for defining, following up, and communicating quality requirements.
3) Evaluating quality throughout the life cycle of the software.
4) Establishing software criticality before starting a project.
5) Using the benefits of software traceability.

Factors that may Adversely Affect Software Quality

1) Not taking quality requirements into account.
2) Not taking software criticality into account.
3) Making excuses to not be concerned with quality (see below).

Following is a list of excuses that you may have heard at times from people who do not believe in the importance of quality [SPM 10]:

- "I do not have to be concerned with quality. My client is only interested in costs and deadlines. Quality has nothing to do with these things."
- "My project did not specify quality objectives."
- "Quality cannot be measured. We never know the number of bugs that we did not find."

- "Relax... it is not software that controls a nuclear plant or a rocket."
- "A client might require greater productivity or quality, but not both at the same time."
- "To us, quality is important. We do process audits constantly."
- "We make quality software since we follow the ISO international standards."
- "That's not an error... it's a bug"
- "We must deliver a very high quality software product, so we set aside lots of time to test it."
- "This project is of high quality because the QA guys look at our documents."
- "This software is of high quality because we reused 90% of it from another project"
- "What is quality... how do you measure quality?"
- "Complexity of code has nothing to do with quality."
- "Quality is the number of bugs in the software shipped to the client."
- "We must not do more than is asked for in the contract. The contract said nothing about quality."
- "If this does not work now, we will repair it later at the client's."
- "The quality of this software is very high. We found 1,000 bugs during testing."
- "We have no more time to test... we have to deliver."
- "We will let the client find the bugs"
- "We have a very tight schedule. We do not have time to do inspections."

3.7 FURTHER READING

GALIN D. *Software Quality: Concepts and Practice.* Wiley-IEEE Computer Society Press, Hoboken, New Jersey, 2017, 726 p.

3.8 EXERCISES

3.1 Describe the definition of a quality model for the software field, as well as what it must allow the user to easily do.

3.2 Consider the McCall quality model:
   a) What do you think about removing the *Testability* attribute from the model?
   b) Should it be included in *Maintainability*?

3.3 It has been determined that the *expandability* and *survivability* attributes are equivalent to *flexibility* and *reliability*.
   a) Is there added value to keeping both?
   b) Can they be integrated given the equivalence?

3.4   Software quality—Compromises:

    a) In the following space, explain that the optimization of one quality factor can be done to the detriment of another quality factor. Explain your reasoning for the following factors:

	Quality factor	**Quality factor**	**Reasoning**
1	Maintainability	Efficiency (of execution)	
2	Reusability	Integrity	

    b) It is necessary to establish the complementary and opposing links between quality factors. In the following table, for each possible correspondence, specify whether it is complementary (C) or opposing (X), and provide an example to support your choice.

	Integrity	Reliability	Performance	Testability	Security	Maintainability
Integrity						
Reliability						
Performance						
Testability						
Security						
Maintainability						

    c) Establish the links between software classes and quality factors. Use the following table. For each software class, give a specific example. For example, for sensitive data, you could use a bank system database. Then for each link between software class and quality factor, assign an importance criterion (indispensable (I) or desirable (D)); provide an argument in favor of this association in the appropriate box. Again, using the example of the bank database, data integrity is an indispensable quality factor, since data must always be correct, regardless of the operation performed or external actions, such as a server crash. Note: All classes do not necessarily have a relationship with all quality factors.

	Integrity	Reliability	Ergonomics	Testability	Security	Maintainability
Human life in danger						
e.g.:						
Long life						
e.g.:						
Experimental system						
e.g.:						
Real-time application						
e.g.:						
Embedded application						
e.g.:						
Sensitive data						
e.g.:						
Embedded systems						
e.g.:						
Belonging to a range						
e.g.:						

3.5   The request for a proposal to develop a laboratory management software for a medical laboratory network includes non-functional requirements for quality factors specifications. In the following table, you will find articles taken from the requirements document. For each section, fill in the name of the element that best corresponds to the requirement (choose only one factor for the requirements section). Use the following definitions of quality factors to fill in the table.

- Reliability: Ability of a program to carry out all functions specified in a reference document, in an operating environment, without failure for a given amount of time.
- Security: Quality attribute for software characterized by the absence of events that endanger the integrity of property or human lives during its operation.
- Integrity: Level of protection of the system and data that it handles against unauthorized or malicious access.
- Ergonomics: Ability of the system to be used with a minimum level of effort.
- Efficacy: Ability of software to optimally use the available physical resources (memory space, central unit time).
- Testability: Ability of software to be verified appropriately, while the system is in operation.
- Maintainability: Ability of software to facilitate operations required to locate and correct an error when the system is in operation.
- Flexibility: Ability of software to adapt to a change in specifications.
- Reusability: Ability of a software component to be reused in different applications.
- Portability: Ability of software to adapt to an environment different from that of the previous applications.
- Compatibility: Quality of several software applications (or components) related to a given function, based on certain criteria, for example, standardization of data structures, internal communications.

No.	Description of the quality requirement	Quality factor
1.	The probability that the "Super-lab" system software will fail during normal operating hours (9 A.M. to 4 P.M.) must be less than 0.5%.	
2.	The "Super-lab" system must transfer the laboratory analysis results to the patient file software.	
3.	The "Super-lab" system includes a function that prepares a detailed report of the test results throughout a hospital stay (a copy of this report will also be used for the family doctor). The execution time (preparation and printing) must be less than 60 seconds; the level of precision and completeness of the report must be 99% or greater.	
4.	The "Super-lab" software functionality is targeted for public hospital laboratories. It must allow for easy adaptation to the private laboratory market.	

No.	Description of the quality requirement	Quality factor
5.	The training of a technician to use the software must be done in less than three days in order for the technician to reach the "C" skill level for using the software. This skill level must allow him to process more than 20 patient files in an hour.	
6.	The "Super-lab" stores access data. As well, it must report unsuccessful access attempts by unauthorized personnel. This report must include the following information: the identification of the access terminal network, the system code used, the date and time of the access attempt.	
7.	The "Super-lab" system includes a patient billing function for laboratory tests. This sub-system will be reused in the physiotherapy center software.	
8.	The "Super-lab" system will process all monthly reports for each hospital department, as listed in Appendix B of the contract.	
9.	The system must serve 12 workstations and 8 pieces of automatic test equipment from an AS250 centralized server, and a CS25 communications server, which will support 25 lines of communication. The system (software + hardware) must meet or exceed the availability specifications described in Appendix C.	
10.	The "Super-lab" software developed on Linux must also be able to work on Windows.	

3.6   Calculate the MTBF for software that must be available during work hours, that is, 37.5 hours a week, and cannot fail more than 15 minutes per week.

3.7   Describe the steps proposed by the IEEE 1061 standard to determine a good definition of non-functional requirements.

3.8   Explain the two different perspectives, the internal and external point of view, for software maintainability, and give an example.

3.9   Describe the three categories used to classify requirements.

3.10   What is the difference between a business requirement and a functional requirement?

3.11   Describe the steps recommended for defining non-functional requirements.

3.12   Explain what is meant by bidirectional traceability of requirements.

Chapter 4

Software Engineering Standards and Models

After completing this chapter, you will be able to:

- understand the evolution and the importance of software engineering standards for the SQA specialist;

- understand the standards for software quality management systems: ISO 9001, ISO/IEC 90003, ISO/IEC 20000, and the TickIT certification process [TIK 07];

- understand the software engineering standards: ISO/IEC/IEEE 12207 and IEEE 730 that govern the content of the SQA plan;

- include other improvement models, norms, standards, and quality processes: CMMI® maturity models of software processes and S^{3m}, ITIL, the CobiT IT governance approach, the ISO/IEC 27000 information security standard, and the ISO/IEC 29110 standards for very small organizations;

- understand that there are also repositories and standards specific to certain application domains: DO-178 and ED-12 for aeronautics, EN 50128 for railways and ISO 13485 for the field of medical devices that contain software;

- understand the importance of standards in terms of SQA.

4.1 INTRODUCTION

Other engineering domains such as mechanical, chemical, electrical, or physics engineering are based on the laws of nature as discovered by scientists. Figure 4.1 illustrates some of the many laws of nature.

Unfortunately, software engineering, unlike other engineering disciplines, is not based on the laws of nature. This explains, in part, some of the setbacks discussed in Chapter 1. The number of defective software, accidents and deaths, projects over

Software Quality Assurance, First Edition. Claude Y. Laporte and Alain April.
© 2018 the IEEE Computer Society, Inc. Published 2018 by John Wiley & Sons, Inc.

Hooke's Law

$$\sigma = E \cdot \varepsilon$$

Newton's Law

$$x(t) = \frac{1}{2}a \cdot t^2 + v_0 \cdot t + x_0$$

Boyle–Mariotte's Law

$$p_1 x V_1 = p_2 x V_2$$

Gravitational Law

$$\vec{F}_{A \to B} = -G\frac{M_A M_B}{AB^2}\vec{u}_{AB}$$

Ohm's Law

$$V = RI$$

Curie's Law

$$E = -\vec{\mu} \cdot \vec{B}$$

Refraction Law

$$\eta_1 \cdot \sin(\theta_1) = \eta_2 \cdot \sin(\theta_2)$$

Coulomb's law

$$F_{12} = \frac{q_1 q_2}{4\pi\epsilon_0}\frac{r_2 - r_1}{|r_2 - r_1|^3}$$

Figure 4.1　A few laws of nature used by some engineering disciplines.

budget and that are delivered past deadlines, and frustrated users are some examples of these setbacks.

The development of software is based only on the laws of logic and mathematics. Software engineering, like other disciplines, is based on the use of well-defined practices for ensuring the quality of its products. In software engineering, there are several standards, which are actually guides for management practices. A rigorous process is the framework for the way standards are developed and approved including, among others, international ISO standards and standards from professional organizations such as IEEE.

The four principles for the development of ISO standards are:

– ISO standards meet a market need.

– ISO standards are based on worldwide expertise.

– ISO standards are the result of a multi-stakeholder process.

– ISO standards are based on consensus.

The ISO standards are developed by consensus, as defined below. Consensus is required to produce a standard that will be accepted by the community of interest.

Consensus
General agreement characterized by the absence of sustained opposition to substantial issues by any important part of the concerned interests and by a process that involves

seeking to take into account the views of all parties concerned and to reconcile any conflicting arguments.

Note: Consensus need not imply unanimity.

ISO Directive, Part 1

Consensus means (adapted from Coallier (2003) [COA 03]):

– That all parties were able to express their views;
– The best effort has been made to take into account all opinions and solve all problems (i.e., all the submissions in a vote of the draft of a standard).

Standard

A set of mandatory requirements established by consensus and maintained by a recognized body to prescribe a disciplined and uniform approach, or to specify a product, with respect to mandatory conventions and practices.

ISO 24765 [ISO 17a]

The first condition to bear in mind is that standards, which differ from other written guidelines, are quasi-legal documents. Standards can be used either to prove or refute elements in a court of law. Standards often become legal requirements when they are adopted by governments and regulatory agencies. When this occurs, the content of a standard is important, since organizations then use it to develop products and services that can greatly impact human life, the environment, or business.

"Organizations formalize behavior to reduce its variability, and, ultimately, to predict it and control it."

Mintzberg (1992) [MIN 92]

The right-hand side of Figure 4.2 shows the development of standards as of the 1970s when the American Department of Defense (DoD) created the "DoD-STD-1679A" military standard [DOD 83]. At that time, a contract was given to a supplier

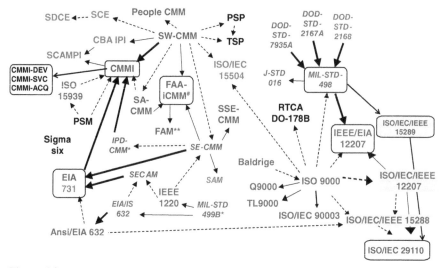

Figure 4.2 The development of standards and models.
Source: Adapted from Sheard (2001) [SHE 01] with the permission of Systems and Software Consortium, Inc.

to develop a piece of software, and it took months, if not a year before receiving the final product. Since the client did not see any development with this process, and, in the end, received boxes containing documents and magnetic tape, this approach was called the "*Big Bang*." The DoD then decided that the entire software development process had to become more transparent to allow documents produced throughout the development cycle to be evaluated. The military standard required that the supplier write and have a number of documents approved. These approvals allowed clients to review, comment, and approve documents instead of waiting until the end and receiving software that did not meet their needs. Review and approval activities for documents were related to certain project management activities: the approval of these documents led to the payment of a large sum of money as agreed to in the contract with the supplier. This enabled the client to remotely control the development of the requested software. Over the years, other organizations such as the IEEE, the International Organization for Standardization (ISO), and the European Space Agency (ESA) developed standards. During the late 1980s, the American DoD decided to use commercial standards instead of military standards, such as ISO/IEC/IEEE 12207 [ISO 17], to develop its software. Military software engineering standards were then removed.

On the left-hand side of the figure, we note the "*Capability Maturity Model*" (CMM®). It was developed at the request of the American DoD, by the Software Engineering Institute (SEI) in order to provide a road map of engineering practices to improve the performance of the development, maintenance and service provisioning processes. This model is described in a later section.

The Continuous Evolution of Standards

The standards presented in this book are constantly evolving. They are reviewed periodically and, if necessary, updated. Some standards are updated about every 5 years. Others are updated if major changes are needed. In an organization, it is possible that different versions of the same standard are used. For example, when an agreement or a contract is signed, we normally quote the latest version of the standard in the contract. In some organizations, such as those working in military defense and aerospace, a development or maintenance project can span over many years if not several decades. In such cases, it is possible that the customer prefers to continue the project with the version of the standard used at the beginning of the project, while for the same organization, a new project will use the latest version of the same standard. Developers and SQA must then carry out their responsibilities with the two versions of the same standard and use templates and checklists specific to each version.

Figure 4.3 illustrates the evolution of standards that are maintained and published under the responsibility of the appointed subcommittee for standardized processes, tools, and supporting technologies for software engineering and systems: ISO/IEC JTC1 SC7.

The ISO Name

Because the name "International Organization for Standardization" would have resulted in different abbreviations in different languages (e.g. "IOS" in English and "OIN" in French), the founders opted for a short name "ISO". This name is derived from the Greek "ISOS", meaning "equal". Whatever the country, whatever the language, the short form of the name of the organization is therefore always ISO.

ISO website (https://www.iso.org/about-us.html)

In the 1980s, there were only five standards; in 2016, more than 160 standards make up the portfolio of sub-committee 7 (SC7). This rapid increase is due, among others, to the fact that more software engineering practices have matured and acquired a broad consensus since the late 1980s.

More than 39 countries actively participate in the development of SC7 standards and 20 countries participate as observers. For countries that actively participate and have the right to vote, the meaning of the Greek word "ISOS" implies that the vote of any ISO member country is equal to the vote of any other country, no matter its size, economic, or political influence.

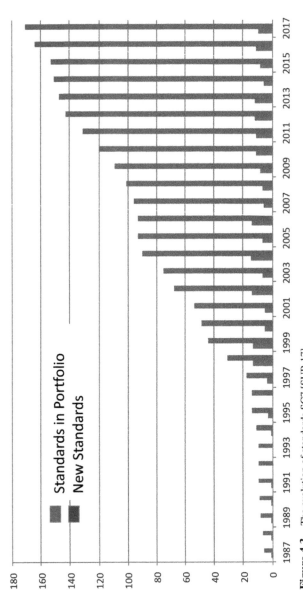

Figure 4.3 The evolution of standards SC7 [SUR 17].

The majority of software engineering standards of SC7 describe proven practices such as configuration management and quality assurance (QA) practices, while a small number of standards, such as ISO 25000 presented in the previous chapter, describe product requirements.

The ISO/IEC/IEEE 24765 [ISO 17a] glossary will be used as the reference for most of the definitions in this book. When a term is not defined in the ISO 24765, the definition of another standard will be used, such as ISO 9001, the IEEE standard or CMMI® for Development [SEI 10a].

ISO/IEC/IEEE 24765—Systems and Software Engineering Vocabulary [ISO 17a].

An online glossary can be consulted and downloaded at the following address: www.computer.org/sevocab

In recent years, some software engineering standards in the portfolio of SC7 saw their scope expanded as the practices described can be applied to a wider area than software engineering alone. For example, the scope of software engineering standards of verification and validation, risk management, and configuration management has been extended to cover the field of systems engineering that develop products which often include hardware (e.g., electronic, mechanical, and optical) and software. Thus, a greater number of engineers and developers use the same standards. This facilitates communication between different domains.

System

Combination of interacting elements organized to achieve one or more stated purposes.

Note 1 to entry: A system is sometimes considered as a product or as the services it provides.

Note 2 to entry: In practice, the interpretation of its meaning is frequently clarified by the use of an associative noun, e.g. aircraft system. Alternatively, the word "system" is substituted simply by a context-dependent synonym, e.g. aircraft, though this potentially obscures a system principles perspective.

Note 3 to entry: A complete system includes all of the associated equipment, facilities, material, computer programs, firmware, technical documentation, services and personnel required for operations and support to the degree necessary for self-sufficient use in its intended environment.

ISO 15288 [ISO 15c]

4.2 STANDARDS, COST OF QUALITY, AND BUSINESS MODELS

In a previous chapter, we presented the notions of cost of quality and business models. Regarding cost of quality, standards are an element of prevention costs: in other words, the costs incurred by an organization to prevent errors from happening during different stages in the development or maintenance process. Table 4.1 lists the different prevention cost elements. Purchasing, training, and standards implementation are also prevention costs.

We will briefly review the main business models in the software industry, namely (adapted from Iberle (2002) [IBE 02]):

- Custom systems written on contract: The organization makes profits by selling tailored software development services for clients.
- Custom software written in-house: The organization develops software to improve organizational efficiency.
- Commercial software: The company makes profits by developing and selling software to other organizations.
- Mass-market software: The company makes profits by developing and selling software to consumers.
- Commercial and mass-market firmware: The company makes profits by selling software in embedded hardware and systems.

The standards are commonly used in the following business models: custom systems written on contract, mass-market software and commercial and mass-market firmware. In these business models, standards are used to optimally manage development and minimize errors and risks. As for the "Custom systems written on contract" business model, it is the client who will decide whether or not to impose standards.

Table 4.1 Prevention Costs (Adapted from Krasner (1998) [KRA 98])

Major category	Subcategories	Definition	Typical cost
Prevention costs	Quality basis definition.	Efforts to define quality, set quality goals, standards, and thresholds.Quality trade-off analysis.	Defining release criteria for acceptance testing and related quality standards.
	Project and process-oriented interventions.	Efforts to prevent poor product quality or improve process quality.	Training, process improvements, metrics collection, and analysis.

In this chapter, we present a brief overview of some standards: process standards, product standards, and quality systems. We also present the CMMI model because its widespread use has made it a de facto standard.

4.3 MAIN STANDARDS FOR QUALITY MANAGEMENT

This section describes the main standards related to the management of software quality: ISO 9000 [ISO 15b] and ISO 9001 [ISO 15] and the application guide for software, the ISO/IEC 90003 standard. We also present a brief overview of the quality standard for the medical domain.

4.3.1 ISO 9000 Family

As described on the ISO website, "The ISO 9000 family addresses various aspects of quality management and contains some of ISO's best known standards. The standards provide guidance and tools for companies and organizations who want to ensure that their products and services consistently meet customer's requirements, and that quality is consistently improved." The ISO 9000 family includes the following four standards.

Standards in the ISO 9000 family include:

- ISO 9001:2015 - sets out the requirements of a quality management system
- ISO 9000:2015 - covers the basic concepts and language
- ISO 9004:2009 - focuses on how to make a quality management system more efficient and effective
- ISO 19011:2011 - sets out guidance on internal and external audits of quality management systems.

  Principles of quality management, http://www.iso.org/iso/qmp_2012.pdf

The ISO 9001 standard provides the basic concepts, principles and vocabulary of quality management systems (QMS) and is the basis for other standards for QMSs [ISO 15]. The "Quality Management Principles" (QMP) are a set of values, rules, standards, and fundamental convictions regarded as fair and that could be the basis for quality management. ISO 9001 proposes the following for each QMP:

– A statement that describes the principle.

– A foundation that explains why this principle is important for the organization.

- The main benefits associated with this principle.
- Possible actions to improve the performance of the organization by applying this principle.

The seven QMP of the ISO 9001, presented in order of priority, are [ISO 15]:

- Principle 1: Customer focus
  - ○ Organizations depend on their customers and therefore should understand current and future customer needs, should meet customer requirements and strive to exceed customer expectations.
- Principle 2: Leadership
  - ○ Leaders establish unity of purpose and direction of the organization. They should create and maintain the internal environment in which people can become fully involved in achieving the organization's objectives.
- Principle 3: Involvement of people
  - ○ People at all levels are the essence of an organization and their full involvement enables their abilities to be used for the organization's benefit.
- Principle 4: Process approach
  - ○ A desired result is achieved more efficiently when activities and related resources are managed as a process.
- Principle 5: System approach to management
  - ○ Identifying, understanding, and managing interrelated processes as a system contributes to the organization's effectiveness and efficiency in achieving its objectives.
- Principle 6: Factual approach to decision making
  - ○ Effective decisions are based on the analysis of data and information.
- Principle 7: Mutually beneficial supplier relationships
  - ○ An organization and its suppliers are interdependent and a mutually beneficial relationship enhances the ability of both to create value.

As an example, the first principle of customer focus is described in detail [ISO 15]:

- Statement
  - ○ The main objective of quality management is to satisfy customer requirements and strive to exceed their expectations.
- Basis
  - ○ Sustainable performance is achieved when an organization obtains and retains the confidence of customers and of other interested parties. Every aspect of the interaction with customers provides an opportunity to create more value for the customer. Understanding the current and future needs of customers and other stakeholders contributes to the sustainable performance of the organization.

– Benefits
  o Increased customer value
  o Increased customer satisfaction
  o Improved customer loyalty
  o Improved recurring business activity
  o Improved corporate image
  o Expanded customer base
  o Increased sales and market share

– Possible actions
  o Identify direct and indirect customers for which the organization creates value.
  o Understand the needs and expectations of current and future customers.
  o Link the objectives of the organization to the needs and expectations of its clients.
  o Communicate the needs and expectations of clients at all levels of the organization.
  o Plan, design, develop, produce, deliver, and support products and services to meet the needs and expectations of customers.
  o Measure and monitor customer satisfaction and take appropriate action.
  o Identify the needs and expectations of interested parties that may affect customer satisfaction and take appropriate action.
  o Actively manage relationships with customers to achieve sustainable performance.

The ISO 9001 standard [ISO 15] applies to all organizations regardless of size, complexity, or business model. ISO 9000 specifies requirements for a QMS, as defined in the following text box, for both internal groups and for external partners.

Management System

Set of interrelated or interacting elements of an organization to establish policies and objectives, and processes to achieve those objectives.

Note 1 to entry: A management system can address a single discipline or several disciplines, e.g. quality management, financial management or environmental management.

Note 2 to entry: The management system elements establish the organization's structure, roles and responsibilities, planning, operation, policies, practices, rules, beliefs, objectives and processes to achieve those objectives.

Note 3 to entry: The scope of a management system can include the whole of the organization, specific and identified functions of the organization, specific and identified sections of the organization, or one or more functions across a group of organizations.

Note 4 to entry: This constitutes one of the common terms and core definitions for ISO management standards.

Quality Management System

Part of a management system with regard to quality.

ISO 9001 [ISO 15]

ISO 9001 [ISO 15] is used worldwide in a wide range of organizations. About a million certificates of conformity to ISO 9001 are issued annually in nearly 187 countries.

ISO 9001 uses the process approach, the Plan-Do-Check-Act (PDCA) approach, and a risk-based thinking approach [ISO 15]:

– The process approach allows an organization to plan its processes and their interactions.

– The PDCA cycle allows an organization to ensure that its processes are adequately resourced and appropriately managed and that opportunities for improvement are identified and implemented.

– The risk-based thinking approach allows an organization to determine the factors that may cause deviation from its processes and its QMS in relation to expected results, to implement preventive measures in order to limit negative effects and exploit opportunities when they arise.

The ISO 9001 [ISO 15] standard indicates that an organization can use a complementary improvement approach to that of the continuous improvement approach such as a drastic change, an innovation or a reorganization.

Figure 4.4 shows the elements of a process and the interaction between these elements. Note that the figure shows the links between processes with "Input sources" (e.g., upstream process) and "Target outputs" (e.g., downstream processes) to illustrate that a process does not work in isolation in an organization. The organization must master not only the elements of a process, but also the interactions and interdependencies between them, in order to improve its overall performance, such as reduced rework as presented in the section on the cost of quality.

Also note that one of the output elements in Figure 4.4 is a service. ISO 9000 defines a service as follows: output of an organization with at least one activity necessarily performed between the organization and the customer. For example, in software development, the organization that developed software for its client could offer implementation services and maintenance of the software.

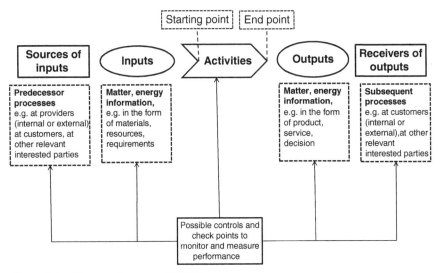

Figure 4.4   Elements of a process [ISO 15].
Source: Standards Council of Canada.

ISO 9001 describes the elements of the PDCA cycle as follows [ISO 15]:

– Plan: establish the objectives of the system, processes and resources to deliver results in accordance with customer requirements and policies of the organization, identify and address risks and opportunities;

– Do: implement what has been planned;

– Check: monitor and measure (if applicable) processes and the products and services obtained against policies, objectives, requirements and planned activities, and report the results;

– Act: take actions to improve performance, as needed.

The requirements of ISO 9001 are presented in the following 10 items:

1) Scope

2) Normative references

3) Terms and definitions

4) Context of the organization

5) Leadership

6) Planning

7) Support

8) Implementation of operational activities

9) Performance evaluation

10) Improvement

We have only briefly described the main articles of the standard. In regards to article 4, ISO 9001 [ISO 15] requires the organization to determine the pertinent external and internal issues. These issues, as for the needs and expectations of the interested parties, the QMS and its scope, have a great influence on the ability to achieve the expected results of its QMS. Article 5 of the standard explains that management must demonstrate leadership and commitment to the QMS and that the customer must ensure the establishment of the quality policy of the organization and must ensure that the responsibilities and authorities of roles have been assigned, communicated, and understood. Article 6 describes the actions to be implemented in regards to the risks and opportunities, the objectives of the QMS, the planning of actions to achieve these objectives, and the planning of changes to the QMS. Article 7 presents the requirements for resources for the establishment of the QMS, its implementation, updating, and the continuous improvement of the QMS: human resources, infrastructure, resources for monitoring, measurement, traceability, the knowledge and skills required of personnel, the needs of internal and external communication and documentation (e.g., creating, updating). Article 8 describes in detail the implementation of operational activities of the organization such as the process planning, the determination and review of the requirements of its products and services, design and development, products and services processes of its external suppliers, production and release of its product and services, and the identification of non-compliant output elements compared with requirements. Article 9 describes the requirements for performance evaluation, such as monitoring the extent of analysis and evaluation, customer satisfaction, internal audits, and management reviews of the QMS. Finally, article 10 provides the requirements for improvement such as improving customer satisfaction, non-compliance and corrective actions, and continuous improvement of the QMS.

Preconceptions
- ISO 9001 is a standard for large organizations.
- ISO 9001 standard is complicated to implement.
- ISO 9001 is expensive to implement.
- ISO 9001 is a standard that applies only to manufacturing.
- ISO 9001 is an administrative burden.

ISO 9001—Debunking the myths [ISO 15]
http://www.iso.org/iso/iso_9001_debunking_the_myths.pdf

The ISO 19011 standard [ISO 11g], which establishes guidelines for internal and external audits of QMSs, will be presented in the chapter on audits. ISO 9004 [ISO 09a], which shows how to increase the efficiency and effectiveness of a QMS, will be presented in the chapter concerning policies and processes.

4.3.2 ISO/IEC 90003 Standard

The ISO/IEC 90003 [ISO 14] standard provides guidelines for the application of the ISO 9001 standard to computer software. It provides organizations with instructions for acquiring, supplying, developing, using and maintaining software. The following text box is part of the introduction to the ISO 90003 standard.

It identifies the issues which should be addressed and is independent of the technology, life cycle models, development processes, sequence of activities and organizational structure used by an organization. The guidance and identified issues are intended to be comprehensive but not exhaustive. Where the scope of an organization's activities includes areas other than computer software development, the relationship between the computer software elements of that organization's quality management system and the remaining aspects should be clearly documented within the quality management system as a whole.

Throughout ISO 9001:2000, "shall" is used to express a provision that is binding between two or more parties, "should" to express a recommendation among possibilities and "may" to indicate a course of action permissible within the limits of ISO 9001:2000. In this International Standard (ISO/IEC 90003), "should" and "may" have the same meaning as in ISO 9001:2000, i.e. "should" to express a recommendation among possibilities and "may" to indicate a course of action permissible within the limits of this International Standard.

Organizations with quality management systems for developing, operating or maintaining software based on this International Standard may choose to use processes from ISO/IEC 12207 to support or complement the ISO 9001:2000 process model.

Where text has been quoted from ISO 9001:2000, that text is enclosed in a box, for ease of identification.

Adapted from ISO 90003 [ISO 14]

An example, Section 7.3.5 of the ISO 9001 standard is provided below as it is presented in the ISO 90003 standard.

ISO 9001:2008, Quality Management Systems—Requirements

7.3.5 Design and Development Verification
Verification shall be performed in accordance with planned arrangements (see 7.3.1) to ensure that the design and development outputs have met the design and development input requirements. Records of the results of the verification and any necessary actions shall be maintained (see 4.2.4).

The explanatory text from the ISO 90003 standard for Section 7.3.5 is shown in the following text box.

Verification of software is aimed at providing assurance that the output of a design and development activity conforms to the input requirements.

Verification should be performed as appropriate during design and development. Verification may comprise reviews of design and development output (e.g. by inspections and walk-throughs), analysis, demonstrations including prototypes, simulations or tests. Verification may be conducted on the output from other activities, e.g. COTS, purchased and customer-supplied products.

The verification results and any further actions should be recorded and checked when the actions are completed.

When the size, complexity or criticality of a software product warrants, specific assurance methods should be used for verification, such as complexity metrics, peer reviews, condition/decision coverage or formal methods.

Only verified design and development outputs should be submitted for acceptance and subsequent use. Any findings should be addressed and resolved, as appropriate. For more information, see ISO 12207.

Adapted from ISO 90003 [ISO 14]

The text of the ISO 90003 standard correctly explains what a software audit is for the organization wishing to set up a QMS as well as for the QMS auditor.

Differences between ISO 9001:2015 and CMMI® for Development Version 1.3

The scope of the ISO 9001 is broader than that of the CMMI-DEV:

– CMMI-DEV applies to development and maintenance activities.

– The ISO 9001 applies to all activities of an organization. Sector-specific applications of ISO 9001 have been developed (e.g., medical devices, petroleum, petrochemical and natural gas industries).

The level of abstraction is different:

– CMMI-DEV is about 470 pages and contains a wealth of practical examples.

– The ISO 9001 is only 29 pages, whereas all the standards of the ISO 9000 family is about 180 pages (www.iso.org).

CMMI-DEV is less subject to interpretation given that each process area in the model is discussed at length. ISO/IEC 90003, a 54-page document, provides guidance for organizations in the application of ISO 9001 to the acquisition, supply, development, operation, and maintenance of computer software.

The assessment is different:

– CMMI: An organization is evaluated by a team comprised of a lead appraiser, licensed by the CMMI Institute (www.cmmiinstitute.com), accompanied by an assessment team that typically consists of members of the organization evaluated and external assessors.

– ISO 9001: The Quality Management System (QMS) of an organization is audited by an audit team, (i.e., one or more persons supported, if needed, by technical experts), authorized by a governmental or non-governmental certification body to perform ISO 9001 audits. An accreditation body is an organization that is usually established by a national government which assesses certification bodies and certifies their technical competence to carry out the certification process.

The CMMI appraisal usually lasts longer and is more in-depth than an ISO 9001 audit:

– CMMI: The main result of the appraisal is a list of strengths and weaknesses for initiating an improvement process. The appraisal team provides, for an appraisal using the staged representation of CMMI, a maturity level rating for the organization assessed.

– ISO 9001: The result of the audit is an ISO 9001 certificate–proof that the audited organization meets the requirements of the standard. A set of audit findings (i.e., results of the collected audit evidence against audit criteria) are documented.

The next section discusses the ISO/IEC/IEEE 12207 [ISO 17] standard for software engineering, which describes all the software life cycle processes (from cradle to grave).

4.4 ISO/IEC/IEEE 12207 STANDARD

The third edition of the ISO/IEC/IEEE 12207 standard [ISO 17] establishes a common framework for software life cycle processes. It applies to the acquisition of systems and software products and services, supply, development, operation, maintenance, and disposal of software products and the development of the software part of a system, whether performed internally or externally to an organization. In this standard, the software also includes the firmware. Each process of the standard is described in a few pages, and includes the following attributes [ISO 17]:

– The title conveys the scope of the whole process.

– The purpose describes the goals of performing the process.

– The outcomes express the observable results expected from the successful performance of the process.

– The activities are sets of cohesive tasks of a process.

– The tasks are requirements, recommendations, or permissible actions intended to support the achievement of the outcomes.

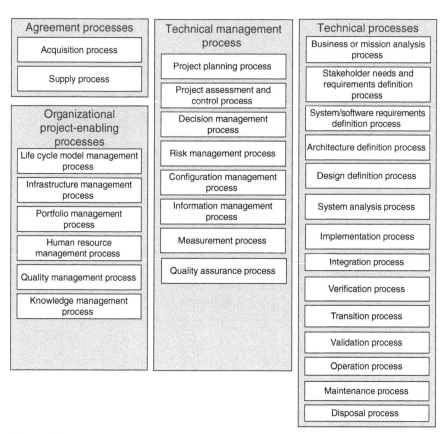

Figure 4.5 The four life cycle process groups of ISO 12207 [ISO 17].
Source: Standards Council of Canada.

ISO 12207 [ISO 17] defines four sets of processes as shown in Figure 4.5:

– Two agreement processes between a customer and a supplier;

– Six organizational project-enabling processes;

– Eight processes for technology management;

– Fourteen technical processes.

As most modern systems are now controlled by software, the ISO 12207 [ISO 17] standard has been updated to interface with the new edition of the standard in engineering systems: ISO/IEC/IEEE 15288 [ISO 15].

Since ISO 12207 [ISO 17] is an important software engineering standard, we briefly describe one process (we will not describe the details of each task): The QA process [ISO 17]:

– Purpose
  o The purpose of the QA process is to help ensure the effective application of the organization's quality management process to the project.

o QA focuses on providing confidence that quality requirements will be fulfilled. Proactive analysis of the project life cycle processes and outputs is performed to assure that the product being produced will be of the desired quality and that organization and project policies and procedures are followed.

– Outcomes
   o As a result of the successful implementation of the QA process:
      □ Project QA procedures are defined and implemented.
      □ Criteria and methods for QA evaluations are defined.
      □ Evaluations of the project's products, services, and processes are performed, consistent with quality management policies, procedures, and requirements.
      □ Results of evaluations are provided to relevant stakeholders.
      □ Incidents are resolved.
      □ Prioritized problems are treated.

– Activities and tasks
   o The project shall implement the following activities and tasks in accordance with applicable organization policies and procedures with respect to the measurement process.
      Note IEEE 730-2014 [IEE 14], software quality assurance (SQA) processes, provides additional detail.
      □ Prepare for QA. This activity consists of the following tasks:
         • Define a QA strategy.
         • Establish independence of QA from other life cycle processes.
      □ Perform product or service evaluations. This activity consists of the following tasks:
         • Evaluate products and services for conformance to established criteria, contracts, standards, and regulations.
         • Monitor that verification and validation of the outputs of the life cycle processes are performed to determine conformance to specified requirements.
      □ Perform process evaluations. This activity consists of the following tasks:
         • Evaluate project life cycle processes for conformance.
         • Evaluate tools and environments that support or automate the process for conformance.
         • Evaluate supplier processes for conformance to process requirements.
      □ Manage QA records and reports. This activity consists of the following tasks:
         • Create records and reports related to QA activities.
         • Maintain, store, and distribute records and reports.
         • Identify incidents and problems associated with product, service, and process evaluations.
      □ Treat incidents and problems. This activity consists of the following tasks:
         • Record, analyze, and classify incidents.

- Identify selected incidents to associate with known errors or problems.
- Record, analyze and classify problems.
- Identify root causes and treatment of problems where feasible.
- Prioritize treatment of problems (problem resolution) and track corrective actions.
- Analyze trends in incidents and problems.
- Identify improvements in processes and products that may prevent future incidents and problems.
- Inform designated stakeholders of the status of incidents and problems.
- Track incidents and problems to closure.

The reader may have noticed the use of the term "shall" in the sentence "The project shall implement the following activities and tasks." In ISO engineering standards terminology, the words "shall", "should" and "may" have been defined as follows: "Requirements of this document are marked by the use of the verb "shall". Recommendations are marked by the use of the verb "should". Permissions are marked by the use of the verb "may". However, despite the verb that is used, the requirements for conformance are selected as described previously".

In other words, unless you, as a supplier of a software, have been granted permission by your customer, all the activities and tasks have to be performed. The implementation of these activities and tasks may be verified through a formal audit conducted by the customer or his representative. Audits will be presented in Chapter 6 of this book.

The ISO 12207 standard can be used in one or more of the following modes [ISO 17]:

- By an organization: to help establish an environment of desired processes. These processes can be supported by an infrastructure of methods, procedures, techniques, tools, and trained personnel. The organization may then employ this environment to perform and manage its projects and progress software systems through their life cycle stages. In this mode this document is used to assess conformance of a declared, established environment to its provisions.

- By a project: to help select, structure and employ the elements of an established set of life cycle processes to provide products and services.

- By an acquirer and a supplier: to help develop an agreement concerning processes and activities.

- By organizations and assessors: to serve as a process reference model for use in the performance of process assessments that may be used to support organizational process improvement.

4.4.1 Limitations of the ISO 12207 Standard

The ISO 12207 standard describes the limitations to its use as follows (adapted from [ISO 17]):

– It does not prescribe a specific system or software life cycle model, development methodology, method, model, or technique.
  o The users of this document are responsible for selecting a life cycle model for the project and mapping the processes, activities, and tasks in this document into that model. The parties are also responsible for selecting and applying appropriate methodologies, methods, models, and techniques suitable for the project.
– It does not establish a management system or require the use of any management system standard.
  o It is intended to be compatible with the QMS specified by ISO 9001, the service management system specified by ISO/IEC 20000-1 [ISO 11h], and the information security management system (ISMS) specified by ISO/IEC 27000.
– It does not detail information items in terms of name, format, explicit content, and recording media.
  o ISO/IEC/IEEE 15289 addresses the content for life cycle process information items (documentation).

4.5 ISO/IEC/IEEE 15289 STANDARD FOR THE DESCRIPTION OF INFORMATION ELEMENTS

With the abandonment of military standards that defined the content and format of information elements, the international community developed a standard, ISO 15289 [ISO 17b], in support, among others, to ISO 12207 [ISO 17], to facilitate the description of the different types of information items to be produced.

Information Item
Separately identifiable body of information that is produced, stored, and delivered for human use.

Note 1 to entry: "Information product" is a synonym. A document produced to meet information requirements can be an information item, or part of an information item, or a combination of several information items.

Note 2 to entry: An information item can be produced in several versions during a project or system life cycle.

ISO 15289 [ISO 17b]

Table 4.2 Generic Types of Information Items Described by the ISO 15289 [ISO 17b]

Type	Purpose	Sample output
Description	Represents a plan or an existing function, design or item.	Description of the design
Policy	Establish an organization's high-level intention and approach to achieve objectives for, and ensuring effective control of, a service, process, or management system.	Quality management policy
Plan	Define when, how, and by whom specific processes or activities are to be performed.	Project plan
Procedure	Defines in detail when and how to perform certain activities or tasks including tools needed.	Problem resolution procedure
Report	Describe the results of activities such as investigations, assessments, and tests. A report communicates decisions.	Problem report Validation report
Request	Record information needed to solicit a response.	Change request
Specification	Provide requirements for a required service, product, or process.	Software specification

The clauses of ISO 12207 [ISO 17] list the artifacts to produce without defining the content. The ISO 15289 standard describes seven types of documents: request, description, plan, policy, procedure, report, and specification. These document types are described in Table 4.2.

For example, the ISO 15289 standard defines what a procedure is and what it must describe. The following text box describes this type of document [ISO 17b].

Procedure

Purpose: Define in detail when and how to perform certain processes, activities or tasks, including tools needed.

A procedure shall include the following elements:

a) Date of issue and the status;

b) Scope;

c) Issuing organization;

d) Approval authority;

e) Relationship with other plans and procedures;

f) Authoritative references;

g) Inputs and outputs;

h) Ordered description of steps to be performed by each participant;

i) Error and problem resolution;

j) Glossary;

k) Change history.

Examples of procedures:

– Audit procedure

– Configuration management procedure

– Enhancement procedure

– Documentation procedure

– Measurement procedure

ISO 15289 [ISO 17b]

The IEEE 730 [IEE 14] standard for the SQA process is presented next. This standard is based on ISO 12207 [ISO 17] and ISO 15289 [ISO 17b].

4.6 IEEE 730 STANDARD FOR SQA PROCESSES

The scope of the IEEE 730-2014 Standard for SQA Processes [IEE 14] is very different from that of previous versions. Unlike previous versions, where the QA plan was the cornerstone of IEEE 730, the new version establishes the requirements for the planning and implementation of SQA activities for a software project. QA according to IEEE is a set of proactive measures to ensure the quality of the software product. The IEEE 730 provides guidance for the SQA activities of products or of services. The QA plan is presented in a clause and an annex of the standard.

The following text box provides the definition SQA of the IEEE 730 [IEE 14].

Software Quality Assurance

A set of activities that define and assess the adequacy of software processes to provide evidence that establishes confidence that the software processes are appropriate for and produce software products of suitable quality for their intended purposes.

IEEE 730 [IEE 14]

Publishers of a standard must use official versions of standards, that is, the latest published version, when producing a new standard or the revision of a published standard. The IEEE 730-2014 [IEE 14] standard is harmonized with the 2008 version of ISO 12207 [ISO 17] and the 2011 version of ISO 15289 [ISO 17b]. After the publication of the IEEE 730 in 2014, a new version of ISO 15289 was published in 2016 and ISO had also initiated a revision of ISO 12207 [ISO 17]. Given that the IEEE 730 used the 2008 version of the ISO 12207 as a normative reference, we present it below.

The IEEE 730 is structured as follows [IEE 14]:

– Clause 1 describes the scope, purpose, and an introduction.

– Clause 2 identifies normative references used by the IEEE 730.

– Clause 3 defines the terms, abbreviations and acronyms.

– Clause 4 describes the context for the SQA processes and SQA activities, and covers expectations for how this standard will be applied.

– Clause 5 specifies the processes, activities and SQA tasks. Sixteen activities grouped into three categories are described: implementation of SQA process, product assurance, and process assurance.

– Twelve informative annexes A–L, where annex C provides guidelines for creating a SQAP.

While the development of ISO standards is done with the participation of member countries and their technical experts, the IEEE standards are developed through the individual contributions of experts.

Along with its standards, the IEEE publishes the list of participants in a working group that developed or updated a norm and who voted (e.g., approve, disapprove, abstain). Professor Laporte actively participated in the working group of the IEEE 730 and voted to approve the revised version. He is also the author of an annex that provides a mapping between the IEEE 730 and ISO 29110, which aims to help very small organizations that develop software or systems. The ISO 29110 will be presented in a section of this chapter.

The following text, taken from clause 7 of ISO 12207-2008, describes the purpose and the outcomes of the SQA process.

Software Quality Assurance Process

Purpose
Provide assurance that work products and processes comply with the provisions and predefined plans.

Outcomes

Following the successful implementation of the software quality assurance process:

    a) a strategy for conducting quality assurance is developed;

    b) evidence of software quality assurance is produced and maintained;

    c) problems and/or non-compliance with the requirements are identified and recorded; and

    d) compliance to standards, procedures and applicable requirements is checked for products, processes and activities.

ISO 12207-2008

The SQA process of the IEEE 730 is grouped into three activities: the implementation of the SQA process, product assurance, and process assurance. Activities consist of a set of tasks.

4.6.1 Activities and Tasks of SQA

IEEE 730 [IEE 14] describes what must be done by a project; it assumes that the organization has already implemented SQA processes before the start of a project. The activities and tasks of the IEEE 730 are presented next.

The standard includes a clause that describes what is meant by compliance. The following text box defines this term.

Compliance

Doing what has been asked or ordered; as required by rule or law (e.g. comply with a regulation).

IEEE 730 [IEE 14]

Compliance with all requirements of IEEE 730 [IEE 14] can be imposed by a client in an agreement (e.g., a contract) with the organization that will develop software. However, a given project may not need to use all the activities of the standard. The implementation of the standard implies the selection of a set of activities adapted to a project. For any activity or task that will not be performed, the standard requires that the SQAP describes them as not applicable and include a justification as to why the activity or task will not be performed. In order to improve, an organization can choose to gradually implement the activities and tasks of the IEEE 730.

Figure 4.6 shows the links between project requirements and artifacts produced during a project. The figure shows two categories of compliance: process assurance

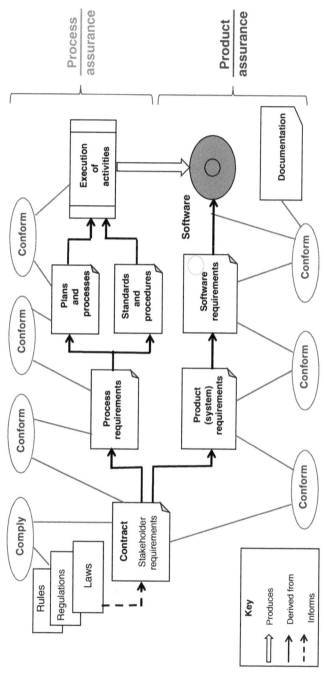

Figure 4.6 The links between requirements and the artifacts of a project [IEE 14].

and product assurance. The figure also describes the transitive relationships for compliance: if process requirements are consistent with the contract and if the process and project plans meet the process requirements, then the project process and plans comply with the contract. This simplifies the work of the SQA as each artifact of the project does not have to be checked against the contract. Each artifact must be verified with respect to its immediate predecessor.

The IEEE 730 [IEE 14] does not require that the activities be performed by any unit of the organization (e.g., an SQA department); it requires that responsibilities are assigned to the SQA function with the resources needed to perform the SQA activities described by the standard.

A basic principle of the IEEE 730 is to first understand the risks of the software product to ensure that the SQA activities are appropriate for the risks of the product. This means that the extent and depth of SQA activities that will be defined in the SQAP are determined by the risk associated with the software product.

4.6.1.1 SQA Process Implementation Activities

These activities aim to develop a strategy for conducting SQA, planning, and executing the activities and producing and maintaining the evidence. The six SQA process implementation activities are [IEE 14]:

– Establish the SQA processes to define and establish documented SQA processes that exist independently of the organization's projects.

– Coordinate with related software processes to coordinate activities and SQA tasks with other software processes such as verification and validation, reviews, audits, and other processes of ISO 12207 [ISO 17] that are relevant to the achievement of project objectives.

– Document SQA planning to document the activities, tasks, and results that are appropriate for the risks of a specific project. Planning SQA also includes adapting generic processes to the specific needs and risks of a project. The result of this planning is documented in the SQAP. Figure 4.7 shows the contents of the SQAP. There are 13 tasks associated to this activity, we have only listed the main tasks of this activity:
  o Use this standard and appendix C to help prepare an SQAP that is appropriate for the project, which meets the needs of all stakeholders and which is appropriate for the risks of the product.
  o Review and update the SQA level as the project evolves.
  o Present information on the status of the project to management in the agreed manner.
  o Estimate the resources of the SQA function (including effort, schedule, people, required skills, tools and equipment) needed to perform the SQA activities, tasks, and outcomes.

```
1 Purpose and scope
2 Definitions and acronyms
3 Reference documents
4 SQA plan overview
        4.1 Organisation and independence
        4.2 Software product risk
        4.3 Tools
        4.4 Standards, practices, and conventions
        4.5 Effort, resources, and schedule
5 Activities, outcomes and tasks
        5.1 Product assurance
                5.1.1 Evaluate plans for conformance
                5.1.2 Evaluate product for conformance
                5.1.3 Evaluate plans for acceptability
                5.1.4 Evaluate product life-cycle support for conformance
                5.1.5 Measure products
        5.2 Process assurance
                5.2.1 Evaluate life cycle processes for conformance
                5.2.2 Evaluate environments for conformance
                5.2.3 Evaluate subcontractor processes for conformance
                5.2.4 Measure processes
                5.2.5 Assess staff skill and knowledge
6 Additional considerations
        6.1 Contract review
        6.2 Quality measurement
        6.3 Waivers and deviations
        6.4 Task repetition
        6.5 Risk to performing SQA
        6.6 Communications strategy
        6.7 Non-conformance process
7 SQA records
        7.1 Analyze, identify, collect, file, maintain and dispose
        7.2 Availability of records
```

Figure 4.7   Table of contents of a SQAP according to the IEEE 730 [IEE 14].

- o Analyze product risks, standards, and assumptions that could impact quality and identify specific SQA activities, tasks, and specific outcomes that could help determine whether these risks are mitigated effectively.
- o Analyze the project and adapt SQA activities accordingly so that they are commensurate with the risks.
- o Define specific measures for the evaluation of projects, software quality, performance of SQA function compared with quality objectives of the project and those of the quality management of the organization.
- o Identify and track changes to the project that require additional planning from SQA function including changes to the requirements, resources, schedule, project scope, priorities, and product risk.

- o If process areas or activities are not adequately addressed by the quality management function of the organization, or if the organization does not have a quality management function, these process areas can be documented in the SQAP.
- Execute the SQAP in coordination with the project manager, the project team, and organizational quality management.
- Manage SQA records to create records of the activities, tasks and results of SQA; manage and control them and make them available to stakeholders.
- Evaluate organizational independence and objectivity to determine if those responsible for SQA occupy a position in the organization that allows them to have direct communication with the organization's management.

We will not describe the remaining 10 activities of IEEE 730 listed below in detail here since they will be described in other chapters of this book.

4.6.1.2 Product Assurance Activities

The five product assurance activities that aim to evaluate adherence to the requirements are [IEE 14]:

- evaluate plans for compliance to contracts, standards, and regulations;
- evaluate product for compliance to established requirements;
- evaluate product for acceptability;
- evaluate the compliance of product support;
- measure products.

4.6.1.3 Process Assurance Activities

The five process assurance activities which verify adherence to standards and procedures are [IEE 14]:

- evaluate compliance of the processes and plans;
- evaluate environments for compliance;
- evaluate subcontractor processes for compliance;
- measure processes;
- assess the skill and knowledge of personnel.

4.7 OTHER QUALITY MODELS, STANDARDS, REFERENCES, AND PROCESSES

This section presents the quality models, standards, reference, and processes specific to the software industry and that are used by many organizations worldwide. First, maturity models for software processes are presented, followed by the ITIL reference and its ISO/IEC 20000-1 standard [ISO 11h]. Next, we will look at the IT governance processes proposed by the CobiT reference. We will then present the family of ISO 27000 standards for information security, followed by the ISO/IEC 29110 standards for very small organizations.

4.7.1 Process Maturity Models of the SEI

The SEI developed several Capability Maturity Models (CMM®). In this section, we will present the CMMI model used to develop products (e.g., software, system) and services.

The CMMI for Development (CMMI-DEV) covers a broader area than its predecessor by adding other practices, such as systems engineering, and the development of integrated processes and products. It was formed from the following model practices: version 2.0 of the SW-CMM, the "*Systems Engineering Capability Model*" of the Electronic Industries Alliance [EIA 98] and version 0.98 of the "*Integrated Product Development CMM*" model.

The CMMI model was developed as two versions: an initial staged version and a continuous version, which is the first CMM model for systems engineering (Systems Engineering CMM, SE-CMM). The CMMI is available in several languages: Germany, English, Traditional Chinese, French, Spanish, Japanese, and Portuguese.

Just like the SW-CMM model, the objective of this model is to encourage organizations to check and continuously improve their development project process and evaluate their level of maturity on a five-level scale as proposed by the staged CMMI model.

Two other CMMI models were developed based on architecture, namely the CMMI for Services (CMMI-SVC) [SEI 10b] and the CMMI for Acquisition (CMMI-ACQ) [SEI 10c]. The CMMI-SVC model provides guidelines for organizations that provide services either internally or externally, whereas the CMMI-ACQ model provides guidelines for organizations that purchase products or services. All three CMMI models use 16 common process areas.

For each level of maturity, a set of process areas are defined. Each area encompasses a set of requirements that must be met. These requirements define which elements must be produced rather than *how* they are produced, thereby allowing the organization implementing the process to choose its own life cycle model, its design methodologies, its development tools, its programming languages, and its documentation standard. This approach enables a wide range of companies to implement this model while having processes that are compatible with other standards.

Table 4.3 describes the maturity levels as well as the process areas for each maturity level in the CMMI-DEV model.

Table 4.3   CMMI Maturity Levels of the Staged Representation [SEI 10a]

CMMI Development Model Maturity Levels [SEI 10a]
Model practices are grouped into 22 process areas, which are further broken down into five maturity levels.

Maturity Level 1: Initial
At maturity level 1, processes are usually ad hoc and chaotic. The organization usually does not provide a stable environment to support processes. Success in these organizations depends on the competence and heroics of the people in the organization and not on the use of proven processes. In spite of this chaos, maturity level 1 organizations often produce products and services that work, but they frequently exceed the budget and schedule documented in their plans.
Maturity level 1 organizations are characterized by a tendency to overcommit, abandon their processes in a time of crisis, and be unable to repeat their successes.

Maturity Level 2: Managed
At maturity level 2, the projects have ensured that processes are planned and executed in accordance with policy; the projects employ skilled people who have adequate resources to produce controlled outputs; involve relevant stakeholders; are monitored, controlled, and reviewed; and are evaluated for adherence to their process descriptions. The process discipline reflected by maturity level 2 helps to ensure that existing practices are retained during times of stress. When these practices are in place, projects are performed and managed according to their documented plans.
Process areas:
• Requirements management

• Project planning

• Project monitoring and control

• Supplier agreement management

• Measurement and analysis

• Process and product quality assurance

• Configuration management

Maturity Level 3: Defined
At maturity level 3, processes are well characterized and understood, and are described in standards, procedures, tools, and methods. The organization's set of standard processes, which is the basis for maturity level 3, is established and improved over time. These standard processes are used to establish consistency across the organization. Projects establish their defined processes by tailoring the organization's set of standard processes according to tailoring guidelines.

Table 4.3 (*Continied*)

Process areas:
- Requirements development
- Technical solution
- Product integration
- Verification
- Validation
- Organizational process focus
- Organizational process definition
- Organizational training integrated project management
- Risk management
- Decision analysis and resolution

Maturity Level 4: Quantitatively managed

At maturity level 4, the organization and projects establish quantitative objectives for quality and process performance and use them as criteria in managing projects. Quantitative objectives are based on the needs of the customer, end-users, organization, and process implementers. Quality and process performance is understood in statistical terms and is managed throughout the life of projects.

Process areas:
- Organizational process performance
- Quantitative project management

Maturity Level 5: Optimizing

At maturity level 5, an organization continually improves its processes based on a quantitative understanding of its business objectives and performance needs. The organization uses a quantitative approach to understand the variation inherent in the process and the causes of process outcomes.

Process areas
- Organizational performance management
- Causal analysis and resolution

Figure 4.8 presents an overview of the CMMI model structure. Each process area has generic and specific goals, practices, and sub-practices. The specific goals, practices, and sub-practices are "specific" to a process area, such as Requirements Management, although the generic goals, practices, and sub-practices apply to all process areas for a given maturity level. Each process area also includes explanatory notes and references for other process areas. As indicated in the legend for Figure 4.8, model components are required, expected, or informative. The SEI defines these three components in the following way [CHR 08]:

    – Required components describe what an organization must achieve to satisfy a process area;

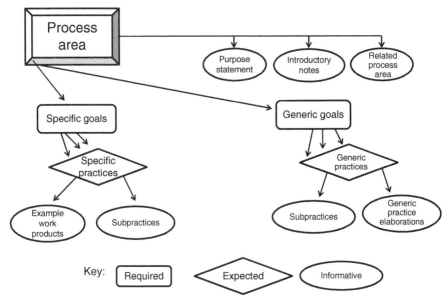

Figure 4.8   Structure of the staged representation of the CMMI [SEI 10a].

– Expected components describe what an organization can implement to achieve a required component. Expected components guide those who implement improvements or perform evaluations;

– Informative components provide details that help organizations initiate the process by specifying the way to understand required and expected components.

The concept of CMM model characteristics was transferred to the CMMI: these are generic goals and practices. For level 2 of the staged representation of the CMMI model, the 10 generic practices are [SEI 10a]:

1) Establish an organizational policy: establish and maintain an organizational policy for planning and performing the process.

2) Plan the process: establish and maintain the plan for performing the process.

3) Provide resources: provide adequate resources for performing the process, developing the work products, and providing the services of the process.

4) Assign responsibility: assign responsibility and authority for performing the process, developing the work products, and providing the services of the process.

5) Train people: train the people performing or supporting the process as needed.

6) Control work products: place selected work products of the process under appropriate levels of control.

Level	Focus	Key process area	
5 Optimizing	*Continuous process improvement*	Organizational performance management causal analysis and resolution	Quality productivity
4 **Quantitatively managed**	*Quantitative management*	Organizational process performance quantitative project management	
3 **Defined**	*Process standardization*	Requirements development Technical solution Product integration Verification Validation Organizational process focus Organizational process definition Organizational training Integrated project management Risk management Decision analysis and resolution	
2 Managed	*Basic project management*	Requirement management Project planning Project monitoring and control Supplier agreement management Measurement and analysis Process and product quality assurance Configuration management	
1 Initial			Risk rework

Figure 4.9 The staged representation of the CMMI® for Development model.
Source: Adapted from Konrad (2000) [KON 00].

7) Identify and involve relevant stakeholders: identify and involve the relevant stakeholders of the process as planned.

8) Monitor and control the process: monitor and control the process against the plan for performing the process and take appropriate corrective action.

9) Objectively evaluate adherence: objectively evaluate adherence of the process and selected work products against the process description, standards, and procedures, and address non-compliance.

10) Review status with higher level management: review the activities, status, and results of the process with higher level management and resolve issues.

Figure 4.9 presents the process areas associated with maturity levels in the staged version of the CMMI development model. The path flow through the CMMI model is carried out gradually. To meet the requirements for a process area, the organization must satisfy all the goals of the given process area. In the same way, in order to move up between maturity levels, all goals must be satisfied for the process areas in question, as well as for all the goals of the process areas of the lower levels of process maturity.

This section presented a maturity model that should help software development organizations: "Given that software maintenance is a specific field of software engineering, it is therefore necessary to focus on the processes and methodologies that take into account the specific characteristics of software maintenance" [BAS 96]. The next section introduces a maturity model designed to improve the software maintenance process.

CMMI® Institute

The CMMI Institute has taken over responsibility from the Software Engineering Institute for the CMMI models. CMMI documents, in several languages, can be downloaded free of charge from http://cmmiinstitute.com.

4.7.2 Software Maintenance Maturity Model (S^{3m})

The software maintenance maturity model S^{3m} (available at www.s3m.ca) proposes a structured approach to address a number of issues that arise in the context of the daily maintenance of software [APR 08]:

- software maintenance is a discipline mainly derived from industrial practices;
- a relatively large gap exists between the academic literature and the industrial practices;
- validation of improvement proposals made by consultants;
- many inconsistencies in terms that are often poorly defined with regard to the different proposals, approaches, presentations, and publications in maintenance;
- scarcity and difficulty in acquiring/adapting a specific methodology for software maintenance;
- there is no universal consensus;
- publications that are often optimistic (i.e., proposing unproven theories and miracle tools);
- problems that take on their full meaning when software and organizations reach a certain size.

Figure 4.10 illustrates the structure of the S^{3m} model.

Some authors have studied the differences and similarities between software development and maintenance activities. The maintenance organizational unit is structured to meet quite different challenges, such as randomly occurring daily events and requests from users, while providing continued service on the software for which it is responsible.

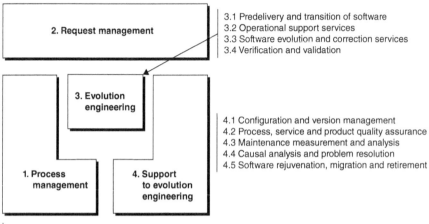

2.1 Event/request management
2.2 Maintenance planning
2.3 Maintenance requests monitoring and control
2.4 SLA and supplier agreement management

2. Request management

3.1 Predelivery and transition of software
3.2 Operational support services
3.3 Software evolution and correction services
3.4 Verification and validation

3. Evolution engineering

4.1 Configuration and version management
4.2 Process, service and product quality assurance
4.3 Maintenance measurement and analysis
4.4 Causal analysis and problem resolution
4.5 Software rejuvenation, migration and retirement

1. Process management

4. Support to evolution engineering

1.1 Maintenance process focus
1.2 Maintenance process/service definition
1.3 Maintenance training
1.4 Maintenance process performance
1.5 Maintenance innovation and deployment

Figure 4.10   Structure of the S^{3m} model [APR 08].

When we refer to software maintenance in the S^{3m} model, we are specifically considering the operational support activities, corrections and software evolution that occur daily. The characteristics that are specific to software maintenance are [ABR 93]:

- Modification requests (MRs) come in on an irregular basis and cannot be accounted for individually in the annual budget planning process.

- MRs are reviewed and prioritized by the developer, often at the operational level. Most do not require senior management involvement.

- The maintenance workload is not managed using project management techniques but rather by using queue management techniques and sometimes supported by Helpdesk software.

- The size and complexity of maintenance requests is such that it can usually be handled by one or two maintainers.

- The maintenance workload is user-services oriented, for the short term, to ensure that the operating software is running smoothly on a daily basis.

- Priorities can be shifted around at any time (sometimes hourly), and problem reports requiring that corrections be made immediately to the software in production take priority over any other work in progress.

Additionally, in many organizations, maintenance is often performed by different organizational groups than the software development groups.

The software maintenance maturity model:

– relates to the daily software maintenance activities rather than large-scale activities that should be managed using proven techniques in project management—in these specific large maintenance projects, it is the CMMI model that should be used;

– is based on the customers' perspective;

– is relevant for the maintenance of application software (a) developed and maintained internally, (b) configured and maintained internally or with a subcontractor's help, and (c) outsourced to an external supplier;

– provides references and details for each exemplary practice;

– offers an improvement approach based on road maps and maintenance categories;

– covers the software maintenance life cycle standards described in ISO 12207;

– covers most ISO 9001 characteristics and practices and relevant parts of the CMMI-DEV that apply to small maintenance;

– integrates references to additional software maintenance practices documented in other software and quality improvement models like ITIL, that we will cover in the next section.

A support team, which includes a technician and programmer for a section of the Canadian Forces, supports some 20 applications, which are all in use. Programmers are constantly in project mode and put the requests they receive from users in a priority order. The projects on which they are working all have delivery deadlines. However, it is very difficult to make time estimates when there may be weeks where the team receives a dozen requests that may require everything come to a halt until everything is fixed and validated.

The team had to set up a stringent daily schedule, where they work 30% of the day on requests and 70% on projects. On top of that, there are also standards for responding to the client in order to ensure better customer service. The team must now respond within 72 hours, and if the request is not completed, it must follow up every 14 days.

The previous sections described models for improving software development and maintenance processes. The next section considers the references and standards for improving processes for IT operations and infrastructure (also known as IT services).

4.7.3 ITIL Framework and ISO/IEC 20000

The ITIL framework was created in Great Britain based on good management practices for computer services. It consists of a set of five books providing advice and recommendations in order to offer quality service to IT service users. It should be pointed out that IT services are typically responsible for ensuring that the infrastructures are effective and running (backup copies, recovery, computer administration, telecommunications, and production data). The ITIL books systematically cover all aspects of the computer operations in a company while at the same time not claiming to have all the answers.

The list of guides illustrated in Figure 4.11 is:

– strategy;

– design;

– transition;

– operation;

– continuous improvement.

Support processes described in the ITIL are focused on daily operations. Their main goals are to resolve the problems when they arise or to prevent them from happening when there is a change in the computer environment or in the way the organization does things.

ITIL describes the support center function and the following five processes:

– Incident management

– Problem management

– Configuration management

– Change management

– Commissioning management

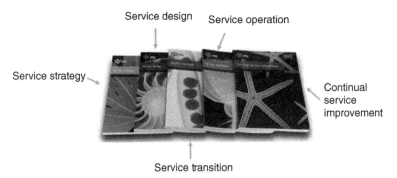

Figure 4.11 The main ITIL guides.

Service operation processes focus more on longer term management than support processes do. The main objective is to ensure that the IT infrastructure meets the business requirements of the organization. ITIL describes the following five processes for service operation:

- Service level management
- Financial management of IT services
- Capacity management
- IT service continuity management
- Availability management

ITIL user group in the USA http://www.itsmfusa.org/

ITIL is therefore a compendium of good practices and a compilation of descriptions of business processes that allow us to benefit from the experience of many organizations. It does not contain any implementation guidelines. ITIL is based on the sharing of operational experiences in IT.

Given the major recognition of ITIL worldwide, an international standard based on ITIL came into being: ISO/IEC 20000-1 [ISO 11h]. The principles of ITIL were successfully conveyed to many companies of all sizes and from all sectors of activity. The three main objectives are:

- to align computer services with the current and future needs of the company and its clients;
- to improve the quality of computer services provided;
- to reduce the long-term cost of service provision.

Just like the CMMI and S^{3m} references, ITIL is a process-based approach founded on the principle of continuous improvement of the Deming Cycle: PDCA.

In the 1980s, the British government wanted to improve efficacy and reduce IT costs in public companies. Initially this meant developing a universal method that could be applied to all public organizations. The project, which was initiated in 1986, really took off in 1988. The conclusions of the study quickly led to the definition of general principles and the development of best practices. These results were also applicable to the private sector. Work groups were struck, bringing together operational managers, independent experts, consultants, and trainers from public organizations and from the private sector. Private companies that participated and accepted to have their work methods studied by competing companies ensured the objectivity of the conclusions drawn, and at the same time, prevented any risk of being influenced by proprietary technologies or systems. ITIL places service at the center of information systems management. This principle was highly innovative in 1980 and promoted the notion of the client of information systems management. In 1989, 10 books were published in a first version of ITIL. The areas covered by this version were the processes of Service Support and Service Delivery. Over 30 books have been produced since, but an update between 2000 and 2011 has reduced the number to five books today.

4.7.3.1 Managing IT Services

The definition of a service, when used in the computing context, is as an organizational unit of the company similar to the accounting department, in that it supports the organization. This concept is also linked to the fact that information systems render services to users; services such as email, desktop support, and others.

ITIL's philosophy is based on the following fundamental concepts:

- taking into account the client's expectations regarding implementing computer services;
- the life cycle of computer projects must incorporate the different aspects of computer services management from the start;
- the implementation of interdependent ITIL processes to ensure service quality;
- the implementation of a way of measuring this quality from the user's point of view;
- the importance of communication between the computer department and the rest of the company;
- ITIL is flexible enough and must remain so in order to adapt to all organizations.

The main subjects handled under ITIL are:

- user support, which includes the management of incidents and is an extension of the concept of a Helpdesk;
- provision of services which involves managing processes that are dedicated to the daily operations of IT (cost control, management of service levels);
- management of the production environment infrastructure which involves implementing the means for network management and production tools (scheduling, backup, and monitoring);
- application management which consists of managing the support of an operational program;
- security management (confidentiality, data integrity, data availability, etc.) of the SI (security process).

Therefore, based on these guidelines, we can help define the processes for IT infrastructure and IT operations groups.

4.7.3.2 ISO/IEC 20000 Standard

The ISO/IEC 20000 standard is the first ISO standard dedicated to managing IT services. Inspired by the former British standard BSI 15000, which is based on ITIL, it was initially published on November 10, 2005, by the ISO. The main contribution of ISO 20000 is an international consensus on ITIL content.

As we show in Figure 4.12, this standard has two parts. The first part, ISO 20000-1 [ISO 11h] represents the certifiable part of the standard. It defines the IT service management requirements. These requirements are:

– specifications for service management;

– planning and implementing service management;

– planning and implementation of new services;

– service delivery process;

– relationship management;

– resolution management;

– control management;

– production management.

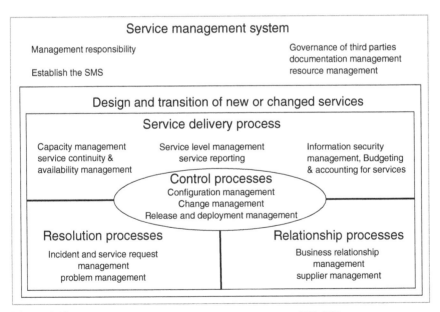

Figure 4.12   Processes of ISO 20000 service management system [ISO 11h].
Source: Standards Council of Canada.

The next section introduces the reference for exemplary practices in IT governance. This guide is used by internal auditors specialized in IT in order to evaluate the quality of controls set up in an organization.

4.7.4 CobiT Process

CobiT [COB 12] is a repository of best practices for IT governance established by ISACA (IT auditors). Oriented on auditing and governance assessment information systems, CobiT provides risk analysis and assessment of the effectiveness of internal controls. This repository of best practices tries to cover several concepts such as the analysis of business processes, technical aspects of IT, control needs in information technology, and risk management.

The CobiT process ensures that technological resources are well aligned with the company's fundamental objectives. It helps to achieve the right level of control to exercise on IT. This process is harmonized with the ITIL reference, the PMBOK® Guide from the Project Management Institute [PMI 13] as well as the ISO 27001 and ISO 27002 standards.

CobiT version 5 covers 34 generic guidance processes and 318 control objectives divided into four process domains:

– planning and organization;
– acquisition and implementation;
– distribution and support;
– monitoring and surveillance.

The framework consists of checklists covering the four process domains, with 34 general control objectives and 302 detailed control objectives. The "planning and organization" domain has eleven goals covering everything related to strategy and tactics. These objectives identify means for IT to contribute most effectively to the achievement of the business goals of the company. The "acquisition and implementation" domain covers six goals for the achievement of the IT strategy: the identification, acquisition, development and implementation of IT solutions and their integration into business processes. The "distribution and support" domain considers thirteen goals, grouping the delivery of IT services required, that is, the operation, security, emergency plans, and training. The "monitoring and surveillance" domain has four objectives that allow management to assess the quality and compliance of process control requirements. The implementation tools include a presentation of cases where companies have implemented processes quickly and successfully using the CobiT methodology. These examples are therefore closely linked to business objectives while focusing particular attention on IT. This will reassure management, standardize work processes, and guarantee the security and controls of IT services.

The CobiT management section of the guide focuses on several aspects: the performance measurement and control of the IT profile and awareness of technological risks. This document provides the key indicator objectives, the key performance indicators, key success factors, and the maturity model. The maturity model evaluates the achievement of one or more general objectives of the process on a scale of 0–5:

- 0: non-existent;
- 1: existing but unorganized (initialized on an ad hoc basis);
- 2: renewable;
- 3: defined;
- 4: managed;
- 5: optimized.

The audit guideline allows an organization to evaluate and justify the risks and shortcomings of general and detailed objectives. Then, once this step is completed, corrective actions can be implemented. This audit guideline respects four principles: developing an in-depth understanding, evaluating controls, verifying compliance, and justifying the risk of not attaining the control objectives.

The next section presents information security standards.

4.7.5 ISO/IEC 27000 Family of Standards for Information Security

The ISO 17799 standard [ISO 05d] initially published in December 2000 by ISO introduces a code of good practices for information security management. The second edition of this standard was published in June 2005, and then obtained a new reference number in July 2007. This standard is now known as ISO/IEC 27002 [ISO 05c].

The ISO 27002 standard comprises 133 practical steps to be used by anyone in charge of implementing or maintaining an ISMS. Information security is defined within the standard as the "preservation of confidentiality, integrity and availability of information."

This standard is not mandatory for companies. However, it may be required under a contract: a service provider could then agree to respect the standardized practices with a client.

The ISO 27002 standard is made up of 11 main sections, which cover security management as well as its strategic and operational aspects. Each section makes up a chapter of the standard:

- security policy;
- information security organization;
- asset management;
- security related to human resources;

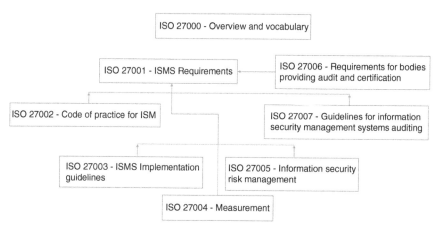

Figure 4.13 Family of ISO 27000 Standards [ISO 05c].
Source: Standards Council of Canada.

- physical and environmental security;

- communications and operations management;

- access control;

- acquisition, development, and maintenance of information systems;

- incident management related to information security;

- activity continuity management;

- legal and regulatory compliance.

The chapters in the standard specify the objectives to reach and list practices that allow for the fulfillment of these objectives. The standard does not provide detailed practices, since each organization is supposed to carry out an evaluation of its own risks in order to determine its needs before choosing the practices that are appropriate in each of the possible cases.

This standard is based on the ISO 27000 family of standards as shown in the Figure 4.13.

Note that the ISO 20000 standard, based on ITIL, connects directly to the ISO 27001 guideline concerning information security.

The next section presents the ISO/IEC 29110 standard and guides that have been developed specifically for small organizations that develop software or systems.

4.7.6 ISO/IEC 29110 Standards and Guides for Very Small Entities

Worldwide there are numerous very small entities (VSEs), namely enterprises, organizations (e.g., public organizations and not-for profit organizations), departments or projects with up to 25 people. For example, in Europe more than 92.2% of companies have between 1 and 9 employees and only 6.5% of firms have between 10 and 49 employees [MOL 13]. In the US, about 95% of companies have 10 employees or less [USC 16]. In Canada, nearly 80% of software companies in the Montreal region have 25 or fewer employees, and 50% of companies have 10 employees or less [GAU 04]. In Brazil, small businesses represent approximately 70% of the total number of companies [ANA 04]. Lastly, in Northern Ireland [MCF 03], a survey showed that 66% of organizations employ fewer than 20 people.

Unfortunately, according to surveys and studies carried out, it is clear that the ISO standards were not developed for VSEs, that they do not meet their needs, and therefore they are difficult to apply in such contexts. In order to help VSEs, an international standardization project developed a set of ISO/IEC 29110 standards and guides [ISO 16f]. ISO 29110 standards were developed by taking relevant information for VSEs from existing standards, such as ISO 12207 and ISO 15289. This "assemblage" is called a profile. We will briefly introduce ISO 15289 at the end of this chapter.

In 2005, Professor Laporte was appointed as project editor of the ISO 29110 standards and guides. Since 2005, the working group has developed a set of ISO 29110 standards and guides to help VSEs that develop software, systems involving hardware and software or who provide maintenance services to their clients. The interest of several countries was such that several ISO 29110 documents have been translated into Czech, French, German, Japanese, Portuguese, Spanish. Some countries, such as Brazil, Japan and Peru, have adopted ISO 29110 as a national standard.

4.7.6.1 ISO 29110 Set of Profiles

The essential characteristic of organizations covered by ISO/IEC 29110 standards is size. However, we know that there are other aspects and characteristics of VSEs that may impact the preparation of profiles, such as the business model, situational factors like the application domain (e.g., medical), uncertainty, criticality, and risk levels. The creation of a profile for each possible combination of characteristics presented above would have resulted in a large and unmanageable number of profiles. Therefore, the profiles were designed so as to be applicable to more than one category. A profile group is a collection of profiles that are related either by process composition (e.g., activities, tasks) or by capacity level, or both [ISO 16f]. For example, the generic profile group was defined as applicable to VSEs that do not develop critical software or critical systems. Critical systems or software are systems or software having the potential for serious impact on the users or environment, due to factors including safety, performance, and security [ISO 16f]. The generic profile group is a roadmap made up of four profiles (Entry, Basic, Intermediate, and Advanced) providing a progressive approach in order to satisfy the greatest amount of VSEs. The main characteristics of these four profiles are described in Table 4.4.

Table 4.4   VSEs Targeted by Generic Profiles [ISO 16f]

Title of profile	VSEs targeted by each profile
Entry	This profile is for VSEs working on small projects (e.g., a project of up to six person-months) and for start-ups.
Basic	This profile is for VSEs developing only one project at a time with a single team.
Intermediate	This profile is for VSEs involved in the simultaneous development of more than one project with more than one team.
Advanced	This profile is for VSEs wishing to significantly improve the management of their business and their competitiveness.

Table 4.5 illustrates the documents that have been developed and the intended recipients. Technical reports are marked "TR" (Technical Report) in their titles whereas the other documents are standards [ISO 16f]:

– ISO/IEC TR 29110-1 [ISO 16f], entitled "Overview," defines commonly used terminology in all documents of VSE profiles. It presents the process, life cycle concepts, and standards as well as all documents constituting the ISO/IEC 29110. It also describes the characteristics and needs of a VSE, and precisely why profiles, documents, standards, and guidelines have been developed for the VSE.

Table 4.5   ISO 29110 Documents and their Targeted Recipients [ISO 16f]

ISO/IEC 29110	Title	Target
Part 1	Overview	The VSE and their customers, evaluators, standards developers, vendors of tools and methodologies.
Part 2	Framework for profile preparation	Standards developers, vendors of tools and methodologies. This document is not intended for VSEs.
Part 3	Certification and Assessment Guide	The VSE and their customers, assessors, accreditation bodies.
Part 4	Profile Specifications	The VSE, standards developers, vendors of tools and methodologies.
Part 5	Management and Engineering and Service Delivery Guides	The VSE and their customers.

- ISO/IEC 29110-2-1, entitled "Framework for profile preparation," presents the concepts of engineering profiles of systems and software for the VSE. It explains the logic behind the definition and application profiles. The document also specifies the elements common to all standardized profiles (structure, conformity assessment) of the ISO/IEC 29110.
- ISO/IEC 29110-3, entitled "Certification and Assessment Guidelines," sets guidelines for the evaluation of the process and the compliance requirements necessary to achieve the objective of the profiles defined for VSEs. The report also contains information which can be useful for developers of methods and assessment tools. ISO/IEC TR 29110-3 is for those who have a direct relation to the assessment process, for example, the assessor and the person requesting the evaluation, which need guidance to ensure the requirements for performing an evaluation are met.
- ISO/IEC 29110-4-m, entitled "Profile Specifications," provides the specifications for all profiles in a profile group based on subsets of relevant standards.
- ISO/IEC 29110-5-m-n provides management and engineering and service delivery guides for the profiles within the generic profile group.

When a new profile is required, only parts 4 and 5 of ISO 29110 are developed without affecting other documents. To facilitate the adoption of ISO 29110 by a large number of VSEs, Working Group 24, which was mandated to develop the ISO 29110, negotiated that the ISO technical reports be available for free.

Technical Reports for the ISO 29110

Technical reports of ISO 29110 are available on the ISO website at the following address: http://standards.iso.org/ittf/PubliclyAvailableStandards/index.html

---

4.7.6.2 ISO 29110 Software Basic Profile

The software basic profile is made up of two processes: the project management process and the software implementation process. The goal of the project management process is to establish a systematic way to carry out the tasks of implementing the software project which will meet the project objectives with regard to quality, schedule, and cost. The purpose of the software implementation process is the systematic performance of the analysis, design, construction, integration, and tests for new or modified software products according to the specified requirements.

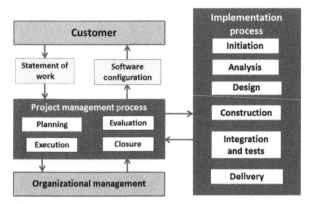

Figure 4.14   Management and implementation processes for the software Basic profile [LAP 16a].

Figure 4.14 shows the project management process and the software implementation process and their activities. Activities consist of tasks. The input to the project management process is a document entitled "Statement of Work (SOW)," the output of the implementation process is the set of deliverables (e.g., documentation, code) that are defined early in the project and called software configuration.

The management and engineering guide of the ISO 29110 software Basic profile is presented next. Remember that a standard describes "what to do", while management and engineering guides describe "how to do it".

4.7.6.2.1 Management Processes for the Software Basic Profile   The goal of the project management process is to establish a systematic way to carry out the tasks of implementing the software project which will meet the project objectives with regard to quality, schedule and cost.

The management process, as illustrated in Figure 4.15, uses the SOW to develop the project plan. Assessment and control tasks for this process will use the project plan to assess project progress. Steps are taken, if necessary, either to eliminate gaps with the project plan or to incorporate changes into the plan. Project closure activities provide the deliverables that are produced during the implementation process to the client for their approval so as to officially close the project. A project repository is established to save work products and control their versions during the project.

The project management process includes four activities that are broken down into 26 tasks.

4.7.6.2.2 Software Implementation Process for the Basic Profile   The purpose of the software implementation process is the systematic performance of

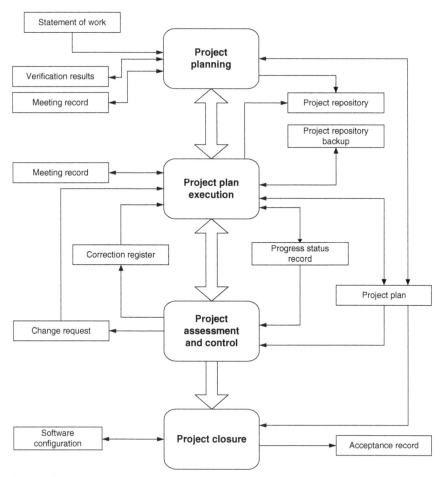

Figure 4.15   Activities for the software Basic profile management process [ISO 11e].
Source: Standards Council of Canada.

the analysis, design, construction, integration, and tests for new or modified software products according to the specified requirements. Carrying out the implementation process, as illustrated in Figure 4.16, is controlled by the project plan. The project plan guides the analysis of software requirements, detailed architecture and design, construction and integration of software, tests and delivery activities for the deliverables. To eliminate the defects in a product, verification and validation activities as well as test tasks are included in this process's activities. The client provides a SOW as input to the project management process and receives the set of deliverables initially identified in the project plan as the result of the implementation process.

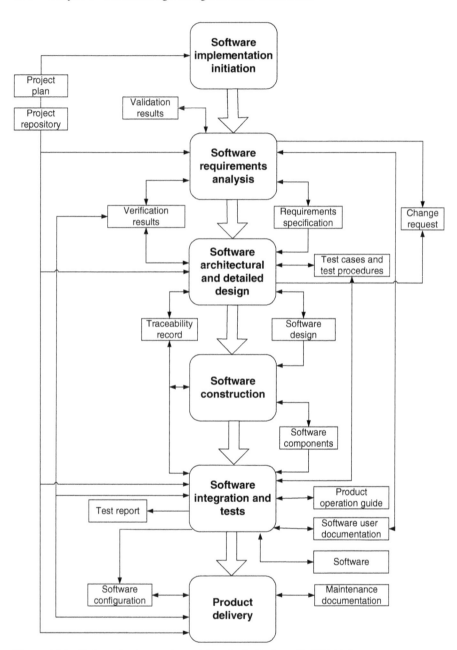

Figure 4.16   Software implementation activities for the Basic profile [ISO 11e].
Source: Standards Council of Canada.

Table 4.6 Description of One Task of the Requirements Analysis Activity [ISO 11e]

Roles	Tasks	Input work products	Output work products
CUS AN	SI.2.4 Validate and obtain approval of the *Requirements Specifications* Validate that requirements specification satisfies needs and agreed upon expectations, including the user interface usability. The results found are documented in a *validation results* and corrections are made until the document is approved by the CUS.	*Requirements Specification [verified]*	*Validation Results* *Requirements Specification [validated]*

The implementation process, as shown in Figure 4.16, includes six activities that are broken down into 41 tasks. To illustrate how management and engineering guides facilitate the implementation of ISO 29110 in a VSE, Table 4.6 provides an example of the requirements analysis task (SI.2.4-Validate and obtain approval of the requirements specification). The left column lists the roles involved in this task: the Analyst (AN) and the Customer (CUS). A second column describes the task as well as some additional information to help execute the task and a third column describes the input work product required for task execution as well as its status (e.g., verified). The last column shows the output work products and their status resulting from the execution of the task.

Table 4.7 shows an example of a document, a change request, and the content proposed by the ISO 29110 management and engineering guide. Note that the guide does not impose specific content by stating that a document "may," not "shall," include the following items, but provides information items that can be grouped. The right-hand column shows the source who requested the document. In this example, a change request is drafted by the client, the development team or by project management. We also note at the end of the description, the applicable status of this document.

4.7.6.3 The Development of Deployment Packages

Although the ISO working group created a management and engineering guide, many VSEs do not have the resources to readily use the two processes described. Therefore, Professor Laporte, as the ISO 29110 project editor, recommended to the delegates for Working Group 24 during a meeting in Moscow in 2007, the development of material that can be used "as is" by VSEs [LAP 08a]. He coordinated the development of a set of documents called the Deployment Package (DP), based on the management and engineering guides.

Table 4.7   Description of the Content of a Document [ISO 11e]

Name	Description	Source
Change request	Identifies a *Software*, or documentation problem or desired improvement, and requests modifications. It may have the following characteristics: – Identifies purpose of change – Identifies request status – Identifies requester contact information – Impacted system(s) – Impact to operations of existing system(s) defined – Impact to associated documentation defined – Criticality of the request, date needed The applicable statuses are: initiated, evaluated, and accepted.	Customer Project management Software implementation

A DP is a set of artifacts that facilitate and accelerate the implementation of the ISO 29110 standard in VSEs by providing them with ready-to-use processes. The table of contents for the package is described in Table 4.8. Note that the tasks from

Table 4.8   Table of Contents of a DP [LAP 08a]

1. Technical description
   Purpose of this document
   Why is this topic important?
2. Relationships with ISO/IEC 29110
3. Definitions
4. Overview of processes, activities, tasks, roles, and products
5. Description of processes, activities, tasks, steps, roles, and products
   Role description
   Product description
   Artifact description
6. Template(s)
7. Example(s)
8. Checklist(s)
9. Tool(s)
10. References to other standards and models (e.g., ISO 9001, ISO/IEC 12207, CMMI)
11. References
12. Evaluation form

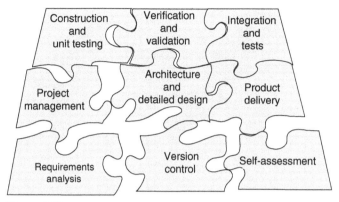

Figure 4.17   DPs for the software Basic profile [LAP 08].

the management and engineering guides were broken down into steps to help the VSE concretely implement the standard.

Figure 4.17 shows the DPs that have been developed for the software Basic profile.

These DPs are available for free in English, French, and Spanish on the Web (www.sqabook.org). This site also provides access to templates, checklists, and examples of the implementation of ISO 29110, such as the example provided below.

4.7.6.4 Example of the Implementation of the Software Basic Profile

Following is a brief description of the implementation of the ISO 29110 basic profile for BitPerfect Inc.—a start-up of four persons located in Lima, Peru [GAR 15].

Peruvian Start-Up VSE - BitPerfect (Adapted from [GAR 15])

The VSE had used agile practices for its first development projects (e.g., Scrum, test driven development, continuous integration). The basic profile of ISO 29110 had been tailored to agile practices already in use. The new process had been successfully applied to a project for a client of the VSE: software that facilitates communication between clients and judicial counselors to the second largest insurance company in Peru.

Project management and software development of the insurance company project took about 900 hours. The table below shows the effort for prevention, execution, evaluation, and correction.

Development phase	Prevention (hours)	Execution (hours)	Evaluation (hours)	Correction (hours)
Preparing the environment (e.g., server)	14			
Developing the project plan		15	3	7
Implementation and project control		108		
Implementation (sprints)		90		
Assessment and control (sprint review)		18		
Software specification		107	28	58
Statement of work		12	3	7
Specification of use cases		95	25	51
Architecture design		35	18	14
Development of the test plan		45	8	11
Coding and testing		253	70	62
Documentation of maintenance and operating guide		14	5	7
Software implementation		6		
Project closure		2		
Total (hours)	**14**	**585**	**124**	**159**

Efforts for prevention, execution, evaluation, and correction [GAR 15]

The percentage of effort invested in the correction of defects (i.e., rework) was only 18% (159 hours/882 hours) for the entire project. This percentage is similar to the performance of a CMM level 3 maturity level. The table below shows the percentage of correction efforts by CMM level of maturity.

Maturity Levels of the CMM Model and Correction Effort [DIA 02])

CMM maturity level	% Correction effort
2	23.2%
3	14.3%
4	9.5%
5	6.8%

It should be noted that this project was the first for which the VSE used the new ISO 29110 based process in a software development project. The correction percentage reflects the learning curve associated with the new process and the new documents to produce.

This Peruvian VSE was subsequently audited by an accredited auditor from Brazil. An ISO 29110 compliance audit was completed and a certificate of compliance for the

Basic profile was issued by the Brazilian standards agency. ISO/IEC 29110 certification has facilitated access to new customers and larger projects. The certificate obtained by the VSE is recognized by the member countries of the International Accreditation Forum (IAF). In 2016, the Peruvian VSE employed more than 23 people.

The certification process for the Peruvian VSE will be presented in the chapter on audits.

4.7.7 ISO/IEC 29110 Standards for VSEs Developing Systems

Since a large number of VSEs develop systems that include hardware and software, members of the SC7 countries mandated Working Group 24 to develop a set of documents similar to those already created for the development of software: a roadmap consisting of a set of four profiles (Entry, Basic, Intermediate, and Advanced). The system development and software development profiles are similar since the strategy was to develop the system profiles using, as the foundation, the software profiles already published.

Figure 4.18 shows the processes and activities for the systems engineering Basic profile. Since this is based on the software Basic profile, there are similarities with

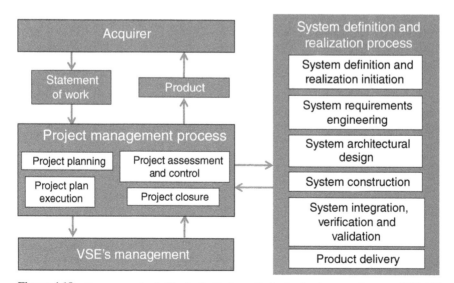

Figure 4.18   Processes and activities for the Basic profile for the development of systems [ISO 16f].

the software Basic profile presented above:

- – a client, called acquirer, presents a SOW for a VSE;
- – a project management process that includes similar activities as the project management process of the software Basic profile. Some tasks were added to the profile for systems engineering such as the management of the purchase of hardware components.
- – a technical process, referred to as System Definition and Realization process, has activities that are similar to the software implementation process. Several tasks were added to the systems engineering profile. It is during the execution of the activity entitled "Construction of the system" that the hardware components are purchased or manufactured by the VSE. It is also at this point that software components are developed. Next, during the system integration activity, software and hardware components are integrated, the system is verified and validated. The ISO 29110 systems engineering guide suggests that a VSE uses the software Basic profile for the development of the software components of the system.

The system and software management and engineering guides are freely available from ISO in French and English. A version in German is also available on the Web.

4.8 SPECIFIC STANDARDS FOR AN APPLICATION DOMAIN

In this section, we provide examples of standards that are specific to an application domain, such as aerospace and railways, to highlight the fact that certain fields comprising critical systems have developed their own standards. Unlike ISO's software engineering standards, which are developed by representatives of member countries, the standards described below have been developed by the organizations involved in a specific field (e.g., aerospace companies and aviation authorities). One notable example is the CobiT guide [COB 12] that is used by internal auditors to audit the information systems organizations.

4.8.1 DO-178 and ED-12 Guidance for Airborne Systems

The development of DO-178 "Software Considerations in Airborne Systems and Equipment Certification," was initiated in 1980 by the Radio Technical Commission for Aeronautics (RTCA). At the same time, the European Organisation for Civil Aviation (EUROCAE) had also developed a similar document. Later, the two organizations decided to develop a common guidance. In 1982, EUROCAE published ED-12 with technical content identical to DO-178. As stated in DO-178C, since

1992, the aviation industry and certification authorities around the world have used the considerations in DO-178B/ED-12B as an acceptable means of compliance for software approval in the certification of airborne systems and equipment. The latest version of these two documents was published in 2011 [EUR 11, RTC 11].

Their purpose is to provide guidance for the production of software for airborne systems and equipment. These systems must provide a high level of safety that complies with airworthiness requirements.

The reader may wonder why this document is called guidance and not a standard. This is the explanation extracted from DO-178 [RTC 11]: "This document recognizes that the guidance herein is not mandated by law, but represents a consensus of the aviation community. It also recognizes that alternative methods to the methods described herein may be available to the applicant. For these reasons, the use of words such as 'shall' and 'must' is avoided."

However, DO-178 provides some text that transforms this guidance to a de facto standard for that industry [RTC 11]: "If an applicant adopts this document as a means of compliance, the applicant should satisfy all applicable objectives. This document should apply to the applicant and any of its suppliers, who are involved with any of the software life cycle processes or the outputs of those processes described herein. The applicant is responsible for oversight of all of its suppliers."

Certification

Legal recognition by the certification authority that a product, service, organization, or person complies with the requirements.

Such certification comprises the activity of technically checking the product, service, organization, or person and the formal recognition of compliance with the applicable requirements by issue of a certificate, license, approval, or other documents as required by national laws and procedures. In particular, certification of a product involves:

a) the process of assessing the design of a product to ensure that it complies with a set of requirements applicable to that type of product so as to demonstrate an acceptable level of safety;

b) the process of assessing an individual product to ensure that it conforms with the certified type design;

c) the issuance of a certificate required by national laws to declare that compliance or conformity has been found with requirements in accordance with items (a) or (b) above.

DO-178 [RTC 11]

The purpose of this guidance includes [RTC 11]:

– objectives for software life cycle processes;

– activities that provide a means for satisfying those objectives;

– descriptions of evidence in the form of life cycle data that indicate that the objectives have been satisfied.

A system safety assessment determines the software level of the software components of a system. The software level may even be specified in the system requirements. The DO-178 divides software into five levels [RTC 11]:

1) Level A: Software whose anomalous behavior, as shown by the system safety assessment process, would cause or contribute to a failure of system function resulting in a catastrophic failure condition for the aircraft.

2) Level B: Software whose anomalous behavior, as shown by the system safety assessment process, would cause or contribute to a failure of system function resulting in a hazardous failure condition for the aircraft.

3) Level C: Software whose anomalous behavior, as shown by the system safety assessment process, would cause or contribute to a failure of system function resulting in a major failure condition for the aircraft.

4) Level D: Software whose anomalous behavior, as shown by the system safety assessment process, would cause or contribute to a failure of system function resulting in a minor failure condition for the aircraft.

5) Level E: Software whose anomalous behavior, as shown by the system safety assessment process, would cause or contribute to a failure of system function with no effect on aircraft operational capability or pilot workload. If a software component is determined to be Level E and this is confirmed by the certification authority, no further guidance contained in this document applies.

 RTCA

RTCA, Incorporated is a not-for-profit corporation formed to advance the art and science of aviation and aviation electronic systems for the benefit of the public. The organization's recommendations are often used as the basis for government and private sector decisions as well as the foundation for many Federal Aviation Administration Technical Standard Orders.

www.rtca.org

EUROCAE

The European Organisation for Civil Aviation Equipment is a non-profit organisation dedicated to aviation standardisation.

www.eurocae.net

4.8.2 EN 50128 Standard for Railway Applications

The EN 50128 standard describes the application domain in the following way [CEN 01]: by specifying the procedures and technical requirements applicable to the development of programmable electronic systems used in railway control and protection applications. It is meant to be used in any field having safety implications. It applies to all software used in developing and implementing railway control and protection systems, including application programming, operating systems, support tools, and micro-programming. It also focuses on the requirements applicable to the systems configured using application data.

The standard was developed for the following reasons [CEN 01]:

– To define a process for the specification and demonstration of dependability requirements for the railway industry;

– To foster an understanding and common approach to managing dependability;

– To provide the railway authorities and industry with a process that allows for a coherent management of reliability, availability, maintainability, and dependability.

The EN 50128 standard introduces five software safety integrity levels (SWSIL) where each level is associated with a degree of risk when using a software system. Level 0 is assigned to software that is non-safety-related, whereas level 4 means a very high risk. A special feature of this standard is that it imposes an organizational structure on organizations that develop risky software, that is, software with safety issues. Figure 4.19 illustrates the different organizational structures based on the SWSIL.

The standard describes the sharing of roles and responsibilities of stakeholders. For level 0, there is no constraint, the designer/implementer, verifier, and validator can all be the same person.

For SWSIL levels 1 and 2, the verifier and validator can be the same person, but they cannot be the designer/implementer. However, the designer/implementer, verifier, and validator can all report through the project manager. At SWSIL levels 3 and 4, there are two acceptable arrangements:

– The verifier and the validator can be one and the same person, but they must not also be the designer/implementer. In addition, the verifier and validator cannot report to the project manager, as the designer/implementer does, and they must have the authority to prevent a product from being released.

– The designer/implementer, verifier and validator must all be different people. The designer/implementer and verifier can report to the project manager, whereas the validator cannot. The validator must have the authority to prevent a product from being released.

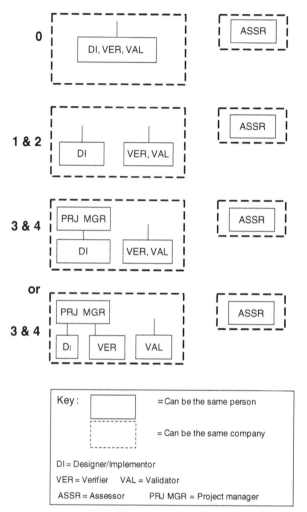

Figure 4.19 Independence versus SWSIL [CEN 01].

Moreover, in this standard, there are requirements about using or not using certain development techniques. The classification table has five levels, from mandatory to not recommended and appears in Table 4.9.

Table 4.10 illustrates the classification of static analysis techniques for SWSIL from 0 to 4. Here, we see the requirement for performing design reviews is "HR" for all levels of criticality.

In Table 4.11, we list requirements made as to whether or not to use certain programming languages, such as C or C++ (without restriction), for the five criticality levels.

Table 4.9   Example of a Technique Classification Table (Translated from [CEN 01])

Classification	Explanation
The use of a technique is mandatory (M).	
The technique is highly recommended (HR) for this safety integrity level.	If this technique or measure is not used then the rationale behind not using it should be detailed in the SQAP or in another document referenced by the SQAP.
The technique or measure is recommended (R) for this safety integrity level.	
The technique or measure has no recommendation (-) for or against being used.	
The technique or measure is positively not recommended (NR) for this safety integrity level.	If this technique or measure is used then the rationale behind using it should be detailed in the SQAP or in another document referenced by the SQAP.

Table 4.10   Description of Classification of Static Analysis Techniques [CEN 01]

Technique	SW IL 0	SW IL 1	SW IL 2	SW IL 3	SW IL 4
Boundary value analysis	-	**R**	**R**	**HR**	**HR**
Checklists	-	**R**	**R**	**R**	**R**
Data flow analysis	-	**HR**	**HR**	**HR**	**HR**
Fagan inspections	-	**R**	**R**	**HR**	**HR**
Walk-through/design reviews	**HR**	**HR**	**HR**	**HR**	**HR**

Table 4.11   Description of Classification of Programming Languages [CEN 01]

Technique	SW IL 0	SW IL 1	SW IL 2	SW IL 3	SW IL 4
Ada	**R**	**HR**	**HR**	**R**	**R**
C or C++ (unrestricted)	**R**	-	-	**NR**	**NR**
Subset of C or C++ with coding standard	**R**	**R**	**R**	**R**	**R**

4.8.3 ISO 13485 Standard for Medical Devices

The ISO 13485 standard focuses on QA of products, customer requirements, and various items related to the management of quality systems [ISO 16d]. This standard specifies the requirements of a QMS that can be used by an organization for design and development, production, installation, and related services for medical devices, as well as the design, development, and supply of related services. The following definition of medical device illustrates the wide spectrum of devices that could be impacted by poor quality software.

Medical Device

Instrument, apparatus, implement, machine, appliance, implant, reagent for *in vitro* use, software, material or other similar or related article, intended by the manufacturer to be used, alone or in combination, for human beings, for one or more of the specific medical purpose(s) of:

– diagnosis, prevention, monitoring, treatment or alleviation of disease;

– diagnosis, monitoring, treatment, alleviation of or compensation for an injury;

– investigation, replacement, modification, or support of the anatomy or of a physiological process;

– supporting or sustaining life;

– control of conception;

– disinfection of medical devices;

– providing information by means of *in vitro* examination of specimens derived from the human body;

and does not achieve its primary intended action by pharmacological, immunological or metabolic means, in or on the human body, but which may be assisted in its intended function by such means.

   Note 1 to entry: Products which may be considered to be medical devices in some jurisdictions but not in others include:

– disinfection substances;

– aids for persons with disabilities;

– devices incorporating animal and/or human tissues;

– devices for *in vitro* fertilization or assisted reproduction technologies.

ISO 13485 [ISO 16d]

Although this standard is independent of ISO 9001, it is based upon it. Given that certain requirements of ISO 9001 were excluded from ISO 13485, organizations whose QMS comply with ISO 13485 can claim compliance with the ISO 9001 standard only if their QMSs comply with all the requirements of the ISO 9001 standard. Appendix B of ISO 13485 provides correspondence between these two standards.

4.9 STANDARDS AND THE SQAP

Standards have a central place in a project's SQAP (Software Quality Assurance Plan). In order to conduct product and process assurance activities, the project team as well as the SQA function need to assess the adherence of project processes and products to the applicable agreements (e.g., contracts), regulations and laws, organizational standards, and procedures. Effectiveness of organizational software processes need to be constantly evaluated and improvements suggested. Problems and non-conformance are identified and recorded. The IEEE 730 standard defines the requirements regarding standards, practices, and conventions that must be described in the SQAP of a project. The SQAP identifies all applicable standards, practices, and conventions used for the project such as:

- documentation standards;
- design standards;
- coding standards;
- standards for comments;
- testing standards and practices.

Once the standards are identified and staff are trained on how to use them, SQA has a duty to conduct process and product assurance evaluations based on organizational QA policies and processes. Both process and product assurance evaluations have to be done on all projects. Process assurance consists of the following activities [IEE 14]:

- evaluate life cycle processes and plans for conformance;
- evaluate environments for conformance;
- evaluate subcontractor processes for conformance;
- measure processes;
- assess staff skills and knowledge.

In many organizations, the requirement of the standards mandated by a customer in an agreement is embedded in the processes of a project before the project is started and developers are trained on these processes.

It is up to the organization to decide if it will inform their developers about the standards that have been embedded in their processes. In the reference section of a process, an organization usually lists the standards used, such as ISO 12207 or ISO 9001, to document the process.

We have seen that organizational standards used by software and system engineers are adapted locally to fit each project specificity. To ensure consistency, high quality and reduce project risks, the SQA is asked to:

- identify the standards and procedures established by the project or organization;
- analyze, for identified projects, the product risks, standards, and assumptions that could impact quality and identify specific SQA activities, tasks, and outcomes that could help determine whether those risks can be effectively mitigated by the project;
- determine whether the proposed product measurements are consistent with standards and procedures established by the project.

Note that standard conformance, regulatory concerns, or customer requirements are to be included in any tailoring for projects that use an agile methodology. As well, the SQA cannot replace the project team's responsibility for the product quality. The IEEE 730 standard insists that resources and information necessary for performing SQA are identified, made available, allocated, and used in projects. The project team should ask the following questions, regarding standards, practices and conventions, at the project planning stage:

- what government regulations and industry standards are applicable to this project? Have all laws, regulations, standards, practices, conventions, and rules been identified?
- what specific standards are applicable to this project? Have specific criteria and standards against which all project plans are to be evaluated been identified and shared within the project team?
- what organizational reference documents (such as standard operating procedures, coding standards, and document templates) are applicable to this project?
- have specific criteria and standards against which software life cycle processes (e.g., supply, development, operation, maintenance, and support processes including QA) are to be evaluated been identified and shared with the project team?
- what reference documents are appropriate to include in the SQAP?
- is the SQA expected to assess the project's compliance with applicable regulations, standards, organizational documents, and project reference documents?

During the project execution, the project team will have to verify:

- the project level of adherence to plans, product quality, processes (such as the life cycle processes), and activities with regards to the applicable standards, procedures, and requirements. Additionally, they need to keep quality records of these activities in case of additional verification;

- that appropriate coding standards and conventions, identified during planning, are applied;
- that the criteria, standards, and contractual requirements against which software engineering practices, development environment, and libraries are to be reviewed have been identified and documented;
- deviations to the SQAP are reported, authorized, and documented.

4.10 SUCCESS FACTORS

The implementation of standards into practice may be facilitated or hindered depending on the factors at play in the organization. The following text box lists a few of these factors.

Factors that Foster Software Quality

1) Understanding the standards used;
2) Educating users about the benefits and hazards of the standards chosen;
3) Promoting standards to be used by upper management;
4) Using the right standards related to the application domain;
5) Using the standards only if they provide a benefit; otherwise, they will not be used or be useful;
6) Adopting standards although they are not mandatory.

Factors that may Adversely Affect Software Quality

1) Using standards to the letter and not according to the essence;
2) Using too many standards.

4.11 FURTHER READING

GLAZER H., DALTON J., ANDERSON D., KONRAD M., and SHRUM S. *CMMI or Agile: Why Not Embrace Both!*. Software Engineering Institute, Carnegie Mellon University, Pittsburgh, USA, 2008, 41 p.

MOORE J. *The Road Map to Software Engineering: A Standards-Based Guide*. Wiley-IEEE Computer Society Press, Hoboken, New Jersey, USA, 2006, 440 p.

PFLEEGER S.L., FENTON N.E., and PAGE P. Evaluating software engineering standards. *IEEE Computer*, vol. 27, issue 9, 1994, pp. 71–79.

4.12 EXERCISES

4.1 Name five advantages and five disadvantages of software engineering standards.

4.2 Provide five advantages and five disadvantages of the CMMI-DEV.

4.3 If an organization implements all the standards needed for the development of its software products, is this sufficient to produce quality software?

4.4 If an organization implements all the standards needed for the development of its software products, is this sufficient to produce software and make a profit or bring other benefits to the organization?

4.5 Can the CMMI-DEV model be used for development in an organization with 10 developers? Explain why or why not.

4.6 Can software be developed in an environment that has adopted agile practices and the CMMI-DEV?

4.7 What could be the benefits of using ISO 29110 for a small organization?

Chapter 5

Reviews

After studying this chapter, you will be able to:

- understand the value of different types of reviews;
- understand the personal review;
- understand the desk-check type of peer review;
- understand the reviews described in the ISO/IEC 20246 standard, the CMMI®, and the IEEE 1028 standard;
- understand the walk-through and inspection review;
- understand the project launch review and project lessons learned review;
- understand the measures related to reviews;
- understand the usefulness of reviews for different business models;
- understand the requirements of the IEEE 730 standard regarding reviews.

5.1 INTRODUCTION

Humphrey (2005) [HUM 05] collected years of data from thousands of software engineers showing that they unintentionally inject 100 defects per thousand lines of code. He also indicates that commercial software typically includes from one to ten errors per thousand lines of code [HUM 02]. These errors are like hidden time bombs that will explode when certain conditions are met. We must therefore put practices in place to identify and correct these errors at each stage of the development and maintenance cycle. In a previous chapter, we introduced the concept of the cost of quality. The calculation of the cost of quality is:

$$
\begin{aligned}
\text{Quality costs} = \ & \text{Prevention costs} \\
& + \text{Appraisal or evaluation costs} \\
& + \text{Internal and external failure costs} \\
& + \text{Warranty claims and loss of reputation costs}
\end{aligned}
$$

Software Quality Assurance, First Edition. Claude Y. Laporte and Alain April.
© 2018 the IEEE Computer Society, Inc. Published 2018 by John Wiley & Sons, Inc.

The detection cost is the cost of verification or evaluation of a product or service during the various stages of the development process. One of the detection techniques is conducting reviews. Another technique is conducting tests. But it must be remembered that the quality of a software product begins in the first stage of the development process, that is to say, when defining requirements and specifications. Reviews will detect and correct errors in the early phase of development while tests will only be used when the code is available. So we should not wait for the testing phase to begin to look for errors. In addition, it is much cheaper to detect errors with reviews than with testing. This does not mean we should neglect testing since it is essential for the detection of errors that reviews cannot discover.

Unfortunately, many organizations do not perform reviews and rely on testing alone to deliver a quality product. It often happens that, given the many problems throughout development, the schedule and budget have been compressed to the point that tests are often partially, if not completely, eliminated from the development or maintenance process. In addition, it is impossible to test a large software product completely. For example, for software that has barely 100 decisions (branches), there are more than 184,756 possible paths to test and for software with 400 decisions, there are $1.38E + 11$ possible paths to test [HUM 08].

"A year of work experience in an environment that requires regular participation in peer reviews is the equivalent of 3 years of work experience in an environment where you do not use these reviews."

Gerald Weinberg, SQTE Magazine, March 2003

In this chapter, we present reviews. We will see that there are many types of reviews ranging from informal to formal.

Informal reviews are characterized as follows:

– There is no documented process to describe reviews and they are carried out in many ways by different people in the organization;
– Participants' roles are not defined;
– Reviews have no objective, such as fault detection rate;
– They are not planned, they are improvised;
– Measures, such as the number of defects, are not collected;
– The effectiveness of reviews is not monitored by management;
– There is no standard that describes them;
– No checklist is used to identify defects.

Formal reviews will be discussed in this chapter as defined in the following text box.

Review

A process or meeting during which a work product, or set of work products, is presented to project personnel, managers, users, customers, or other interested parties for comment or approval.

ISO 24765 [ISO 17a]

A process or meeting during which a software product, set of software products, or a software process is presented to project personnel, managers, users, customers, user representatives, auditors, or other interested parties for examination, comment, or approval.

IEEE 1028 [IEE 08b]

In this chapter, we present two types of review as defined in the IEEE 1028 standard [IEE 08b]: the walk-through and the inspection. Professor Laporte contributed to the latest revision of this standard. We will also describe two reviews that are not defined in the standard: the personal review and the desk-check. These reviews are the least formal of all of the types of reviews. They are included here because they are simple and inexpensive to use. They can also help organizations that do not conduct formal reviews to understand the importance and benefits of reviews in general and establish more formal reviews.

Peer reviews are product activity reviews conducted by colleagues during development, maintenance, or operations in order to present alternatives, identify errors, or discuss solutions. They are called peer reviews because managers do not participate in this type of review. The presence of managers often creates discomfort as participants hesitate to give opinions that could reflect badly on their colleagues and the person who requested the review may be apprehensive of negative feedback from his own manager.

Figure 5.1 shows the variety of reviews as well as when they can be used throughout the software development cycle. Note the presence of phase-end reviews, document reviews, and project reviews. These reviews are used internally or externally for meetings with a supplier or customer.

Figure 5.2 lists objectives for reviews. It should be noted that each type of review does not target all of these objectives simultaneously. We will consider what the objectives are for each type of review in a subsequent section.

The types of reviews that should be conducted and the documents and activities to be reviewed or audited throughout the project are usually determined in the software quality assurance plan (SQAP) for the project, as explained by the IEEE 730 standard

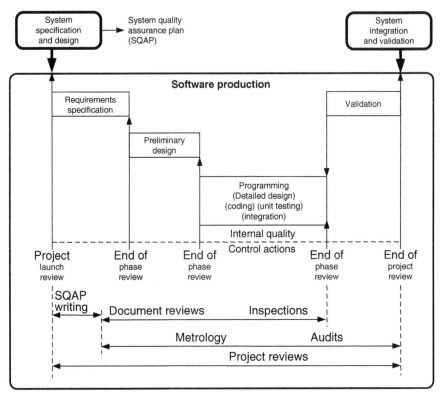

Figure 5.1 Types of reviews used during the software development cycle [CEG 90] (© 1990 – ALSTOM Transport SA).

[IEE 14], or in the project management plan, as defined by the ISO/IEC/IEEE 16326 standard [ISO 09]. The requirements of the IEEE 730 standard will be presented at the end of this chapter.

As illustrated in Figure 5.3, to produce a document, that is, a software product (e.g., documentation, code, or test), source documents are usually used as inputs to the review process. For example, to create a software architecture document, the developer should use source material such as the system requirements document, the software requirements, a software architecture document template, and possibly a software architecture style guide.

A review of just the software product, for example, a requirements document, by its author is not sufficient to detect a large number of errors. As illustrated in Figure 5.4, once the author has completed the document, the software product is compared by his or her peers against the source documents used. At the end of the review, peers who participated in the review will have to decide if the document produced by the author is satisfactory as is, if significant corrections are required or if the document must be

- Identify defects
- Assess / measure the quality of a document (e.g., the number of defects per page)
- Reduce the number of defects by correcting the defects identified
- Reduce the cost of preparing future documents (i.e., by learning the type of defects each developer makes, it is possible to reduce the number of defects injected in a new document)
- Estimate the effectiveness of a process (e.g., the percentage of fault detection)
- Estimate the efficiency of a process (e.g., the cost of detection or correction of a defect)
- Estimate the number of residual defects (i.e., defects not detected when software is delivered to customer)
- Reduce the cost of tests
- Reduce delays in delivery
- Determine the criteria for triggering a process
- Determine the completion criteria of a process
- Estimate the impacts (e.g., cost) of continuing with current plans, e.g. cost of delay, recovery, maintenance, or fault remediation
- Estimate the productivity and quality of organizations, teams and individuals
- Teach personnel to follow the standards and use templates
- Teach personnel how to follow technical standards
- Motivate personnel to use the organization's documentation standards
- Prompt a group to take responsibility for decisions
- Stimulate creativity and the contribution of the best ideas with reviews
- Provide rapid feedback before investing too much time and effort in certain activities
- Discuss alternatives
- Propose solutions, improvements
- Train staff
- Transfer knowledge (e.g., from a senior developer to a junior)
- Present and discuss the progress of a project
- Identify differences in specifications and standards
- Provide management with confirmation of the technical state of the project
- Determine the status of plans and schedules
- Confirm requirements and their assignment in the system to be developed

Figure 5.2   Objectives of a review.
Source: Adapted from Gilb (2008) [GIL 08] and IEEE 1028 [IEE 08b].

corrected by the author and peer reviewed again. The third option is only used when the revised document is very important to the success of the project. As discussed below, when an author makes many corrections to a document, it inadvertently creates other errors. It is these errors that we hope to detect with another peer review.

The advantage of reviews is that they can be used in the first phase of a project, for example, when requirements are documented, whereas tests can only be performed when the code is available. For example, if we depend on tests alone and errors are injected when writing the requirements document, these will only become apparent

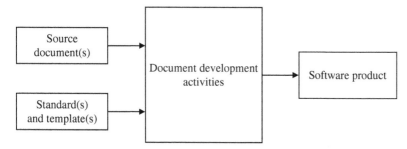

Figure 5.3   Process of developing a document.

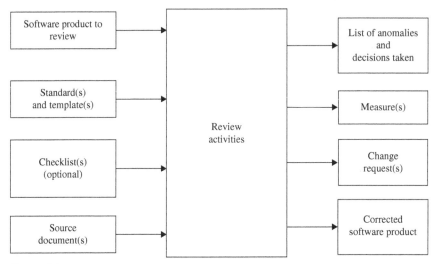

Figure 5.4   Review process.

when the code is available. However, if we use reviews, then we can also detect and correct errors during the requirements phase. Errors are much easier to find and are less expensive to correct at this phase. Figure 5.5 compares errors detected using only tests and using a type of review called inspections.

For illustration purposes, we used an error detection rate of 50%. Several organizations have achieved higher detection rates, that is, well over 80%. This figure clearly illustrates the importance of establishing reviews from the first phase of development.

5.2 PERSONAL REVIEW AND DESK-CHECK REVIEW

This section describes two types of reviews that are inexpensive and very easy to perform. Personal reviews do not require the participation of additional reviewers, while desk-check reviews require at least one other person to review the work of the developer of a software product.

5.2.1 Personal Review

A personal review is done by the person reviewing his own software product in order to find and fix the most defects possible. A personal review should precede any activity that uses the software product under review.

The principles of a personal review are [POM 09]:

- – find and correct all defects in the software product;
- – use a checklist produced from your personal data, if possible, using the type of defects that you are already aware of;

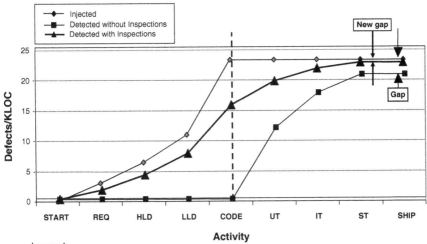

Legend:
- REQ = requirement
- HLD = high-level architecture design
- LLD = detailed design
- CODE = coding and debugging
- UT = unit testing
- IT = integration testing
- ST = system testing
- SHIP = delivery to customer
- KLOC = thousand lines of code

Figure 5.5   Error detection during the software development life cycle [RAD 02].

– follow a structured review process;

– use measures in your review;

– use data to improve your review;

– use data to determine where and why defects were introduced and then change
your process to prevent similar defects in the future.

Checklist

A checklist is used as a memory aid. A checklist includes a list of criteria to verify the
quality of a product. It also ensures consistency and completeness in the development of
a task. An example of a checklist is a list that facilitates the classification of a defect in a
software product (e.g., an oversight, a contradiction, an omission).

The following practices should be followed to develop an effective and efficient personal review [POM 09]:

– pause between the development of a software product and its review;

– examine products in hard copy rather than electronically;

– check each item on the checklist once completed;

– update the checklists periodically to adjust to your personal data;

– build and use a different checklist for each software product;

– verify complex or critical elements with an in depth analysis.

Figure 5.6 outlines the process of a personal review.

As we can see, personal reviews are very simple to understand and perform. Since the errors made are often different for each software developer, it is much more efficient to update a personal checklist based on errors noted in previous reviews.

ENTRY CRITERIA
• None

INPUT
• Software product to review

ACTIVITIES
1. Print:
• Checklist for the software product to be reviewed
• Standard (if applicable)
• Software product to review
2. Review the software product, using the first item on the checklist and cross this item off when the review of the software product is completed
3. Continue review of the software product using the next item on the checklist and repeat until all the items in the list have been checked
4. Correct any defects identified
5. Check that each correction did not create other defects.

EXIT CRITERIA
• Corrected software product

OUTPUT
• Corrected software product

MEASURE
• Effort used to review and correct the software product measured in person-hours with an accuracy of +/–15 minutes.

Figure 5.6   Personal review process.
Source: Adapted from Pomeroy-Huff et al. (2009) [POM 09].

5.2.2 Desk-Check Reviews

A type of peer review that is not described in standards is the desk-check review [WAL 96], sometimes called the Pass around [WIE 02]. It is important to explain this type of peer review because it is inexpensive and easy to implement. It can be used to detect anomalies, omissions, improve a product, or present alternatives. This review is used for low-risk software products, or if the project plan does not allow for more formal reviews. According to Wiegers, this review is less intimidating than a group review such as a walk-through or inspection. Figure 5.7 describes the process for this type of review.

As shown in Figure 5.7, there are six steps. Initially, the author plans the review by identifying the reviewer(s) and a checklist. A checklist is an important element of a review as it enables the reviewer to focus on only one criterion at a time. A checklist is a reflection of the experience of the organization. Then, individuals review the software product document and note comments on the review form provided by the author. When completed, the review form can be used as "evidence" during an audit.

In this book, several checklists are presented. Here is a list of some important features of checklists:

– each checklist is designed for a specific type of document (e.g., project plan, specification document);

– each item of a checklist targets a single verification criteria;

– each item of a checklist is designed to detect major errors. Minor errors, such as misspellings, should not be part of a checklist;

– each checklist should not exceed one page, otherwise it will be more difficult to use by the reviewers;

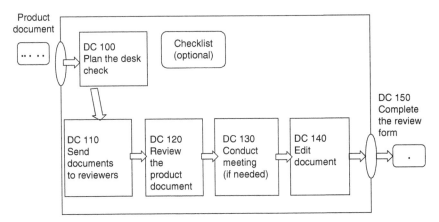

Figure 5.7   Desk-check review.

– each checklist should be updated to increase efficiency;
– each checklist includes a version number and a revision date.

During a course on peer reviews given in Sweden by Professor Laporte, an employee of an organization proudly shared the code checklist he had developed. His list included more than 250 items!

The employee was kindly asked to share some elements of his list with other course participants who quickly concluded that a large number of items on the list could be detected by a page-formatting tool. A large number of elements could also be detected by a good compiler. Participants also identified a list of items that detected minor errors. Following this discussion, there remained a list of about one page of detection criteria for major errors.

The following text box presents a generic checklist, that is, a checklist that can be used for almost any type of document to be reviewed (e.g., project plan, architecture). For each type of software product (e.g., requirements or design), a specific checklist will be used. For a list designed to facilitate the detection of errors in requirements, we could add the EX identifier and include the following element: EX 1 (testable)—the requirement must be testable. For a list of verifications for a test plan, one might use the TP identifier.

 Generic Checklist

LG 1 (COMPLETE). All pertinent information should be included or referenced.
LG 2 (RELEVANT). All information must be relevant to the software product.
LG 3 (BRIEF). Information must be stated succinctly.
LG 4 (CLEAR). Information must be clear to all reviewers and users of the document.
LG 5 (CORRECT). Information does not contain errors.
LG 6 (COHERENT). Information must be consistent with all other information in the document and its source document(s).
LG 7 (UNIQUE). Ideas must be described once and referenced afterward.
Adapted from Gilb and Graham (1993) [GIL 93]

In the third step of the desk-check process, the reviewers verify the document and record their comments on the review form. The author reviews the comments as part of step 4. If the author agrees with all the comments, he incorporates them into his document. However, if the author does not agree, or if he believes the comments have a major impact, then he should convene a meeting with the reviewers to discuss the comments. After this meeting, one of three options should be considered: the comment is incorporated as is, the comment is ignored, or it is incorporated with modifications. For the next step, the author can make the corrections and note the effort spent reviewing and correcting the document, that is, the time spent by the reviewers as well as the time spent by the author to correct the document and conduct the meeting if this is the case. The activities of the desk-check (DC) review are described in Figure 5.8. In the final step, the author completes the review form illustrated in Figure 5.9.

ENTRY CRITERIA
- The document is ready for a review

INPUT
- Software product to review

DC 100. Plan the Desk-Check
Author:
- Identifies reviewers
- Chooses the checklist(s) to use
- Completes the first part of the review form

DC 110. Send documents to reviewers
Author:
- Provides the following documents to the reviewers:
  - Software product to review
  - Review form
  - Checklist(s)

DC 120. Review the software product
Reviewers:
- Check the software product against the checklist
- Complete the review form with
  - comments
  - effort to conduct the review
- Sign and return the form to the author

DC 130. Call a meeting (if needed)
Author:
- Reviews the comments
  - If the author agrees with all the comments, they are incorporated in the software product
  - If the author does not agree with all the comments, or believes some comments have a significant impact, then the author:
    - Convenes a meeting with the reviewers
    - Leads the meeting to discuss the comments and determine course of action:
      - Incorporate the comment as is
      - Ignore the comment
      - Incorporate the comment with modifications

DC 140. Correct the software product
The author incorporates the comments received.

DC 150. Complete the review form
Author:
- Completes the review form with:
  - Total effort (i.e., by all the reviewers) required to review the software product
  - Total effort required to correct the software product
- Signs the review form

EXIT CRITERIA
- Corrected software product

OUTPUT
- Corrected software product
- Completed and signed review form

MEASURE
- Effort required to review and correct the software product (person hours).

Figure 5.8   Desk-check review activities.

Desk check review form

Comment No.	Document page	Line # / location	Comments	Disposition of comments*	Remarks
1					
2					
3					
4					
5					
6					
7					
8					
9					
10					
11					
12					
13					
14					
15					
16					
17					
18					
19					
20					
21					
22					
23					
24					

Name of author: _____
Document title: _____
Document version: _____

Review date (yyyy-mm-dd): _____
Reviewer name: _____

Disposition of comment: Inc: Incorporate as is;   NOT: Not incorporate,   MOD: Incorporate with modification

Effort to review document (hour): _____
Effort to correct document (hour): _____

Signature of reviewer: _____

Signature of author: _____

Figure 5.9   Desk-Check review form.

Figure 5.9 illustrates a standard form used by reviewers to record their comments and the time they devoted to the revision of the document. The author of the document collects these data and adds the time it took him to correct the document. The forms will be retained by the author as "evidence" for an audit by the SQA of the organization the author belongs to, or by the SQA of the customer.

As an alternative to the distribution of hard copies to reviewers, one can place an electronic copy of the document, the review form and the checklist in a shared folder on the Intranet. Reviewers are invited to provide comments as annotations to documents over a defined period of time. The author can then view the annotated document, review the comments, and continue the Desk-Check review as described above.

In the next sections, we describe more formal reviews.

Support site for Weigers Peer Reviews http://www.processimpact.com/pr_goodies.shtml

5.3 STANDARDS AND MODELS

In this section, we present the ISO/IEC 20246 standard on work product reviews, the Capability Maturity Model Integration (CMMI) model, and the IEEE 1028 standard, which lists requirements and procedures for software reviews.

5.3.1 ISO/IEC 20246 Software and Systems Engineering: Work Product Reviews

The purpose of ISO/IEC 20246 Work Product Reviews is [ISO 17d]: "to provide an International Standard that defines work product reviews, such as inspections, reviews and walk-throughs, that can be used at any stage of the software and systems life cycle. It can be used to review any system and software work product. ISO/IEC 20246 defines a generic process for work product reviews that can be configured based on the purpose of the review and the constraints of the reviewing organization. The intent is to describe a generic process that can be applied both efficiently and effectively by any organization to any work product. The main objectives of reviews are to detect issues, to evaluate alternatives, to improve organizational and personal processes, and to improve work products. When applied early in the life cycle, reviews are typically shown to reduce the amount of unnecessary rework on a project. The work product review techniques presented in ISO/IEC 20246 can be used at various stages of the generic review process to identify defects and evaluate the quality of the work product."

Work Product

Artefact produced by a process.

  Example: Project plan, requirements specification, design documentation, source code, test plan, test meeting minutes, schedules, budgets, and incident reports.

  Note 1 to entry: A subset of the work products will be baselined to be used as the basis of further work and some will form the set of project deliverables.

ISO/IEC 20246

ISO 20246 includes an annex that describes the alignment of the activities of the ISO 20246 standard and the procedures of the IEEE 1028 standard presented below.

5.3.2 Capability Maturity Model Integration

The CMMI® for Development (CMMI-DEV) [SEI 10a] is widely used by many industries. This model describes proven practices in engineering. In this model, a part of the "Verification" process area is devoted to peer reviews. Other verification activities will be considered in more detail in a later chapter. Figure 5.10 is an extract of the staged representation of the CMMI-DEV which describes peer reviews.

VERIFICATION
A process area in the engineering category of Maturity Level 3

Purpose
The purpose of the process area "Verification" is to ensure that selected work products meet their specified requirements.

Peer reviews are an important part of verification and are a proven mechanism for effective defect removal. An important corollary is to develop a better understanding of the work products and the processes that produced them so that defects can be prevented and process improvement opportunities can be identified.

Peer reviews involve a methodical examination of work products by the producers' peers to identify defects and other changes that are needed.

Example of Peer review methods include the following:
        Inspections;
        Structured walk-throughs
        Deliberate refactoring
        Pair programming

Specific Objective 2 - Perform Peer Reviews
        Specific practice 2.1 Prepare for Peer reviews
        Specific practice 2.2 Conduct Peer reviews
        Specific practice 2.3 Analyse Peer review data

Figure 5.10 Peer reviews as described in the process area "Verification" of the CMMI-DEV. *Source*: Adapted from Software Engineering Institute (2010) [SEI 10a].

The process and product quality assurance process areas provide the following list of issues to be addressed when implementing peer reviews [SEI 10a]:

- Members are trained and roles are assigned for people attending the peer reviews.
- A member of the peer review who did not produce this work product is assigned to perform the quality assurance role.
- Checklists based on process descriptions, standards, and procedures are available to support the quality assurance activity.
- Non-compliance issues are recorded as part of the peer review report and are tracked and escalated outside the project when necessary.

According to the CMMI-DEV, these reviews are performed on selected work products to identify defects and to recommend other changes required. The peer review is an important and effective software engineering method, applied through inspections, walk-throughs or a number of other review procedures.

Reviews that meet the CMMI requirements listed in Figure 5.10 are described in the following sections.

5.3.3 The IEEE 1028 Standard

The IEEE 1028-2008 Standard for Software Reviews and Audits [IEE 08b] describes five types of reviews and audits and the procedures required for the completion of each type of review and audit. Audits will be presented in the next chapter. The introductory text of the standard indicates that the use of these reviews is voluntary. Although the use of this standard is not mandatory, it can be imposed by a client contractually.

The purpose of this standard is to define reviews and systematic audits for the acquisition, supply, development, operation and maintenance of software. This standard describes not only "what to do" but also how to perform a review. Other standards define the context in which a review is performed and how the results of the review are to be used. Examples of such standards are provided in Table 5.1.

Table 5.1 Examples of Standards that Require the Use of Systematic Reviews

Standard identification	Title of the standard
ISO/IEC/IEEE 12207	Software Life Cycle Processes
IEEE 1012	IEEE Standard for System and Software Verification and Validation.
IEEE 730	IEEE Standard for Software Quality Assurance Processes

The IEEE 1028 standard provides minimum acceptable conditions for systematic reviews and software audits including the following attributes:

– team participation;
– documented results of the review;
– documented procedures for conducting the review.

Conformance to the IEEE 1028 standard for a specific review, such as an inspection, can be claimed when all mandatory actions (indicated by "shall") are carried out as defined in this standard for the review type used.

This standard provides descriptions of the particular types of reviews and audits included in the standard as well as tips. Each type of review is described with clauses that contain the following information [IEE 08b]:

a) Introduction to review: describes the objectives of the systematic review and provides an overview of the systematic review procedures;

b) Responsibilities: defines the roles and responsibilities needed for the systematic review.

c) Input: describes the requirements for input needed by the systematic review;

d) Entry criteria: describes the criteria to be met before the systematic review can begin, including the following:
  1) Authorization,
  2) Initiating event;

e) Procedures: details the procedures for the systematic review, including the following:
  1) Planning the review;
  2) Overview of procedures;
  3) Preparation;
  4) Examination/evaluation/recording of results;
  5) Rework/follow-up;

f) Exit criteria: describe the criteria to be met before the systematic review can be considered complete;

g) Output: describes the minimum set of deliverables to be produced by the systematic review.

5.3.3.1 Application of the IEEE 1028 Standard

Procedures and terminology defined in this standard apply to the acquisition of software, supply, development, operation, and maintenance processes requiring systematic reviews. Systematic reviews are performed on a software product according to the requirements of other local standards or procedures. The term "software product" is used in this standard in a very broad sense. Examples of software products include specifications, architecture, code, defect reports, contracts, and plans.

Anomaly
Any condition that deviates from expectations based on requirements specifications, design documents, user documents, standards, etc., or from someone's perceptions or experiences.
    Note: Anomalies may be found during, but not limited to, the review, test, analysis, compilation, or use of software products or applicable documentation.

IEEE 1028 [IEE 08b]

The IEEE 1028 standard differs significantly from other software engineering standards in that it does not only enumerate a set of requirements to be met (i.e., "what to do"), such as "the organization shall prepare a quality assurance plan," but it also describes "how to do" at a level of detail that allows someone to conduct a systematic review properly. For an organization that wants to implement these reviews, the text of this standard can be adapted to the notation of the processes and procedures of the organization, adjusting the terminology to that which is commonly used by the organization and, after using them for a while, improve the descriptions of the review.

This standard concerns only the application of a review and not their need or the use of the results. The types of reviews and audits are [IEE 08b]:

– management review: a systematic evaluation of a software product or process performed by or on behalf of the management that monitors progress, determines the status of plans and schedules, confirms requirements and their system allocation, or evaluates the effectiveness of the management approaches used to achieve fitness for purpose;

– technical review: a systematic evaluation of a software product by a team of qualified personnel that examines the suitability of the software product for its intended use and identifies discrepancies from specifications and standards;

– inspection: a visual examination of a software product to detect and identify software anomalies including errors and deviations from standards and specifications;

– walk-through: a static analysis technique in which a designer or programmer leads members of the development team and other interested parties through a software product, and the participants ask questions and make comments about any anomalies, violation of development standards, and other problems;

– audit: an independent assessment, by a third party, of a software product, a process or a set of software processes to determine compliance with the specifications, standards, contractual agreements, or other criteria.

Table 5.2 summarizes the main characteristics of reviews and audits of the IEEE 1028 standard. These features will be discussed in more detail in this chapter and in the following chapter on audits.

In the following sections, walk-through and inspection reviews are described in detail. These reviews are described to clearly demonstrate the meaning of a "systematic review" as opposed to improvised and informal reviews.

5.4 WALK-THROUGH

"The purpose of a walk-through is to evaluate a software product. A walk-through can also be performed to create discussion for a software product" [IEE 08b]. The main objectives of the walk-through are [IEE 08b]:

– find anomalies;
– improve the software product;
– consider alternative implementations;
– evaluate conformance to standards and specifications;
– evaluate the usability and accessibility of the software product.

Other important objectives include the exchange of techniques, style variations, and the training of participants. A walk-through can highlight weaknesses, for example, problems of efficiency and readability, modularity problems in the design or the code or non-testable requirements. Figure 5.11 shows the six steps of the walk-through. Each step is composed of a series of inputs, tasks, and outputs.

5.4.1 Usefulness of a Walk-Through

There are several reasons for the implementation of a walk-through process:

– identify errors to reduce their impact and the cost of correction;
– improve the development process;
– improve the quality of the software product;
– reduce development costs;
– reduce maintenance costs.

Table 5.2 Characteristics of Reviews and Audits Described in the IEEE 1028 Standard

	Management review	Technical review	Inspection	Walk-through	Audit
Objective	Monitor progress	Evaluate conformance to specifications and plans	Find anomalies; verify resolution; verify product quality	Find anomalies, examine alternatives; improve product; forum for learning	Independently evaluate conformance with objective standards and regulations
Recommended group size	Two or more people	Two or more people	3–6	2–7	1–5
Volume of material	Moderate to High	Moderate to High	Relatively low	Relatively low	Moderate to High
Leadership	Usually the responsible manager	Usually the lead engineer	Trained facilitator	Facilitator or author	Lead auditor
Management participates	Yes	When management evidence or resolution may be required	No	No	No; however management may be called upon to provide evidence
Output	Management review documentation	Technical review documentation	Anomaly list, anomaly summary, inspection documentation	Anomaly list, action items, decisions, follow-up proposals	Formal audit report; observations, findings, deficiencies

Source: Adapted from IEEE 1028 [IEE 08b].

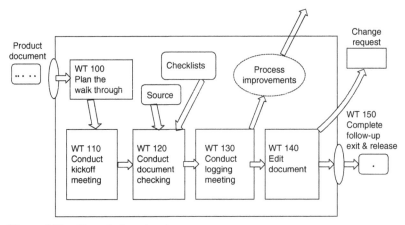

Figure 5.11   The walk-through review.
Source: Adapted from Holland (1998) [HOL 98].

5.4.2 Identification of Roles and Responsibilities

Four roles are described in the IEEE 1028: leader, recorder, author, and team member. Roles can be shared among team members. For example, the leader or author may play the role of recorder and the author could also be the leader. But, a walk-through shall include at least two members.

The standard defines the roles as follow (adapted from IEEE 1028 [IEE 08b]):

- Walk-through leader
  o conduct the walk-through;
  o handle the administrative tasks pertaining to the walk-through (such as distributing documents and arranging the meeting);
  o prepare the statement of objectives to guide the team through the walk-through;
  o ensure that the team arrives at a decision or identified action for each discussion item;
  o issue the walk-through output.

- Recorder
  o note all decisions and identified actions arising during the walk-through meeting;
  o note all comments made during the walk-through that pertain to anomalies found, questions of style, omissions, contradictions, suggestions for improvement, or alternative approaches.

- Author
  o present the software product in the walk-through.

- Team member

- o adequately prepare for and actively participate in the walk-through;
- o identify and describe anomalies in the software product.

The IEEE 1028 standard lists improvement activities using data collected from the walk-throughs. These data should [IEE 08b]:

- − be analyzed regularly to improve the walk-through process;
- − be used to improve operations that produce software products;
- − present the most frequently encountered anomalies;
- − be included in the checklists or in assigning roles;
- − be used regularly to assess the checklists for superfluous or misleading questions;
- − include preparation time and meetings; the number of participants should be considered to determine the relationship between the preparation time and meeting and the number and severity of anomalies detected.

To maintain the efficiency of walk-throughs, the data should not be used to evaluate the performance of individuals.

IEEE 1028 also describes the procedures of walk-throughs.

5.5 INSPECTION REVIEW

This section briefly describes the inspection process that Michael Fagan developed at IBM in the 1970s to increase the quality and productivity of software development.

The purpose of the inspection, according to the IEEE 1028 standard, is to detect and identify anomalies of a software product including errors and deviations from standards and specifications [IEE 08b]. Throughout the development or maintenance process, developers prepare written materials that unfortunately have errors. It is more economical and efficient to detect and correct errors as soon as possible. Inspection is a very effective method to detect these errors or anomalies.

History of the Inspection Process at IBM

An employee of IBM, Michael Fagan, was working in a computer chip design and manufacturing department. Since the tests were not sufficient to detect errors, he developed a technique to examine the designs before they were transferred to the production department. This approach could detect defects that tests had not detected, thus reducing losses and delays to start the production.

In 1971, Fagan was transferred to the software development department. Upon his arrival, he found that software development was chaotic. Even in the absence of measures, he considered that a large percentage of the development budget was allocated to rework. He estimated the cost of the rework and found that 30–80% of the development budget was used to correct defects unintentionally injected by developers. He decided to use a similar approach to the one he had used for the detection of defects in computer chips, that is, a review, to detect design and coding errors.

The effort required to perform the reviews was low compared to the high amount of rework that would have been needed without the review. It is these results that triggered the development of the inspection process. Fagan proposed to reduce the number of injected defects and to detect and correct errors as close to the task within which they were injected.

The objectives developed by the Fagan Inspection Process are:

– find and fix all defects in the product;

– find and fix all defects in the development process that produces the defects in a product (e.g., remove the causes of defects in the product).

For more than three and a half years, Fagan, hundreds of developers and their managers conducted inspections. During this period, the inspection process had been constantly improved. In 1976, he published an article on the inspection of the design and of code in the *IBM Systems Journal* [FAG 76]. Given the results, IBM asked Fagan to promote the inspection process in other divisions of the company. For helping to save millions of dollars, Fagan was awarded the largest individual corporate award ever awarded by IBM at that time.

Adapted from Broy and Denert (2002) [BRO 02]

According to the IEEE 1028 standard, inspection allows us to (adapted from [IEE 08b]):

a) verify that the software product satisfies its specifications;

b) check that the software product exhibits the specified quality attributes;

c) verify that the software product conforms to applicable regulations, standards, guidelines, plans, specifications, and procedures;

d) identify deviations from provisions of items (a), (b), and (c);

e) collect data, for example, the details of each anomaly and effort associated with their identification and correction;

f) request or grant waivers for violation of standards where the adjudication of the type and extent of violations are assigned to the inspection jurisdiction;

g) use the data as input to project management decisions as appropriate (e.g., to make trade-offs between additional inspections versus additional testing).

Figure 5.12 shows the major steps of the inspection process. Each step is composed of a series of inputs, tasks and outputs.

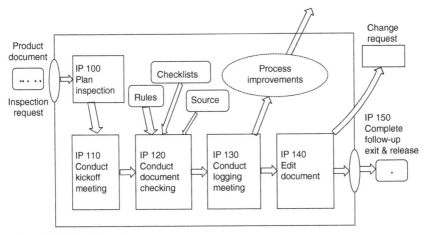

Figure 5.12 The inspection process.
Source: Adapted from Holland (1998) [HOL 98].

The IEEE 1028 standard provides guidelines for typical inspection rates, for different types of documents, such as anomaly recording rates in terms of pages or lines of code per hour. As an example, for the requirements document, IEEE 1028 recommends an inspection rate of 2–3 pages per hour. For source code, the standard recommends an inspection rate of 100–200 lines of code per hour.

An organization that has just started the implementation of formal reviews, such as inspections, could review documents at a higher rate than the inspection rates proposed by IEEE 1028. As more reliable measures are collected, such as defect detection rate and defect removal effectiveness, the organization can decide to reduce the review rate in order to achieve higher detection rates and therefore reduce the defect escape rate. Section 5.8 presents measures and an example of the defect detection rate calculation.

Finally, IEEE 1028 also describes the procedures of inspection.

5.6 PROJECT LAUNCH REVIEWS AND PROJECT ASSESSMENTS

In the SQAP of their projects, many organizations plan a project launch or kick-off meeting as well as a project assessment review, also called a lessons learned review.

5.6.1 Project Launch Review

The project launch review is a management review of: the milestone dates, requirements, schedule, budget constraints, deliverables, members of the development team, suppliers, etc. Some organizations also conduct kick-off reviews at the beginning of each of the major phases of the project when projects are spread over a long period of time (as in several years).

Before the start of a project, team members ask themselves the following questions: who will the members of my team be? Who will be the team leader? What will my role and responsibilities be? What are the roles of the other team members and their responsibilities? Do the members of my team have all the skills and knowledge to work on this project?

The following text box describes the kick-off review meeting used for software projects at Bombardier Transport.

Case Study at Bombardier Transport (Adapted from Laporte et al. (2007) [LAP 07b])

A project launch session is usually done at the beginning of a new project or at the beginning of a project phase. It can also be done for an iterative development project to prepare for the next iteration. In this case, it is called a project relaunch session. This type of session is also well suited in cases where the performance of a project and/or process must be improved and when a project has to be rectified.

Depending on the size, complexity, and type of project (e.g., new development or re-use of critical software development), a typical project launch session will last 1–2 days in one location. During a project launch session, it is important that all team members are fully dedicated to this activity. To reduce disturbances (e.g., telephone calls), the project launch meeting may be held outside of the project office or building. The following table shows a typical schedule for a one-day project launch session. As the table shows, the theme of the Software Management Project (SMP), processes, roles and responsibilities are first discussed in agenda item 4 and then in item 8. Roles and responsibilities (R&R) are discussed under the heading Software Quality Assurance and Verification & Validation.

A project launch review is a workshop, usually led by a facilitator, during which the project team members define the project plan, including activities, deliverables, and schedule. The project launch workshop can last between one and three days. But for a typical project at Bombardier Transport, a one-day session is normally sufficient.

To illustrate the roles of team members, an example of project planning performed during the project launch session is described below. The objectives of the project launch review at Bombardier Transport are:

- define the project plan using an integrated team approach;

- ensure common understanding of objectives, processes, deliverables, and the role and responsibilities (R&R) of all team members;
- facilitate the exchange of information and provide "just in time" training to project members.

Typical Agenda of a Project Launch Meeting at Bombardier Transport

Time	Agenda
08h30	Welcome, agenda review, and discussions regarding participant expectations Meeting roles to assign: secretary and time-keeper
09h00	Overview of the software engineering process at Bombardier Transport
10h30	Software project management process:
	1. Project inputs
	2. Project scope, constraints and assumptions
	3. Project iterations and their associated objectives (e.g., imposed milestones)
	4. Structure of the project team and role assignment
	5. High-level architecture
	6. Tailoring of deliverables (e.g., for each iteration)
	7. Personnel requirements
	8. Relationships with other groups and associated roles/responsibilities
	9. Identification and analysis of risks
12h00	Lunch break
13h00	Software project management process (continued)
14h30	Break
14h45	Software development process:
	• Define requirements and their attributes
	• Description of traceability
15h00	Software Configuration management process:
	• Process overview
	• Identification of configuration items
	• Identification of the Baseline for each iteration
	• Audits and version control
15h45	Software Quality Assurance and Verification and Validation processes:
	• Identification of roles and activities
16h00	Software infrastructure and training:
	• Development environment
	• Test and validation environment
	• Qualification environment
	• Project training needs
16h30	Summary and conclusion
17h00	Adjournment

5.6.2 Project Retrospectives

If the poor cousin of software engineering is quality assurance, the poor cousin of quality assurance reviews is the project retrospective. It is ironic that a discipline, such as software engineering, which depends as much as it does on the knowledge of the people involved, dismisses the opportunity to learn and enrich the knowledge of an organization's members. The project retrospective review is normally carried out at the end of a project or at the end of a phase of a large project. Essentially, we want to know what has been done well in this project, what has gone less well and what could be improved for the next project. The following terms are synonymous: lessons learned, post mortem, after-action-review.

Post Mortem

A collective learning activity that can be organized for projects or at the end of a phase or at project completion. The main motivation is to reflect on what has happened in the project to improve future practices for individuals who participated in the project and for the organization as a whole. The result of a post-mortem is a report.

Dingsøyr (2005) [DIN 05]

Lessons Learned

The knowledge gained during a project which shows how project events were addressed or should be addressed in the future with the purpose of improving future performance.

PMBOK® Guide [PMI 13]

Basili et al. (1996) [BAS 96] published the first controlled experiments that captured experience. This approach, called Experience Factory, where experience is gathered from software development projects, is packaged and stored in a database of experience. The packaging refers to the generalization, adaptation, and formalization of the experience until it is easy to reuse. In this approach, experience is separate from the organization that is responsible for capturing the experience.

A post mortem review, conducted at the end of a phase of a project or at the end of a project, provides valuable information such as [POM 09]:

– updating project data such as length, size, defects, and schedule;
– updating quality or performance data;
– a review of performance against plan;
– updating databases for size and productivity;

– adjustment of processes (e.g., checklist), if necessary, based on the data (notes taken on the proposal process improvement (PIP) forms, changes in design or code, lists of default controls indicated and so on).

There are several ways to conduct project retrospectives; Kerth (2001) lists 19 techniques in his book [KER 01].

Support sites for lessons learned:

Karl Wiegers: http://www.processimpact.com/

Norm Kerth: http://www.retrospectives.com/

Some techniques focus on creating an atmosphere of discussion in the project, others consider past projects, still others are designed to help a project team to identify and adopt new techniques for their next project, and some address the consequences of a failed project. Kerth recommends holding a 3-day session to make a lasting change in an organization [KER 01]. This section presents a less stringent and less costly approach to capturing the experience of project members.

The facilitator of a project retrospective session should not be the project manager. It is best, to preserve neutrality, that it be a person who was not directly involved in the project.

Since a retrospective session may create some tension, especially if the project discussed has not been a total success, we propose rules of behavior so that the session is effective. The rules of behavior at these sessions are:

– respect the ideas of the participants;
– maintain confidentiality;
– not to blame;
– not to make any verbal comment or gesture during brainstorming;
– not to comment when ideas are retained;
– request more details regarding a particular idea.

The following quote outlines the basis of a successful assessment session.

Rule of Thumb for a Successful Lessons Learned Session
"Regardless of what we discover, we truly believe that everyone did their best, given his qualifications and abilities, resources and the project context."

Kerth (2001) [KER 01]

The main items on the agenda during a project retrospective review are:

- list the major incidents and identify the main causes;
- list the actual costs and the actual time required for the project, and analyze variances from estimates;
- review the quality of processes, methods, and tools used for the project;
- make proposals for future projects (e.g., indicate what to repeat or reuse (methodology, tools, etc.), what to improve, and what to give up for future projects).

In many organizations, the transfer of knowledge and lessons learned is not necessarily done from one project to another. For example, one of the co-authors conducted lessons learned sessions in a division of an international transport company. Several times during the same lessons learned session, participants raised problems encountered during the project that had just come to an end and other participants said they had the same problems in previous projects!

A retrospective session typically consists of three steps: first, the facilitator explains, along with the sponsor, the objectives of the meeting; second, he explains what a retrospective session is, the agenda and the rules of behavior; lastly, he conducts the session.

A retrospective session takes place as follows:
Step One

- presentation of the facilitators by the sponsor;
- introduction of participants;
- presentation of the assumption:

- o Regardless of what we discover, we truly believe that everyone did the best job, given his qualifications and abilities, resources, and project context.
- – Presentation of the agenda of a typical retrospective session lasting approximately three hours:
  - o introduction;
  - o brainstorm to identify what went well and what could improve;
  - o prioritize items;
  - o identify the causes;
  - o write a mini action plan.

Step Two—Introduction to the retrospective session

- – what is a retrospective session?
- – when is a lesson really learned?
  - o what is individual learning, team learning?
  - o what is learning in an organization?
- – why have a retrospective session?
- – potential difficulties of a retrospective session;
- – session rules;
- – what is brainstorming? The rules of brainstorming are:
  - o no verbal comments or gestures;
  - o no discussion when ideas are retained.

Step Three—Conducting the retrospective session

- – chart a history (timeline) of the project (15–30 minutes);
- – conduct brainstorming (30 minutes);
  - o individually, identify on post-it notes:
    - □ what went well during the project (e.g., what to keep)?
    - □ what could be improved?
    - □ were there any surprises?
  - o collect ideas and post them on the project history chart
- – clarify ideas (if necessary);
- – group similar ideas;
- – prioritize ideas;
- – find the causes using the "Five Why" technique:
  - o what went well during the project?
  - o what could be improved?
- – final questions;
  - o for this project, name what you would have liked to change?
  - o for this project, name what you wish to keep.

- write a mini action plan;
  - what, who, when?
- end the session;
  - ensure the commitment to implement the action plan;
  - thanks to the sponsor and the participants.

"Those who do not learn from history are doomed to repeat it."

Santayana (1905)

"Insanity is to do the same thing again and expect different results."

Einstein

"The desire and the ability of an organization to continuously learn from any source and rapidly convert this learning into action is its ultimate competitive advantage."

Jack Welch, former CEO of General Electric

"The fastest way to succeed is to double failures!"

Thomas Watson, former President of IBM

Even if logic dictates that conducting project retrospective or lessons learned sessions are beneficial for the organization, there are still some factors that affect these types of sessions:

- leading lessons learned sessions takes time and often management wants to reduce project costs;
- lessons learned benefit future projects;
- a culture of blame (finger pointing) can significantly reduce the benefits of these sessions;
- participants may feel embarrassed or have a cynical attitude;
- the maintenance of social relationships between employees is sometimes more important than the diagnosis of events;
- people may be reluctant to engage in activities that could lead to complaints, to criticism, or blame;
- some people have beliefs that predispose them to the acceptance of lessons learned; beliefs such as "Experience is enough to learn" or "If you do not have experience, you will not learn anything";
- certain organizational cultures do not seem able or willing to learn.

Case Study at the Ministry of Justice of Quebec

In 1999, the Ministry of Justice of Quebec decided to group the management activities of the fines and offences departments. More than 700,000 cases were processed and $110 million was collected annually. However, an increase in accounts receivable was noted and a decline in revenues. These sectors were supported by two computer systems. The system for the management of the offences was designed in the early 1990s. The revenue control system, designed in 1983, was used by collectors to track the payment of fines. The project involved the development of a new system, i.e. the offence management and collection of fines system, to assist the activities of the new department. The project had an overall net savings of $46.7 million and a decrease in costs of 35.9%. This project has received several awards.

Project Retrospective

In December 2006, the project director held meetings to conduct a series of project retrospectives. Three, three-hour sessions were conducted. Three groups of people participated in the project retrospective sessions: users and project leaders (12 people), managers (5 people) and developers (6 people). The agenda for the retrospective activities was as follows: the director presented the objectives of the meeting and the facilitators; he then announced that this was a session which aimed to improve the way we work on future projects; it was not a session designed to blame the people who have made mistakes; the session should help identify what went well during the project and what could be improved. With that said, the director wished all a good session and withdrew.

After describing the project timeline, a first session to collect "what went well" ideas began. Early in the session, participants seemed somewhat shy to express their ideas. Then, when they grasped that the facilitators would not allow criticism of the ideas, they started to open up and express themselves freely. Then, a session to collect ideas about "what could be improved" was conducted. At the end of the retrospective sessions, participants were very enthusiastic, and expressed their satisfaction about the sessions held. Also, they had a greater understanding of the total picture of the completed multi-year project and had many ideas for future projects.

Translated from Laporte (2008) [LAP 08]

5.7 AGILE MEETINGS

For several years, agile methods have been used in industry. One of these methods, "SCRUM," advocates frequent short meetings. These meetings are held every day or every other day for about 15 minutes (no more than 30 minutes). The purpose of these meetings is to take stock and discuss problems. These meetings are similar to management meetings described in the IEEE 1028 standard but without the formality.

In my team, we work according to the agile methodology and at the end of each Sprint, we have a review session where each member identifies what, in his opinion, was good or not so good during the Sprint. We identify the list of things to improve that each member would like to really consider. Afterward, problems that have the most votes are inserted into backlogs according to their priority.

During these meetings, the "Scrum Master" typically asks three questions of the participants:

– What have you accomplished, in the "to do" list of tasks (Backlog), since the last meeting?
– What obstacles prevented you from completing the tasks?
– What do you plan to accomplish by the next meeting?

At the Acme Company, which develops systems that manage promotional campaigns and their return on investment calculations, adopting the Scrum development methodology provided many benefits. With the recruitment of new employees, knowledge transfer was easier thanks to "Daily Scrum" meetings and code reviews. In the last week of an iteration, all developers, novices or experts, are involved in the code review. The goal was not to underestimate the skills of older developers, but to generalize principles and also to train new developers. At the end of an iteration, a retrospective session is organized to check the elements to retain and improve for the next iteration. The application of these techniques reduces project deadlines and develops better software.

These meetings allow all participants to be informed on the status of the project, its priorities, and the activities that need to be performed by members of the team. The effectiveness of these meetings is based on the skills of the "Scrum Master." He should act as facilitator and ensure that the three questions are answered by all participants without drifting into problem-solving.

"None of us is as smart as all of us!"

Gerald Weinberg

5.8 MEASURES

An entire chapter is devoted to measures. This section describes only the measures associated with reviews. Measures are mainly used to answer the following questions:

- How many reviews were conducted?
- What software products have been reviewed?
- How effective were the reviews (e.g., number of errors detected by number of hours for the review)?
- How efficient were the reviews (e.g., number of hours per review)?
- What is the density of errors in software products?
- How much effort is devoted to reviews?
- What are the benefits of reviews?

The measures that allow us to answer these questions are:

- number of reviews held;
- identification of the revised software product;
- size of the software product (e.g., number of lines of code, number of pages);
- number of errors recorded at each stage of the development process;
- effort assigned to review and correct the defects detected.

"If you have just started to do inspections, you should detect about 50% of defects in your software product (this figure varies from a minimum of close to 20% to 90-95%). If you support the inspection with good project management, design and coding standards, and the process measures are publicized, it is possible to systematically obtain a defect detection rate of approximately 90%."

Ed Weller

Tables 5.3 and 5.4, presented at a meeting of software practitioners, show the data that can be collected. Table 5.3 shows the number of reviews, the type of documents, and the errors documented during a project.

The data collected allow us to estimate the number of residual errors and the defect detection efficiency for the development process as illustrated in Table 5.4. For example, for the requirements analysis activity, 25 defects were detected, two defects during the development of the high-level design, one defect during the detailed

Table 5.3 A Company's Peer Review Data [BOU 05]

Product type	Number of inspections	Number of lines inspected	Operational defects detected	Average OP defect density/1000 lines	Minor defects detected	Average minor defect density/1000 lines
Plans	18	5903	79	13	469	79
System requirements	3	825	13	16	31	38
Software requirements	72	31476	630	20	864	27
System design	1	200	–	–	1	5
Software design	359	136414	109	1	1073	8
Code	82	30812	153	5	780	25
Test document	30	15265	62	4	326	21
Process	2	796	14	18	27	34
Change request	8	2295	56	24	51	22
User document	3	2279	1	0	89	39
Other	72	29216	186	6	819	28
Totals	**650**	**255481**	**1303**	**5**	**4530**	**18**

Table 5.4 Error Detection Throughout the Development Process [BOU 05]

Attributed activity	Detection activity							Activity escape	Post-activity escape
	RA	HLD	DD	CUT	T&I	Post-release	Total		
System design	6	1	1	0	3	2	13		15%
RA	25	2	1	0	1	1	30	17%	3%
HLD		32	7	2	8	3	52	38%	6%
DD			43	15	5	7	70	39%	10%
CUT				58	21	4	83	30%	5%
T&I					8	2	10	20%	20%
Total	**31**	**35**	**52**	**75**	**46**	**19**	**258**		**7%**

Legend: attributed activity, project phase where the error occurred; detection activity, phase of the project where the error was found; RA, requirements analysis; HLD, preliminary design; DD, detailed design; CUT, coding and unit testing; T&I, test and integration; post-release, number of errors detected after delivery; activity escape, percentage of errors that were not detected during this phase (%); post-activity escape, percentage of errors detected after delivery (%).

design, zero defects in the coding and debugging activities, one failure during testing activities and integration, and one failure after delivery.

We can calculate the defect detection efficiency of the review conducted during the requirements phase:

$$(30 - 5)/30 \times 100 = 83\%.$$

We can also calculate the percentage of defects that originate from the requirements phase:

$$30/258 \times 100 = 12\%.$$

It is therefore possible, given these data, to make different decisions for a future project. For example:

- to reduce the number of defects injected during the requirements phase, we can study the 25 defects detected and try to eliminate them;
- you can reduce the number of pages inspected per unit of time in order to increase defect detection;
- there is a large number of defects that were not detected during preliminary and detailed design activities: 38% and 39%, respectively. A causal analysis of these defects could reduce these percentages.

Case of Implementing Measures Without Validating the Effects or Adapting Them to the Context

In a Montreal hospital, a renowned doctor from France became manager of a clinical research project. He tried to introduce verification measures throughout the development process. Although his idea was not bad, he used data from the aviation industry and from a much more experienced group than his team to make assessments of software products and of his employees.

Unfortunately, many developers were laid off before the measures were adapted to the context of a small team in a clinical research environment which did not have a CMMI maturity level 2 and the doctor returned to the practice of medicine!

5.9 SELECTING THE TYPE OF REVIEW

To determine the type of review and its frequency, the criteria to be considered are: the risk associated with the software to be developed, the criticality of the software,

Table 5.5   Example of a Matrix for the Selection of a Type of Review

Product	Technical drivers—complexity		
	Low	Medium	High
Software requirements	Walk-through	Inspection	Inspection
Design	Desk-check	Walk-through	Inspection
Software code and unit test	Desk-check	Walk-through	Inspection
Qualification test	Desk-check	Walk-through	Inspection
User/operator manuals	Desk-check	Desk-check	Walk-through
Support manuals	Desk-check	Desk-check	Walk-through
Software documents, for example, Version Description Document (VDD), Software Product Specification (SPS), Software Version Description (SVD)	Desk-check	Desk-check	Desk-check
Planning documents	Walk-through	Walk-through	Inspection
Process documents	Desk-check	Walk-through	Inspection

software complexity, the size and experience of the team, the deadline for completion, and software size.

Table 5.5 is an example of a support matrix for selecting a type of review. The column "document review" shows a list of products to review. The column "complexity" shows the classification criteria and type of review to be used. In this example, the degree of complexity is measured as low, medium, and high. Complexity is defined as the level of difficulty for understanding a document and verifying it. A low complexity level indicates that a document is simple or easily checked while the high complexity level is defined for a product that is difficult to verify. Table 5.5 is only an example. The criteria for choosing the type of review and the product to review should be documented in the project plan or the SQAP.

In Chapter 1, we briefly introduced an example of the software quality for the aircraft engine manufacturer Rolls-Royce. Following is a concrete example of the application of code inspections at Rolls-Royce.

***Code Inspections at the Aircraft Engine Manufacturer
Rolls-Royce***

A group from the Rolls-Royce Company develops embedded software for aircraft engine controllers. Obviously, these programs are critical, and software and development is subject to the DO-178 de facto standard presented in the previous chapter.

This company has significantly increased the efficiency of the inspection process. Developers have considerably reduced the number of errors in their software before the software is delivered to the systems engineering team.

The engine manufacturer has managed this without modifying either the process or the tools and without adding additional effort for inspections. In this company, over 52% of verification efforts are tests and 24% of total efforts are peer reviews.

The method developed at Rolls-Royce is based on measuring the ability of developers and those who participate in inspections and to select, using data (which is confidential), the people who will inspect the product of a particular developer.

The company measured the effectiveness of each author and reviewer. The effectiveness of the authors is measured in number of errors injected per 1000 lines of code. The company found a large variation in defect injection between authors (i.e., 0.5–18 defects injected per 1000 lines of code). The company also measured the detection efficiency for each reviewer. Some reviewers detected only 36% of errors, while the best detected 90%. Another development team noted a factor of 10 between the reviewers. The company noticed that the best authors are not necessarily the best reviewers.

Rolls-Royce has developed the following table to describe the effectiveness of authors and reviewers. The left side shows the data for the injection rate of errors for developers A to F. For example, developer A typically injects 0.5 errors per 1000 lines of code and developer F injects 18. The right side of the table shows the effectiveness of detection of reviewers A to F. Note that reviewer A has a 75% detection rate and reviewer F has a rate of 30%.

			Reviewer effectiveness Defect detection rate					
			C	B	A	E	D	F
			94%	80%	75%	50%	45%	30%
	A	0.5	0.0	0.1	0.1	0.3	0.3	0.4
Author effectiveness	B	1.0	0.1	0.2	0.3	0.5	0.6	0.7
	C	3.0	0.2	0.6	0.8	1.5	1.7	2.1
	D	4.0	0.2	0.8	1.0	2.0	2.2	2.8
Defects introduced per 1000 lines	E	10.0	0.6	2.0	2.5	5.0	5.5	7.0
	F	18.0	1.1	3.6	4.5	9.0	9.9	12.6

"Effectiveness of authors and reviewers."

This table is used to assign one or more effective reviewers to detect errors of an author who makes many errors. Rolls-Royce says the best organizations (world class organization) produce software with one residual defect per 1000 lines of code (defect escape rate). At Rolls-Royce, they have managed to reduce the number of residual defects to 0.03 per 1000 lines of code.

While this approach has been used for code inspections, it can also be used for other artifacts (e.g., requirements, architecture) of a project.

Adapted from Nolan et al. (2015) [NOL 15]

5.10 REVIEWS AND BUSINESS MODELS

In Chapter 1, we presented the main business models for the software industry [IBE 02]:

- – Custom systems written on contract: The organization makes profits by selling tailored software development services for clients.
- – Custom software written in-house: The organization develops software to improve organizational efficiency.
- – Commercial software: The company makes profits by developing and selling software to other organizations.
- – Mass-market software: The company makes profits by developing and selling software to consumers.
- – Commercial and mass-market firmware: The company makes profits by selling software in embedded hardware and systems.

Each business model is characterized by its own set of attributes or factors: criticality, the uncertainty of needs and requirements (needs versus expectations) of the users, the range of environments, the cost of correction of errors, regulation, project size, communication, and the culture of the organization.

Business models help us understand the risks and the respective needs in regards to software practices. Reviews are techniques that detect errors and thus reduce the risk associated with a software product. The project manager, in collaboration with SQA, selects the type of review to perform and the documents or products to review throughout the life cycle in order to plan and budget for these activities.

The following section explains the requirements of the IEEE 730 standard with regard to project reviews.

5.11 SOFTWARE QUALITY ASSURANCE PLAN

The IEEE 730 standard defines the requirements with respect to the review activities to be described in the SQAP of a project. Reviews are central when it comes time to assess the quality of a software deliverable. For example, product assurance activities may include SQA personnel participating in project technical reviews, software development document reviews, and software testing. Consequently, reviews are to be used for both product and process assurance of a software project. IEEE 730 recommends that the following questions be answered during project execution [IEE 14]:

– Have periodic reviews and audits been performed to determine if software products fully satisfy contractual requirements?

– Have software life cycle processes been reviewed against defined criteria and standards?

– Has the contract been reviewed to assess consistency with software products?

– Are stakeholder, steering committee, management, and technical reviews held based on the needs of the project?

– Have acquirer acceptance tests and reviews been supported?

– Have action items resulting from reviews been tracked to closure?

The standard also describes how reviews can be done in projects that use an agile methodology. It states that "reviews can be done on a daily basis," which reflects the agile culture of conducting a daily activity.

We know that SQA activities need to be recorded during the course of a software project. These records serve as proof that the project did the activities and can provide these records when asked. Review results and completed review checklists can be a good source of evidence. Consequently, it is recommended that project teams keep a record of the meeting minutes for all technical and management reviews they conduct.

Finally, an organization should base process improvement efforts on the results of in-process as well as completed projects, gathering lessons learned, and the results of ongoing SQA activities such as process assessments and reviews. Reviews can play an important role in organization-wide process improvement of software processes. Preventive actions are taken to prevent occurrence of problems that may occur in the future. Non-conformances and other project information may be used to identify preventive actions. SQA reviews propose preventive actions and identify effectiveness measures. Once the preventive action is implemented, SQA evaluates the activity and determines whether the preventive action is effective. The preventive action process can be defined either in the SQAP or in the organizational quality management system.

5.12 SUCCESS FACTORS

Although reviews are relatively simple and highly effective techniques, there are several factors that can greatly help their effectiveness and efficiency. Conversely, many factors can affect the review to the point of no longer being used in an organization. Some factors related to an organization's culture, which can promote the development of quality software, are listed below.

Factors that Foster Software Quality

1) Visible management commitment
   - provide the resources and time to conduct reviews such as an inspection;
   - ensure that reviews are planned in the project plan or in the quality assurance plan;
   - maintain reviews (e.g., inspections) even when the schedule is tight;
   - occasionally revise the overall results of reviews and consider proposals to improve the process;
   - attend the training session;
   - conduct reviews with colleagues (e.g., inspect a project plan, a software quality assurance plan)

2) Good team spirit
   - reviews (e.g., inspections) are made by team members in order to help each other and increase product quality.

Following are the factors related to an organization's culture that can harm the development of quality software.

Factors that may Adversely Affect Software Quality

1) Using reviews to evaluate the performance of a developer;

2) When the "ego" of authors do not accept anomalies identified in their documents
   - I do not need help, because I'm the best!
   - I do not need help from someone who is junior to me

3) Participants do not properly prepare for the meeting;

4) The team members have not been trained to perform the reviews, especially the inspection;

5) Wanting to get revenge;
   - The lack of a skilled facilitator;
   - The facilitator must ensure that reviewers do not embarrass the author of a document with derogatory remarks or gestures (i.e., body language).

5.13 TOOLS

Following are some tools for effective reviews.

Tools for Conducting Reviews

Several free software tools are available to facilitate code review:

– Idutils: an indexing tool that allows the creation of a database of identifiers used in a program;

– Egrep: a tool to search regular expression patterns in text files;

– Find: allows a system file to be viewed;

– Diff: to compare two files and show differences;

– Cscope: a C-code browser;

– LXR: a web interface to explore the source code online and offer cross-reference.

5.14 FURTHER READING

WIEGERS K. The seven deadly sins of software reviews. *Software Development*, vol. 6, issue 3, 1998, pp. 44–47.

WIEGERS K. A little help from your friends. *Peer Reviews in Software*, Pearson Education, Boston, MA, 2002, Chapter 2.

WIEGERS K. Peer review formality spectrum. *Peer Reviews in Software*, Pearson Education, Boston, MA, 2002, Chapter 3.

5.15 EXERCISES

5.1   Develop a checklist for an architecture document.

5.2   Identify the activities that must be performed by Quality Assurance.

5.3   List the benefits of walk-throughs or inspections from the perspective of these key players:

a) development manager;

b) developers;

c) quality assurance;

d) maintenance personnel.

5.4   Provide some reasons for not carrying out inspections.

5.5 Name some objectives that are not the goal of an inspection.

5.6 Calculate the residual error given the following: 16 errors were identified in a 36-page document. We know our error detection rate is 60% and that we inject 17% of new errors when we make corrections to the errors detected. Calculate the number of errors per page in the document that remain after completing the review. Explain your calculation.

5.7 Develop a checklist from the Java/C++ programming guide.

5.8 What benefits do these key players get from a review?

    a) analysts,

    b) developers,

    c) managers,

    d) SQA,

    e) maintainers,

    f) testers.

5.9 Describe the advantages and disadvantages of formal reviews.

5.10 Describe the advantages and disadvantages of informal reviews.

5.11 Provide criteria for selecting a type of review.

5.12 Why should we do project retrospectives?

5.13 Complete the table on the next page by putting an "X" in the appropriate columns.

Objective of the peer review	Desk-check	Walk-through	Inspection
Find defects/errors			
Verify compliance with the specifications			
Verify compliance with standards			
Check that the software is complete and correct			
Assess maintainability			
Collect data			
Measure the quality of the software product			
Train personnel			
Transfer knowledge			
Ensure that errors were corrected			

Chapter 6

Software Audits

After completing this chapter, you will be able to understand the:

- utility of software audits;
- audit of a management system;
- software audit and problem resolution according to the ISO 12207 standard;
- software audit process recommended by the IEEE 1028 standard;
- assessment of type audit recommended by the CMMI® model;
- corrective and preventive process;
- audit section of the SQA plan recommended by the IEEE 730 standard.

6.1 INTRODUCTION

Different types of reviews were presented in the previous chapter. This chapter is dedicated to the audit, which is one of the most formal types of reviews. We begin by providing definitions that are presented in some standards.

Different types of conformity certificates (e.g., audits) respond to different needs, such as the needs of an organization that develops software products or those of a client of a software product supplier. The independence level of the auditor as well as the cost varies depending on the audit type. The notes of the definition in the following text box indicate that there are internal audits and external audits performed by second and third parties.

Software Quality Assurance, First Edition. Claude Y. Laporte and Alain April.
© 2018 the IEEE Computer Society, Inc. Published 2018 by John Wiley & Sons, Inc.

Management System

System to establish policy and objectives and to achieve those objectives.

ISO 19011 [ISO 11g]

Audit Criteria

Set of policies, procedures or requirements used as references against which objective evidence is compared.

ISO 9000 [ISO 15b]

Audit Evidence

Records, statement of fact or other information, which are relevant to the audit criteria and verifiable.

ISO 9000 [ISO 15b]

Audit

Systematic, independent and documented process for obtaining audit evidence (3.3) and evaluating it objectively to determine the extent to which the audit criteria (3.2) are fulfilled.

Note 1: Internal audits, sometimes called first party audits, are conducted by the organization itself, or on its behalf, for management review and other internal purposes (e.g. to confirm the effectiveness of the management system or to obtain information for the improvement of the management system). Internal audits can form the basis for an organization's self-declaration of conformity. In many cases, particularly in small organizations, independence can be demonstrated by the freedom from responsibility for the activity being audited or freedom from bias and conflict of interest.

Note 2: External audits include second and third party audits. Second party audits are conducted by parties having an interest in the organization, such as customers, or by other persons on their behalf. Third party audits are conducted by independent auditing organizations, such as regulators or those providing certification.

ISO 19011 [ISO 11g]

There are different types of audit:

– audits to verify the compliance to a standard, such as the audits described in International Organization for Standardization (ISO) standards such as ISO 9001 and IEEE 1028;

– compliance audits for a model such as the Capability Maturity Model Integration (CMMI) that are used to select a supplier before awarding a contract or assess a supplier during a contract;

– audits ordered by the management team of the organization to verify the progress of a project against its approved plan.

A mandate can be assigned to an external consultant or to members of the personnel of the organization that are not involved with the project, specifying the questions that the audit should answer, the audit schedule, the audit participants, and the format of the audit report (e.g., findings and recommendations).

9.2 Internal Audit

9.2.1 The organization shall conduct internal audits at planned intervals to provide information on whether the quality management system:

a) conforms to:
   1) the organization's own requirements for its quality management system;
   2) the requirements of this International Standard;
b) is effectively implemented and maintained.

9.2.2 The organization shall:

a) plan, establish, implement and maintain an audit programme(s) including the frequency, methods, responsibilities, planning requirements and reporting, which shall take into consideration the importance of the processes concerned, changes affecting the organization, and the results of previous audits;

b) define the audit criteria and scope for each audit;

c) select auditors and conduct audits to ensure objectivity and the impartiality of the audit process;

d) ensure that the results of the audits are reported to relevant management;

e) take appropriate correction and corrective actions without undue delay;

f) retain documented information as evidence of the implementation of the audit programme and the audit results.

Note: See ISO 19011 for guidance.

ISO 9001 [ISO 15]

We end this section with the definition of an audit as presented by the Project Management Institute (PMI®) given that audits are often ordered by project managers themselves.

Quality Audit

A quality audit is a structured, independent process to determine if project activities comply with organizational and project policies, processes, and procedures. The objectives of a quality audit may include:

– identify all good and best practices being implemented;

– identify all nonconformity, gaps, and shortcomings;

– share good practices introduced or implemented in similar projects in the organization and/or industry;

– proactively offer assistance in a positive manner to improve implementation of processes to help the team raise productivity; and

– highlight contributions of each audit in the lessons learned repository of the organization.

Quality audits can be either planned or random, and can be conducted by internal or external auditors.

Quality audits can confirm the execution of approved change modifications such as defect corrections and corrective actions.

PMBOK® Guide [PMI 13]

In addition to verifying compliance, it should be noted that this audit definition underlines the fact that the audit can also be done to improve organizational performance.

Remember that in the cost of quality model, audit is an evaluation practice also referred to as a detection activity: the verification or evaluation costs of a product or of a service during the different phases of the development life cycle.

Table 6.1 presents the components of detection costs.

In Chapter 1, the main software business models were presented. Audits allow for detecting defects and therefore diminish the software product development risks. Custom systems written on contract, mass market software, commercial, and mass-market firmware use audits extensively.

The software audits can be performed from several perspectives:

– an external registrar who audits the quality system to assess its compliance for ISO 9001 certification;

– an external auditor that audits for compliance to government standards, such as for the medical industry (e.g., standards of the Food and Drug Administration), or to certify the achievement of a CMMI maturity level;

Table 6.1 Software Quality Cost Categories (Adapted from Krasner (1998) [KRA 98])

Major category	Sub-category	Definition	Typical elementary costs
Detection costs	Discover the state of the product	Discover the non-conformity level	Test, software quality assurance (SQA) inspections, reviews
	Ensure meeting the stated quality	Quality control mechanism	Product quality audits, new versions delivery decision

- an internal auditor that audits a project or software process to ensure that internal controls are adequate. This perspective, for example, can focus on security and fraud detection or simply the identification of inefficiencies. It is a way to ensure that the mandatory processes of the information system are effectively applied. These audits may also prepare an organization to comply with a law or an external regulation such as the Sarbanes-Oxley law of 2002 [SAR 02];

- SQA that audits projects and processes on behalf of management to ensure that teams follow the mandated life cycle processes. This perspective is mainly concerned with the effectiveness and improvement of processes;

- the management team or a project manager may also request that an audit be conducted to determine if a specific internal activity or project assigned to an external provider is complying with the contract specifications and agreed/prescribed clauses. This audit perspective is performed at a specific time or milestone to verify if the work is progressing according to plans;

- a professional designation, such as the board of professional engineers of California, may decide to audit a professional to determine if his work meets its commitments to the code of ethics and that the engineer correctly applies, through his services, the concerns for safeguarding life, health, property, economic interests, the public interest, and the environment.

For more information on the application of the Sarbanes-Oxley Act (SOX) to internal auditing controls for an ISO 9001 quality management system, visit the following website: http://www.asq.org and type SOX in the search box.

Why audit? Whatever the business model of the organization, that is, contracted or in-house software development, commercial software development, mass-market software, or embedded software, all organizations, even public organizations, want to meet their objectives. One way to ensure meeting their objectives is to ensure the constant compliance and improvement of processes. Audit activities are usually targeted to the most important projects of the organization and its external supplier's activities. For example, Professor April [APR 97] describes the results of one year of SQA audits, at Bell Canada, and the findings. In internal audits, SQA promotes defect prevention and encourages teams to always meet their customer commitments while respecting internal rules. Software project audits are usually requested by management to ensure that the software team and contracted suppliers:

- know their duties and obligations toward the public, their employers, their customers, and their colleagues;

- use the processes, practices, techniques, and normalized methods suggested by the company;
- reveal any deficiencies and shortcomings in daily operations and try to identify required corrective actions (CAs);
- are encouraged to develop a personal training plan for their professional skills;
- are monitored in the course of their work on high profile projects of the company.

Any project manager or software developer can expect to be audited at some point in their career. It is therefore important to understand this formal review process and ensure one is always ready to be audited. It is also quite possible that you will have to participate in an audit as a member of the auditing team. Software project audits are described in ISO 12207, ISO 9001, CobiT [COB 12], and the CMMI.

6.2 TYPES OF AUDITS

6.2.1 Internal Audit

An internal audit, also called a first-party audit, can be useful for a software supplier wanting to obtain an ISO 9001 certification. It is the least expensive approach to prepare for conformity to an international standard.

ISO/IEC 17050-1 [ISO 04a] describes the requirements for a supplier conformity statement that indicates that a product (including a service), process, management system, individual, or organization meet the specified requirements originating from a standard, a process model, a law, or regulation. The supplier conformity declaration form shall identify: the issuer of the statement, the subject of the statement, standards or specified requirements, and the person who signed the statement. Figure 6.1, taken from ISO 17050-1, shows an example of a supplier conformity declaration form.

ISO/IEC 17050-2 [ISO 04b] indicates that the organization that declares conformity must make supporting documentation available with this form.

A copy of this declaration can be displayed on the website of the organization. The organization should also have put in place a procedure allowing for the re-evaluation of the validity of the declaration, when necessary, should there be any modifications to the development processes or to the standards used.

6.2.2 Second-Party Audit

We have already defined what a second-party audit is. For example, a client can impose that suppliers demonstrate conformity to a standard. The customer can also audit the supplier to verify conformity. The auditor can be an employee of the

ISO/IEC 17050-1:2004(E)

A.2 Example of form of declaration of conformity

Supplier's declaration of conformity (in accordance with ISO/IEC 17050-1)	
1) **No.**	
2) **Issuer's name:**	....
Issuer's address:	....
	....
3) **Object of the declaration:**	....
	....
	....
4) **The object of the declaration described above is in conformity with the requirements of the following documents:**	

	Documents No.	Title	Edition/Date of issue
5)	....	....	....
	....	....	....
	....	....	....

Additional information:

6) ....

....

....

Signed for and on behalf of:

....

....

(Place and date of issue)

7) ....

(Name, function)   (Signature or equivalent authorized by the issuer)

Figure 6.1   Example of a conformity declaration form [ISO 04a]. *Source*: Standards Council of Canada.

customer, for example a member of the SQA department, or an external consultant that has pertinent knowledge of the customer's business domain.

It is typically the customer that incurs the cost of the audit (e.g., auditor's travelling expenses). The supplier will pay for his employee's costs associated with participating in the audit.

6.2.3 Third-Party Audit

As described earlier in this chapter, a third-party audit is typically conducted by an independent organization. We would like to emphasize that ISO does not offer certification services per say. It does not deliver certificates either. It is the ISO 19011 [ISO 11g] standard entitled "Guidelines for auditing management systems" and the ISO/IEC 17021-1 [ISO 15a] standard, entitled "Conformity assessment-Requirements for bodies providing audit and certification of management systems-Part 1: Requirements" that are used to assess the compliance to a standard such as ISO 9001.

ISO 19011 includes a section that provides guidelines on the necessary competence of auditors and describes their recommended evaluation process.

ISO 17021 states that "the certification organization must be a legal entity or part of a legal entity to ensure that it can be legally held accountable of its certification activities."

It is important to note that a standard certification is not an obligation unless a customer mandates it. An organization that is not accredited can be perfectly reliable.

An International Accreditation Association

The primary purpose of IAF is twofold. Firstly, to ensure that its accreditation body members only accredit bodies that are competent to do the work they undertake and are not subject to conflicts of interest. The second purpose of the IAF is to establish mutual recognition arrangements, known as Multilateral Recognition Arrangements (MLA), between its accreditation body members which reduces risk to a business and its customers by ensuring that an accredited certificate may be relied upon anywhere in the world.

(www.iaf.nu)

6.3 AUDIT AND SOFTWARE PROBLEM RESOLUTION ACCORDING TO ISO/IEC/IEEE 12207

ISO 12207 defines audit requirements and a decision management process.

6.3.1 Project Assessment and Control Process

The purpose of the Project Assessment and Control process is [ISO 17]: to assess if the plans are aligned and feasible; determine the status of the project, technical and process performance; and direct execution to help ensure that the performance is according to plans and schedules, within projected budgets, to satisfy technical objectives. One important mandatory task of this process is the conduct of management and technical reviews, audits, and inspections.

6.3.2 Decision Management Process

The decision management process follows the audit and the report distribution to stakeholders which recommend the start of CAs.

The purpose of the Decision Management process is [ISO 17]: to provide a structured, analytical framework for objectively identifying, characterizing, and evaluating a set of alternatives for a decision at any point in the life cycle and select the most beneficial course of action. One of the main activities of this process is named: "Make and manage decisions." This activity includes the following mandatory tasks [ISO 17]:

- determine preferred alternatives for each decision;
- record the resolution, decision rationale, and assumptions;
- record, track, evaluate, and report decisions
  o Note 1: This includes records of problems and opportunities and their disposition, as stipulated in agreements or organizational procedures and in a manner that permits auditing and learning from experience.

ISO 12207 describes other types of audits to perform such as configuration audits that will be presented in the configuration management chapter.

6.4 AUDIT ACCORDING TO THE IEEE 1028 STANDARD

In the previous chapter, we have seen that the IEEE 1028 describes the many different types of reviews and audits as well as the procedures for the execution of these reviews. Note that this standard explains how to conduct audits and not on their need or how to use the audit reports.

Audit

An independent examination of a software product, software process, or set of software processes performed by a third party to assess compliance with specifications, standards, contractual agreements, or other criteria.

Note: An audit should result in a clear indication of whether the audit criteria have been met.

IEEE 1028 [IEE 08b]

Table 6.2 summarizes the audit characteristics according to the IEEE 1028 standard. These characteristics will be explained in more detail in this section.

The purpose of the IEEE 1028 standard is to define systematic reviews and audits that apply to user's software acquisition, supplier, development, operation, and maintenance processes. The standard describes how to conduct an audit. It also describes the minimum acceptable requirements for software audits, explaining:

- the audited team participation;
- the audit documented results;
- the documented procedure to conduct the audit.

To perform an audit, the auditor will have to read a set of documents (e.g., software process, products) that will need to be available at audit time. The IEEE 1028 standard lists software products that are most likely to be audited. Table 6.3 shows a partial list of these software products.

Table 6.2 Audit Characteristics According to IEEE 1028 [IEE 08b]

Characteristic	Audit
Objective	Independently evaluate conformance with objective standards and regulations
Decision making	Audited organization, initiator, acquirer, customer, or user
Change verification	Responsibility of the audited organization
Recommended group size	One to five people
Group attendance	Auditors; the audited organization may be called upon to provide evidence
Group leadership	Lead auditor
Volume of material	Moderate to high, depending on the specific audit objectives
Presenter	Auditors collect and examine information provided by audited organization

Table 6.3   Examples of Software Project Products that can be Audited (Adapted from IEEE 1028 [IEE 08b])

Contracts	Backup recovery plans	Software design descriptions	Software requirements specifications	Software test documentation
Software project management plans	Customer or user representative complaints	Unit development folders	Walk-through reports	Request for proposal
Operation and user manuals	Installation procedures	Risk management plans	Applicable standards, regulations, plans, and procedures	Disaster plans
Maintenance plans	Contingency plans	Development environment	System build procedures	Software architecture descriptions
Reports and test data	Source code	Software verification and validation plans	Software configuration management plans	Software user documentation

The standard also addresses the topic of auditing software project processes. The organization should prepare checklists that are specific to each process or product that is audited.

6.4.1 Roles and Responsibilities

The software audit roles and responsibilities are (adapted from IEEE 1028 [IEE 08b]:

- – an initiator: shall be responsible for the following activities:
  a) decide upon the need for an audit;
  b) decide upon the purpose and scope of the audit;
  c) decide the software products or processes to be audited;
  d) decide the evaluation criteria, including the regulations, standards, guidelines, plans, specifications, and procedures to be used for evaluation;
  e) decide who will carry out the audit;
  f) review the audit report;
  g) decide what follow-up action will be required;
  h) distribute the audit report.
     The initiator may be a manager in the audited organization, a customer or user representative of the audited organization, or a third party.

- a lead auditor: shall be responsible for the audit. He must ensure that the activity is done according to the agreed upon audit rules and that it has met its objective. He is responsible for the following activities:
  a) prepare an audit plan;
  b) choose and manage the audit team;
  c) lead the team and note observations;
  d) prepare the audit report;
  e) note all deviations;
  f) recommend CAs.

- the recorder: shall document anomalies, action items, decisions, and recommendations made by the audit team.

- auditor(s): shall examine products, as defined in the audit plan. They shall document their observations. All auditors shall be free from bias and influences that could reduce their ability to make independent, objective evaluations, or they shall identify their bias and proceed with acceptance from the initiator.

- the audited organization: shall provide a liaison to the auditors and shall provide all information requested by the auditors. When the audit is completed, the audited organization should implement CAs and recommendations.

6.4.2 IEEE 1028 Audit Clause

As with other IEEE 1028 reviews, audits are described by a clause that contains the following information [IEE 08b]:

- introduction to review: describes the objectives of the systematic review and provides an overview of the systematic review procedures;

- responsibilities: defines the roles and responsibilities needed for the systematic review.

- input: describes the requirements for input needed by the systematic review;

- entry criteria: describes the criteria to be met before the systematic review can begin, including the following:
  o authorization;
  o initiating event;

- procedures: details the procedures for the systematic review, including the following:
  o planning the review;
  o overview of procedures;
  o preparation;
  o examination/evaluation/recording of results;
  o rework/follow-up;

– exit criteria: describe the criteria to be met before the systematic review can be considered complete;

– output: describes the minimum set of deliverables to be produced by the systematic review.

6.4.3 Audit Conducted According to IEEE 1028

The IEEE 1028 standard insists that before an audit can start, the initiator has appointed an auditor and clarified the scope of the audit. The audited organization should have communicated the audit procedure and the rules against which each project team is audited. When the auditor has confirmed this situation, the auditor will have to demonstrate that he is experienced, trained, and certified for this audit scope.

The audit process, described in Figure 6.2, shows the recommended audit process activities (adapted from IEEE 1028 [IEE 08b]):

– plan the audit: an audit plan is prepared and approved by the initiator;

– prepare and lead an opening meeting: an opening meeting, involving the auditors and the audited organization, has the objective of explaining the scope, audit process, and the targeted software process and products that will be audited, the audit schedule, the expected contributions from individuals that will be interviewed, resources involved (e.g., the meeting rooms, access to quality records) as well as the information and documents that will be audited;

– prepare the audit: this activity reviews the audit plan to be completed, the audited organization readiness, the access and preliminary review of the

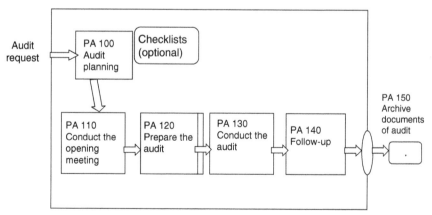

Figure 6.2   Audit process activities according to IEEE 1028.
Source: Adapted from Holland (1998) [HOL 98].

processes and products that will be audited, the organization life cycle and standards, and the evaluation criteria before the audit start;

– collect the objective evidence: this is the heart of the audit and requires the collection and analysis of evidence. The key findings will be presented to the audited organization. Then a final report is produced and communicated to the audit sponsor;

– audit follow-up: the audit sponsor communicates with the audited organization to determine the required CAs.

In this figure, one activity is added to the IEEE 1028 current activity list: "Archive documents of audit" to ensure that all the documentation associated with the audit is kept. In fact, the audit documentation is useful as evidence that the organization is conducting ISO 9001 audits and improvements, or for a future external audit for example. Archived documents can include:

– the approved audit mandate;

– the audit plan;

– identification of participants;

– the minutes, opening and closing document;

– the auditor's checklist;

– the audit report;

– the list of improvements needed;

– the CAs and their follow-up to closure.

Finally, note that audit activities, like all other software life cycle activities, can be audited as well. In large organizations, many auditors conduct software project audits according to a yearly audit plan.

The following text box presents an example of an extract from an audit report.

Acme Corporation
Audit Report

Date (YY-MM-DD):_____
Sponsor: _____
Auditor: _____

To support process improvement at Acme, the software quality assurance department has conducted a process audit of the software development life cycle processes.

Intention

To verify if the software activities and the results obtained comply with the mandated documented processes and procedures and assess their effectiveness when executed.

Scope

The audit was conducted in the engineering department on the 15th of January 2017. The audited project was entitled "Sustainable Environment". The audited development life cycle methodology was version 1.6.

Auditor(s): Mary Peters and John McCullen

Involved Personnel: Michael Phillips, Phillip Cordingley, Julia Perth (software engineering department)

   The auditors are pleased to announce that personnel were extremely cooperative during this audit.

Definitions

Finding: situation or condition that is a non-conformity towards a quality standard, a design, a process, a procedure, a policy, a work order, a contract or other standard and that would require an improvement request.

   Note: the information presented by the auditors to the interviewees as preventive measures do not require a formal improvement request. These topics can be verified, in detail, in a follow-up audit.

Summary

In general, auditors have observed that the project did not have a structured plan. This situation created the need for last minute interventions. The customer requirements were not clearly defined.

   After completing this audit, one (1) process change request was documented, four (4) notes were generated and eight (8) improvement requests were raised. The conformity percentage of this audit was 75.8 %.

Improvement Requests

Request number	Description	Assigned to department
ETS-2017-015	The requirements traceability matrix was not updated during the development process. It needs to be updated frequently.	Software engineering
ETS-2017-016	Source code inspections were not done during the project.	Software engineering

Findings

– Development process - Step SD-120 – write software requirements and test plan. The requirements traceability matrix was not updated during the software development project execution. This matrix will only be completed at the end of the project. This matrix should be updated frequently to ensure that all the requirements have been allocated and satisfied.

– Development process - Step SD-130 – specification review and approval. The specification document is still not approved since November 21, 2016. This augments the project risks since the baseline of this artefact is not formalized.

Note

– Development process - Step SD-160 – test procedure development.

The test plan was finalised too late in the process and was presented to the customer at the test readiness review only. The test plan should have been finalized at the design review time. The causes reported were tight time constraints and inappropriate planning.

6.5 AUDIT PROCESS AND THE ISO 9001 STANDARD

The ISO 9001 standard has made quality system audits popular. Never before has a standard had so much attention and acceptance worldwide. Production plants were the early adopters of the ISO 9001 standard. They were aware of the benefit of having an independent quality certification as a competitive advantage. This popularity grew when suppliers started asking for the certification before awarding contracts.

Although there is a growing popularity of the use of ISO 9001 for production organizations, it is not the case for service organizations such as software developers. These organizations have a more difficult time separating the final product from the development life cycle processes [MAY 02]. IT requires an additional effort to interpret ISO 9001 clauses in this context.

To ease the use of ISO 9001 for the service industry, including the software industry, interpretation guides have been produced. For software, two sources are popular for the interpretation of ISO 9001 [ISO 15]:

– ISO/IEC 90003 is the interpretation guide for software for the ISO 9001 standard [ISO 14]; with respect to internal audits, ISO 90003 provides the following information:
   o When software development organizations plan their development program, it usually coordinates with the audit planning when projects and quality management systems are selected. Audit planning tries to select projects in order to cover all the development life cycle processes for each phase;
   o This could require an audit of different projects, at different development life cycle phases, or the audit of only one project throughout its evolution across life cycle phases. When the targeted project modifies its schedule, the

internal audit calendar should also be reviewed to adjust audit dates or to choose another project.

– the interpretive material (TickIT guide version 5.5 [TIK 07]) developed to train and certify auditors of the ISO 9001 software quality system.

In the United Kingdom, TickITplus is a certification program that can allow multiple IT standards to be covered by one certification arrangement—www.tickitplus.org/

These ISO 9001 interpretation guides clarify how each of the ISO 9001 clauses apply to an organization that develops, maintains, and operates software products. It is interesting to note that the ISO 90003 [ISO 14] interpretation guide is aligned with ISO 12207.

We have discussed that improvement and conformity assessments will always consider the assessment of the current processes used by an organization. The software processes used by the organization will be compared to the standard clauses or to process model practices like those of the CMMI. Capers Jones compares assessment benchmarking to a "medical checkup" of the software organization; this is unavoidable for the identification of what is going right and what has to change. This "medical checkup" is done using a list of the best practices of the industry. In the software industry, this examination is used for benchmarking or to kick start a process improvement program.

We have seen that in order to be ISO 9001 certified, an independent audit will need to be performed. The certifier will need to use an audit process for this.

6.5.1 Steps of a Software Audit

Figure 6.3 presents a model of the key steps that have to be successfully executed in order to obtain an objective assessment of the state of the software processes of an organization.

The first step consists of identifying and interviewing the individuals that will be part of the audit team. These individuals may need training and certifications to join the assessment team.

The second step, once the team is in place, is to have an initial contact with the audited organization either with a meeting or through a preliminary questionnaire. This is followed by a meeting to agree to the scope of the audit, the processes that will be reviewed, a proposed schedule, and the number of individuals involved. The main objective of this step is to prepare everyone for the audit so it is not a surprise. The project manager is then asked to send back the preliminary conformity assessment and provide the security clearance to access the project documentation drives. Questionnaire answers and a preliminary review of the project documents (e.g.,

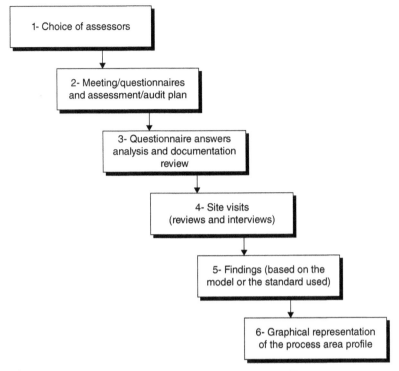

Figure 6.3   Major steps of a software audit or assessment [APR 08].

project management and technical documents) will allow the auditor to tailor the choice of interviewees' and solidify the audit schedule.

Next, we need to prepare for the audit that typically lasts 1 or 2 days. This fourth step consists of visiting the site, making presentations, interviewing individuals, and reviewing artifacts. During these 2 days, the inventory of interviews and documents is kept up to date and findings are captured and rated.

Tips for Your Conduct when Being Audited and Interviewed

1) **Answer directly and honestly.**
   Your responsibility is to provide the requested information. If you do not understand the question just say so. If the question does not apply to your role or your project just say so. Being honest is the best approach in audits.

2) **Do not volunteer information that is not questioned.**
   Answer directly to the question asked. Resist discussing all kinds of other topics that are not pertinent to the specific issue.

3) **Bring and show examples of the work.**
Bring examples of the review topic to show how the process was executed, communicated, tracked.

4) **Ask for help from others when necessary.**
The project is being audited, not you. If you do not know the answer to a question, refer to another person.

[OBR 09]

ISACA is the Information Systems Audit and Control Association. ISACA publishes the CobiT guidelines that can be used to audit a software project: www.isaca.org

Using the information collected, initial findings are made and validated with the participants. Professional judgment is required to determine if the execution of the practices satisfies the reference standards or model. Once findings are validated and ranked, the final report is produced. This takes approximately 10 days. It is sent to the distribution list identified earlier in the audit plan. A last and optional step is to help the team achieve compliance, if asked.

Finally, all the audit information is packaged and archived for future reference. All the compliance data are also analyzed and saved.

Audit Findings
Results of the evaluation of the collected audit evidence against audit criteria.

Note 1: Audit findings indicate either conformity or non-conformity.

Note 2: Audit findings can lead to improvement opportunities or identification or recording good practices.

Note 3: If the audit criteria are selected from legal or other requirements, the audit finding is termed compliance or non-compliance.

Note 4: Adapted from ISO 9000:2005, definition 3.9.5.

ISO 19011 [ISO 11g]

Conformity
Fulfilment of a requirement.

ISO 9000 [ISO 15b]

Nonconformity
Non-fulfilment of a requirement.

ISO 9000 [ISO 15b]

All types of audits should publish their process and make it available.

Process audits should be conducted on a regular basis for two main reasons: first, to ensure that practitioners use the processes of the organization and secondly to discover errors, omissions, or misunderstandings in the application of a process. Process audits are also used to assess the degree of use and understanding of practitioners. For example, a new document management process was introduced and practitioners were invited to produce and update documents using this new process. It is well known that engineers are not very inclined to document their work. They often see documentation as a "necessary evil." Following is an example of an audit conducted to evaluate the conformity of developers to the documentation management process of an organization.

Document Management Process Audits

In a defense industry organization, one of the authors coordinated the documentation of the document management process. A few months after its deployment to the engineers, the QA department conducted an audit to measure the level of compliance to this process.

As presented in the table below, results were not very good. After the published audit report, management decided to send an instruction that all personnel should follow this guideline. It also stated that a second audit would be planned to check on conformity again. As we can see, the second audit showed a higher conformity rate.

The auditor, when conducting the second audit, gathered feedback from engineers. This information was used by the owner of this process to improve it and to increase the level of compliance to reach at least 80% of conformity of all activities in a future audit.

Results of Two Process Conformity Audits

Document process activities	First audit results	Second audit results
Reviewer comments	38%	78%
Document approval matrix completed	24%	67%
Effort checklist completed	18%	33%
Document review checklist completed	5%	44%
Configuration management checklist completed	5%	27%
Document distribution list completed	38%	39%
Document formally approved	100%	100%

Laporte and Trudel (1998) [LAP 98]

6.6 AUDIT ACCORDING TO THE CMMI

The Software Engineering Institute (SEI) has developed assessment methods to be used in conjunction with its process models. A Software Process Assessment and a Software Capability Evaluation (SCE) [BYR 96] are available for use. The SCE is used for supplier selection to verify that contractual clauses are met.

Using the SCE for Supplier Evaluation

A USA agency had specified, in their subway transit contract with its main supplier, that all the contractors involved, such as for the propulsion sub-system, be assessed and demonstrate a minimum of CMM maturity level 2 using the SCE. An initial assessment was done and action plans were developed so that all non-conforming suppliers achieve this level within 24 months of contract signature. Each action plan was also reviewed and approved by the customer. After this period, a second assessment was done to verify the supplier's conformity.

When the SEI updated its model to launch the CMMI for Development in early 2000, the SCE audits became less used. In the CMMI, audits are activities described as part of the "process and product quality assurance." This process area aims at providing management with an objective state of the project's software processes and products. For the CMMI, audits are one of the many techniques used, similar to inspections and walk-throughs, to conduct objective assessments. The CMMI states what an objective assessment is in the Process and Product Quality Assurance process area. To ensure objectivity, the following issues must be addressed [SEI 10a]:

- description of the reporting structure of the SQA to show its independence;
- establish and maintain clearly formulated assessment criteria:
  o what is going to be assessed?
  o when, or, how will the process be assessed?
  o how will the assessment be conducted?
  o who must be involved in the assessment?
  o what product of an activity will be assessed?
  o when and how will the product of an activity be assessed?
- use the formulated criteria to assess the conformity of the process descriptions, standards, and procedures executed and for the product of an activity assessment.

6.6.1 SCAMPI Assessment Method

The SEI developed its own assessment method named SCAMPI (Standard CMMI Appraisal Method for Process Improvement) to support its CMMI model implementation. This method can be used for improving your processes internally, for the selection of external suppliers or for the monitoring of processes. Concerning the improvement of processes, this assessment method can be used for [SEI 06]:

- establishing a baseline of the strengths and weaknesses of the current processes;
- determining the maturity level of the current processes;
- measuring the progress with regards to the last process assessment;
- generating inputs for the process improvement plan;
- preparing the organization for a customer assessment;
- auditing the life cycle processes.

To learn more about the SEI SCAMPI appraisal method, visit the following site: http://resources.sei.cmu.edu/library/asset-view.cfm?assetid=5325

The following text box shows an extract of an evaluation report, using the SEI software assessment method.

Excerpt from a Formal Evaluation Report of the ACME Corporation
Findings
Concerning the requirements management process area:

- an inconsistent process is used to transpose the system requirements to software requirements;
- the design and coding are often started before the software requirements are defined;
- software requirements are inconsistently defined.

Explanation of the Findings

The requirements definition is the first step in a software development project. The whole project depends on the quality of the requirements and how they were transmitted. If the wrong requirements are used in the life cycle (e.g., system requirements, software requirements, high-level design, detailed design, code, unit tests, integration, and system testing and maintenance), the cost of corrective action increases at each stage of the life cycle.

The evaluation team learned that the ACME Corporation's engineers did not follow a standard process to define requirements and their transmission to other stages of the life cycle are not well communicated. The evaluation team also learned that the process should ensure that the documentation is also transmitted in the development cycle to avoid over-reliance on the software personnel. Such a process should also ensure the completeness, clarity and accuracy of requirements with a level of detail appropriate to the level of the document in which a requirement is defined. This means that the software requirement defined in a system performance specification should be verifiable at the system level, a software requirement defined in the system's design specification must be verifiable at the system integration and software requirements defined in the detailed design specification should be verifiable in the code.

Design and coding activities are often carried out before the requirements are defined. The evaluation team heard of situations where the requirements were developed after the code was written. The way the process is monitored is highly dependent on the project manager and project engineers. They must be aware of the importance of following the process that must be taken into account when planning the project.

The following text box shows another example of an improvement cycle in a telecommunications company.

Example of a Mentoring Process Used at Bell Canada

As shown in the figure below, the audit activity, when it is aimed at improvement, is rarely performed in isolation. Here is an example of the process, introduced at Bell Canada in the 1990s, which describes all the interactions between learning, improving the process and ensuring the compliance of the project to the local methodology that is supported by tools. Project teams could rely on a mentoring service to help them better understand how to properly use the internal methodology and tools.

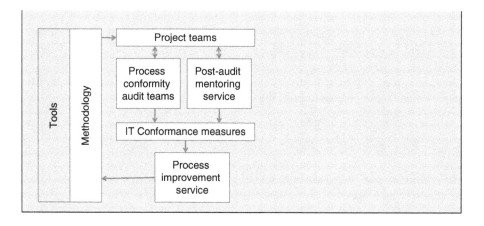

The next section presents the CAs that are implemented after a software audit.

6.7 CORRECTIVE ACTIONS

After an internal or external audit, an organization must perform CAs to correct the observed deficiencies. It is also possible to treat the preventive actions, an incident report, and customer complaints using the CA process.

Corrective Action

Action to eliminate the cause of a nonconformity and to prevent recurrence.

Note 1: There can be more than one cause for a nonconformity.

Note 2: Corrective action is taken to prevent recurrence whereas preventive action is taken to prevent occurrence.

ISO 9000 [ISO 15b]

An intentional activity that realigns the performance of the project work with the project management plan.

PMBOK® Guide [PMI 13]

Preventive Action

An intentional activity that ensures the future performance of the project work is aligned with the project management plan.

PMBOK® Guide [PMI 13]

A CA aims at eliminating potential causes of non-conformity, a defect, or any other adverse event to prevent their recurrence. Although rectifying a problem focuses on the correction of a very specific case, a CA eliminates the root cause of a problem. CMMI has a process area named "Causal analysis and resolution" to identify and eliminate root causes. The purpose of this process area is to identify causes of selected outcomes and take action to improve process performance [SEI 10a].

6.7.1 Corrective Actions Process

The problems encountered when developing systems that include software, or that occur during their operation, can come from defects in the software, in the development process itself, or in the hardware of the system.

To facilitate the identification of problem sources and apply appropriate CAs, it is desirable that a centralized system is developed to track issues through to resolution and to determine their root cause.

Since the validation, monitoring, and problem resolution may require coordination of different groups within an organization, the SQAP should specify the groups authorized to report/raise incident reports and CAs as well as for the submission of unresolved issues to management.

The following text box is an excerpt from the IEEE 730 standard that describes the requirements regarding this process.

Corrective and Preventive Action Process

A process for resolving software problems defined in the project plan or a separate process documented in either the SQA plan or the organizational quality management plan. Non-conformances are addressed by the project team using a defined Corrective Action (CA) process which may be documented in the project plan, the SQA Plan, or the organizational quality management plan. In response to a non-conformance, the project team proposes a corrective action. SQA reviews each proposed corrective action to determine whether it addresses the associated non-conformance. If the proposed corrective action does address the non-conformance, SQA identifies appropriate effectiveness measures that determine whether a proposed corrective action is effective in resolving the non-conformance. Once a corrective action is implemented, SQA evaluates the related activity and determines whether the implemented corrective action is effective.

IEEE 730 [IEE 14]

An organization must implement a CA process following the conduct of internal or external audits. This process should cover software products, agreements and software development plans.

To facilitate the management of many non-compliances, it would be desirable to use a software tool such as a spreadsheet or a database. In very small organizations, free issue-tracking tools like Bugnet can do the job. In large organizations, you either need to develop your own database or purchase a commercial tool (http://www.bugnetproject.com/).

A CA process, in a closed loop, may include the following elements:

– inputs: for example, an audit report, a non-compliance, or a problem report;
– activities:
  o register non-conformities in the issue tracking tool used by the organization,
  o analyze and validate the problem to ensure that the organization's resources are not wasted,
  o classify and prioritize the issue/problem,
  o analyze the problems to conduct trend analysis that can be identified and addressed,
  o propose a solution to the problem,
  o solve the problem and ensure that the resolution does not cause other problems, or if the non-compliance issue cannot be resolved within the project, refer it to the appropriate management level,
  o verify the problem resolution,
  o inform stakeholders of the problem resolution,
  o archive the problem documentation,
  o update the information in the issue tracking tool;
– outputs: for example, the resolution file, a corrected version of the software.

Figure 6.4 describes a problem resolution process.

Some organizations include a section in their problem report template where the person responsible for the service, as the head of the development organization, must propose a possible solution to correct the problem as well as a planned resolution date (see Figure 6.5).

Once this information is stored in the tool, the SQA can track every issue through to resolution and may sometimes have to intervene to remind the person responsible that he has unresolved issues that have exceeded the registered deadline.

As an independent body, the SQA has an escalation mechanism in the organization for raising the issue to a higher level in the event that the problem is not resolved. Here is a three-level escalation mechanism:

– escalate to the first level: if CAs are not carried out in accordance with the commitments, the SQA representative meets with the head of the software project

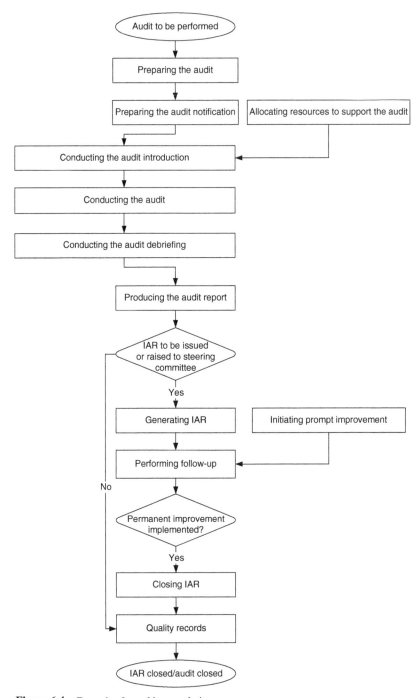

Figure 6.4 Example of a problem resolution process.

```
┌──────────────────────────────────────────────────────────────────┐
│ Problem report                                                     │
│                                                                    │
│                                                                    │
│ Priority:_____        Project name: _____   Date: _____       │
│                                                                    │
│ Process name: _____   Phase number: ___Raised by: _____   │
│                                                                    │
│ Number of days to answer: _____   Close date: _____         │
│                                                                    │
│ Number of days to fix this problem: _____                        │
│                                                                    │
│ Finding: _____         │
│                                                                    │
│ Requirement/Standard impacted:                                     │
│ _____           │
│                                                                    │
│ Immediate solution proposed:                                       │
│ _____           │
│                                                                    │
│ Root cause: _____           │
│                                                                    │
│ Permanent solution proposed:                                       │
│ _____           │
│                                                                    │
│ Acceptance date of permanent solution: _____   │
│                                                                    │
│ Follow-up action (if necessary):                                   │
│ _____           │
└──────────────────────────────────────────────────────────────────┘
```

Figure 6.5   Problem report and resolution proposal form.

to examine the plan to be implemented to ensure CAs, status of CAs and the risk of not completing CAs. The parties negotiate and agree on CAs and new deadlines for the implementation of CAs. The SQA representative documents this agreement and obtains the signature of the head of the software project;

– escalate to the second level: if the head of the software project is not receptive (to CA or the time limit), the SQA manager meets with the software project manager to review the plan to implement CA, the status of CAs and the risk of not completing CAs. The SQA manager documents decisions and obtains the signature of the software project manager;

– escalate to third level: if the software project manager is not receptive (to CA or the time limit), the SQA manager meets with senior management to discuss a plan to initiate CAs. The SQA manager documents decisions and obtains the signature of senior management.

6.8 AUDITS FOR VERY SMALL ENTITIES

Very small entities (VSEs) have modest means and little time to devote to an audit. However, many VSEs wish to be audited or assessed either to meet the requirements of a client or to increase their brand both nationally and internationally. Thus, a VSE could stand out from other VSEs and become an organization with which a client could develop a business relationship.

When the ISO workgroup was mandated to develop the ISO 29110, a survey was addressed to VSEs located in more than 32 countries. A large number (74%) of VSEs that responded to the survey said that it was very important to them to obtain a certification to improve their profile. A formal certification was requested by 40% of VSEs that answered this survey [LAP 08].

In the chapter on standards, we introduced the ISO 29110 standard. The VSE can, as described in this chapter, conduct an internal audit in accordance with ISO 17050. If the organization complies with ISO 29110, it may therefore declare itself as conforming to this standard, as long as the auditor's independence can be demonstrated by the absence of a parallel activity to be audited, divergence or conflict of interest.

VSEs also have access to external audits of second and third parties. An audit by a second party can be performed by external auditors. In this way, a VSE can demonstrate that an external auditor confirmed its conformity to ISO 29110. This type of audit can be performed at a low cost to a VSE.

Recall that third-party audits are conducted by independent auditing organizations, such as regulatory authorities or bodies granting registration or certification. The certification process is illustrated in Figure 6.6. It begins when a company contacts a certification body to start the certification process. Once the auditor has determined that the VSE is willing to be audited, the auditor launches the audit: the preparation of the audit (e.g., document review, planning, and preparing the audit), the implementation of the audit (e.g., conducting the opening meeting, reviewing documents, collecting and verifying information, producing findings and conclusions, and conducting the closing meeting), the preparation and distribution of the audit report and the end of the audit. Following the issuance of the audit certificate, the certification body will perform, typically on an annual basis, a surveillance audit to ensure

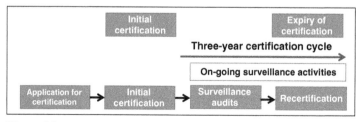

Figure 6.6 ISO/IEC 29110 certification approach.

continued conformance. An audit for the renewal of the certification is conducted to confirm continued conformance.

ISO 29110 was implemented in a Peruvian start-up of four people. The VSE was created in 2012. After implementing the Basic profile, two processes were executed in a 900-hour customer project [GAR 15]. Only 18% of the total effort consisted of rework. The following text box describes the ISO 29110 third-party audit. The VSE had more than 10 employees at the time of certification in 2014.

ISO 29110 Conformity Audit in Peru

The certification process was executed in two steps. During step 1, the existing documentation of the software development process was assessed. During the second step, the implementation and execution of the ISO/IEC 29110 Project Management (PM) and Software Implementation (SI) processes of the Basic profile were audited. At the end of each step, the Brazilian registrar published an observation report.

The VSE received the auditors' observations and corrective measures were taken. The VSE technical team implemented the recommendations and procedures were updated and distributed to the development team members.

The effort dedicated by the VSE for the first step of the audit, which does not include the auditor's travel expenses, was about $1000. The VSE invested 22 hours of work for step 1.

For step 2 of the audit process, the auditor cost was $1200. The VSE invested 63 hours for step 2 of the audit.

Steps 1 and 2 of this audit were conducted in April 2014. In July 2014, a Brazilian registrar delivered a 3-year conformity certificate for the PM and SI Basic profile of ISO/IEC 29110. A first surveillance audit was done in 2015, a second in 2016, and the recertification will be launched in 2017.

Garcia et al. (2015) [GAR 15]

After the successful audit of the Peruvian VSE, a local paper reported on this event. Following this article, Peruvian companies contacted the VSE to know more about the process and discuss business opportunities. Later, a large Peruvian insurance company attributed a software development contract to this VSE. In 2017, this VSE had 23 employees.

6.9 AUDIT AND THE SQA PLAN

The IEEE 730 standard [IEE 14] demands that the SQA activities of a project need to be coordinated with audits and other life cycle processes required to ensure the conformity and quality of the process and the product. It requires SQA to ensure that the audit concerns have been well explained to the project teams and that they audit the software development activities periodically to determine consistency with defined software life cycle processes. For this to happen, the organizational processes must be published to the organization beforehand.

It also imposes that SQA, independent from the project teams, audit projects periodically to determine conformance to defined project plans and assess the skill and knowledge needs of the project and compare them to the skill and knowledge of the organization's staff to identify any gaps. Where projects involve external suppliers, it is a good practice to conduct at least one compliance audit and include it in the contract terms.

For the audit activities of the project, the IEEE 730 standard asks that the project team be ready to answer the following questions:

- is a subcontractor or external supplier required by the contract? If yes, have periodic reviews and audits been performed to determine if software products fully satisfy contractual requirements?
- have any issues raised as part of these supplier audits been reviewed and assessed for impact?
- have corrective/preventive action plans been developed for any non-conformities identified during the supplier audit?
- have project non-conformances been recorded and appropriately resolved?
- have project CAs plans been established for items that did not meet the system requirements?

For SQA, the following questions should be answered for each project identified in the audit plan of the organization:

- has an appropriate and effective audit strategy been developed for the project?
- has an appropriate and effective audit strategy been implemented for the project?
- has compliance of selected software work products, services or processes with requirements, plans, and agreements been determined according to the audit strategy?
- are audits conducted by an appropriate independent party?
- have the audit results been documented?
- have all issues detected during an audit been documented as non-conformances?
- have all non-conformances been considered for CA?
- have all CAs that were implemented proven to be effective as determined by effectiveness measures?
- has an appropriate justification been provided for each non-conformance not requiring CA?

6.10 PRESENTATION OF AN AUDIT CASE STUDY

In this section, the result of project conformity audits completed at a US system development location of Bombardier Transport, where Professor Laporte was involved, is presented.

Software Engineering Performance Improvement at Bombardier Transport
[LAP 07a], [LAP 07b]
The performance of a subway signaling software development project was assessed, by external assessors, twice between 2003 and 2006. The 2003 assessment created a baseline or reference point for the progress observed in 2006. During these two visits, the same assessment method was used to evaluate the software processes, the project performance, and the organizational change management. This text describes the assessed organization, explains the multidimensional methodology used to conduct the assessments as well as the business objectives and the quantitative improvements achieved.

Context Description
At that time, Bombardier Transport had more than 30 software development locations and more than 950 software engineers. The Total Transit Systems (**TTS**) division offers transport solutions for cities and airports. The product portfolio contains a range of automated transport systems, monorails, light trains, and subways. The Pittsburgh division had close to 100 software engineers locally with 30 located at the Bombardier Transport in Hyderabad, India.

The Software Engineering Competence Center
Competence centers were aimed at lowering corporate technical risks and costs. The software engineering competence center supported strategic initiatives. It conducted product reviews and proposed actions to reduce risks. The competence center was asked to assess the Pittsburgh software development organization.

Evaluation Methodology
The process dimension reuses a tailored version of the industry-proven CMM evaluation methods. Depending on business needs (organizational and project list) and the scope of the evaluation, the process areas of the CMM are prioritized (high/medium/low). Then, an evaluation agenda is created using Bombardier SWE Process role names. The agenda is then updated with the individuals involved in the project who are associated with those roles. Communication is conducted in advance to ensure smooth participation and to manage expectations. During the collecting evidence step, an evaluation sheet is used to log the gathered/analyzed data. This evaluation sheet is also used to establish the maturity indicators employed in the site findings.

Site Evaluation

Three phases were used to evaluate the site:

Planning phase:
– establishment of organizational scope
– visit preparation (agenda)
– information gathering (extended version only)
– team build-up

On-site phase:
– opening presentation
– collection of evidence (interviews, documentation reviews)
– documentation of the findings (strengths and weaknesses)
– site debriefing with management representatives

Assessment report writing phase:
– prepare site findings and recommendations report
– prepare interim and final reports

Process Evaluation

The table below summarizes the key process areas of the CMM® and how they relate to a maturity level. To achieve a maturity level, an organization must have successfully completed the practices for lower levels. For this evaluation, the six process areas located at CMM level 2 were evaluated.

			Result
Level	**Characteristic**	**Key process areas**	**Productivity & quality**
Optimizing (5)	Continuous process capability improvement	Process change management Technology change management Defect prevention	
Managed (4)	Product quality planning; tracking of measured software process	Software quality management Quantitative process management	
Defined (3)	Software process defined and institutionalized to provide product quality control	Peer reviews Intergroup coordination Software product engineering Integrated software management Training program Organization process definition Organization process focus	
Repeatable (2)	Management oversight and tracking of project; stable planning and product baselines	Software configuration management Software quality assurance Software subcontract management Software project tracking & oversight Software project planning Requirements management	
Initial (1)	Ad hoc (success depends on heroes)	"People"	Risk

"The CMM model [PAU 95]."

The evaluation criteria provided the following results: a process area can either conform, partially conform of not conform to the CMM. The results of the first evaluation are presented in the following table.

Result of the First Assessment

Process area	Conformity level
Requirements management	Conform
Software project planning	Partially conform
Software project tracking and oversight	Partially conform
Software configuration management	Conform
Software quality assurance	Conform
Software subcontractor management	Partially conform

The team did not, during the second evaluation, assess processes since the software processes at the Pittsburgh location had already formally been evaluated a few months before as CMM level 3 compliant.

Evaluation of Performance

Performance measurement of organizational software processes, using the earned value technique, contributes to achieving the business goals. The first step in implementing this technique was to identify the performance measures already in use in the organization. Then, the techniques were validated to assess their applicability, validity, and precision. Finally, the collected data were used to assess performance. The elements considered during this assessment were:

– the Cost Performance Index (CPI); measures performance and takes corrective action when required, as well as compares performance with that of past projects. This indicator is calculated by dividing the budget cost of work performed by the actual cost of work performed;
– the Schedule Performance Index (SPI); measures performance and takes action to realign the project schedule if needed. It is calculated by dividing the budget cost of work performed by the budget cost of work scheduled;
– the Critical Ratio (CR) is the result of the following calculation: SPI × CPI;
  ○ a CR < 0.9 or <1.2 means that the project is under control;
  ○ a CR > 0.8 or <0.9 or >1.2 or <1.3 means that corrective measures are necessary;
  ○ a CR < 0.8 or >1.3 means that both the scope and the estimate of this project need to be completely reviewed.

Results of the First Performance Evaluation

Project ID	CPI	SPI	CR
Project A	0.72	0.97	0.7
Project B	0.93	0.86	0.8
Project C	1.0	1.0	1.0

Results of the Second Performance Evaluation

Project ID	CPI	SPI	CR
Project D	0.88	0.67	0.59
Project E	1.22	0.96	1.17
Project C	0.93	1.01	0.94
Project F	1.04	0.75	0.78

Evaluating Organizational Change Management

The next table describes the questionnaires used from the American company IMA (www.imaworldwide.com) to evaluate the change management practices in projects that involve technological changes.

Change Management Questionnaire

Questionnaire name	Description
Culture assessment	Assessment of the fit between the desired change and the actual organizational culture in order to identify potential barriers and to leverage actual cultural strengths
Organizational stress level	Evaluation of the priorities for resources in the organization
Implementation history	Assessment of barriers and lessons learned from previous change projects (since past problems are likely to recur, this tool allows identification of the issues that need to be managed for the change project to be successful)
Sponsor assessment	Evaluation of the resources, reinforcement (e.g., motivation) and communications commitments made and demonstrated by the sponsor(s) of a change project
Change agent skills	Evaluation of the skills and motivation of those responsible for facilitating the implementation of organizational changes
Individual readiness	Evaluation of the reasons why people may resist an organizational change

This next table describes the results obtained following two on-site evaluations.

Results of Two Evaluations

Questionnaire name	Ideal result	First evaluation results	Second evaluation results
Culture assessment	100	54	65
Organizational stress level	Less than 200	752	700
Implementation history	100	62	64
Sponsor assessment	100	Not evaluated	90
Change agent skills	100	Not evaluated	66
Individual readiness	100	70	68

The next table describes the positive trend of the organization's performance indicators.

Organizational Performance Indicators

Indicators	2003	2006	Delta
Income	19.6M US$	35.9M US$	+84 %
Productivity	194K US$	264K US$	+36 %
Profitability	2.7M US$	5.5M US$	+104 %
Number of employees	101	136 136 + 32 (India)	+34 % +66 %

The evaluation team concluded that between 2003 and 2006 the organization developed and deployed many software processes, improved its capacity to manage change and demonstrated that its process performance had a significant impact on the business performance of the organization.

Some Recommendations:

– *Deploy an Institutionalized Software Sizing Approach for New Projects*
A shared software sizing technique is indispensable. We know that counting lines of code is not always appropriate, but the site needs a software size measure to compare project performance. This information can be used later to better estimate project efforts, improving the predictability and profitability of future projects.

– *Assimilate Lessons Learned in the Organizational Processes*
Lessons learned are captured and stored on the organization intranet to allow managers to browse through them when needed. Sadly, lessons learned were rarely used by other projects. To try to promote their use, it is recommended that processes, procedures and checklists be updated during the final project review where lessons learned are generated. Additionally, it would be possible to use lessons learned as part of peer reviews.

– *Improve the Peer Review Process*
To improve the peer review effectiveness and defect detection rate, it is recommended that an inspection process be deployed. Inspections are well known to identify defects. It is recommended that the Bombardier Transport procedure entitled "BES Software Peer Reviews" be adopted in the Pittsburgh site. This procedure conforms to IEEE 1028: Software Reviews and Audits standard. The adoption of an inspection process should be relatively easy since the Pittsburgh site already conducts a less formal form of peer review. Inspections also conform to the "Six Sigma" scheme of Bombardier Transport and allows for a non-conformity cost reduction.

6.11 SUCCESS FACTORS

The following text boxes list factors that affect quality with regards to the software audit.

Factors that Foster Software Quality

1) An organization that places quality before schedules and budgets.
2) A documented and public auditing process.
3) Prior training of personnel regarding audits.
4) Trained and certified auditors.
5) A professional audit approach for long-term improvement.
6) An organization that acts on audit recommendations.
7) Resources available to conduct corrective actions.

Factors that may Adversely Affect Software Quality

1) Surprise audits in projects that are already in trouble.
2) Unclear audit rules that are not thought out or understood by anyone.
3) Audits used for personal gain and politics.
4) Unclear development life cycle processes and untrained staff (e.g., on organizational processes).

5) Managers that, by their decisions and statements, send the message that real delivery work has to be done and that we will take care of audit recommendations when we have time.

6) Management that edits reports or pressures auditors to lessen the impacts of audit report results.

7) Internal audits used to surprise and catch employees off-guard with a "QA police" squad instead of teaming up to improve the processes.

8) Management that pays attention to audit reports the month before the annual ISO 9001 audit.

6.12 FURTHER READING

APRIL A., ABRAN A., AND MERLO E. Process assurance audits: Lessons learned. In: Proceedings of ICSE 98, Kyoto, Japan, April 19–25, 1997.

CRAWFORD S. G. AND FALLAH M. Software development process audits—A general procedure. In: Proceedings of the 8th international conference on Software engineering, London, UK, 1985, pp. 137–141.

HELGESON J.W. *The Software Audit Guide*. ASQ Quality Press, Milwaukee, WI, 2010.

OUANOUKI R. AND APRIL A. IT process conformance measurement: A Sarbanes-Oxley requirement. In: Proceeding of the IWSM Mensura, Palma de Mallorca, Spain, November 4–8, 2007. Available at: http://s3.amazonaws.com/publicationslist.org/data/a.april/ref-197/1111.pdf

VAN GANSEWINKEL V. Making quality assurance work (Audits), Professional Tester, April, 2003.

6.13 EXERCISES

6.1 A manager just heard that auditors will come and visit a software package acquisition project team. How can you adequately prepare?

6.2 You have been promoted as the SQA specialist at the Acme Corporation. The manager asks you to explain how to assess the quality of software projects quantitatively. Explain what needs to be implemented first and then the assessment approach that can be used.

6.3 List the deliverables that can be reviewed during an audit.

6.4 The IEEE 1028 standard provides guidelines for the audit of software processes and products. The organization should have ready-made checklists for each process and product audited. Develop a checklist for the following products and processes:

a) a SQA plan;

b) an inspection process;

c) a design document;

d) a walk-through report;

e) source code.

6.5   Describe the different characteristics between an audit and an inspection.

6.6   What training, education, and experience are necessary to play the lead quality auditor role in a software audit?

6.7   Name the most popular ISO 9001 software interpretation guides. Why are they necessary?

6.8   Draw a diagram which describes the typical steps of a software assessment and audit.

6.9   List the necessary support tools for software audits. Explain the two most complex tools.

6.10   List the key success factors of an audit.

6.11   In a quantitative assessment of a deliverable, explain what a quality attribute is.

6.12   Your manager asks you to assess the quality of the requirements phase. Explain how you can use the concepts of conformity evaluation for this life cycle phase.

Chapter 7

Verification and Validation

After completing this chapter, you will be able to:

- – understand what is meant by verification and validation (V&V);
- – understand the benefits and costs of applying V&V techniques;
- – understand the traceability technique and its usefulness;
- – learn about the IEEE 1012 V&V standard and models used in industry;
- – understand the processes and activities of V&V;
- – understand the activities of the software validation phase;
- – learn how to develop and use a V&V checklist for your project;
- – understand how to write a V&V plan for your project;
- – learn about the V&V tools available;
- – understand the relationship between V&V and the software quality assurance plan.

7.1 INTRODUCTION

In an article about safety, Leveson [LEV 00] brilliantly explains the dangers of modern software-based systems. The following text box summarizes her thinking.

> The introduction of new technologies, coupled with the increasing design complexity, is starting to produce a change in the nature of accidents. Although accidents related to equipment failure are reduced, system crashes are increasingly occurring. System accidents occur during interactions between components (e.g., electromechanical, digital, and human) rather than by the failure of an individual component. The increasing use of software is closely linked to the increasing frequency of system crashes, since software

Software Quality Assurance, First Edition. Claude Y. Laporte and Alain April.
© 2018 the IEEE Computer Society, Inc. Published 2018 by John Wiley & Sons, Inc.

usually controls the interactions between the components allowing virtually unlimited complexity in the interactions between components.

These accidents often involve software that correctly implements the specified behavior, but there was a misunderstanding about what this behavior should be. Software related accidents are usually caused by deficient software requirements and not by coding errors or software design problems.

Ensuring that the software meets its requirements or trying to make it more reliable will not make it safer as the requirements at the outset could be deficient. Software can be highly reliable and correct and be unsafe when:

– the software correctly implements its requirements, but its behavior is not safe from a systemic point of view;
– the requirements do not specify certain behaviors required for system safety (consequently the requirements are incomplete);
– the software has unforeseen behaviors (that are dangerous) beyond what has been specified in the requirements.

[LEV 00]

The introduction of innovative products associated with software that contain an increasing number of functions often involves increasing the number of computer processing units and size of software. For example, in the automobile sector, over 60 small computers (named electronic control units (ECUs)) with more than 100 million lines of code from different suppliers are used in many car models [REI 04]. These networked ECUs control, amongst others, ignition, braking, the entertainment system, and now autopilot and self-parking. Failure of some units will result in recalls, dissatisfaction of customers, or deterioration in the performance of the vehicle, whereas a failure with the autopilot, direction, acceleration, or braking system could cause an accident, injury, or death.

"In a typical commercial development organization, the cost to ensure (which is the assurance that the system works satisfactorily in terms of functional and non-functional requirements in its operation environment) prototyping, testing and verification activities can vary from 50% to 75% of the total development cost."

Hailpern and Santhanam (2002) [HAI 02]

According to the IEEE 1012 standard for V&V [IEE 12], the goal of V&V in software projects is to help the organization incorporate quality in the software throughout the software life cycle. The V&V process provides an objective evaluation of software products and processes. It is a simple question of addressing quality during development as opposed to trying to add quality to a product after its been built.

Often it is not possible, and perhaps seldom reasonable, to inspect all the fine details of all the software products created during the development and maintenance life cycle, particularly because of a lack of time and budget. Therefore, all organizations have to make certain compromises and this is where the software engineer is expected to follow a rigorous process of justification and selection. Software teams must establish V&V activities during the project planning phase, so as to choose the techniques and approaches that will allow the products to have a proper level of V&V. The choice of these activities and their priorities is based on assessing risk factors and their potential impact. These V&V activities need to be added to the development process of the project in order to reduce risks to an acceptable level.

In this chapter, we present V&V for software. First, we explain the concepts of V&V and the intended benefits and costs. We then present the international standards and models that define V&V activities or impose them in certain situations. We continue with the presentation of an inventory of the various V&V techniques available as well as their utility to address the particular concerns of the software engineer. We then present the typical contents of a V&V process. This is followed by a brief discussion on the importance of independent V&V (IV&V) in critical projects. We then provide more detail for one of the V&V activities often required for safety: traceability. Finally, we explain how to develop and use checklists.

Verification

Confirmation, through the provision of objective evidence, that specified requirements have been fulfilled

ISO 9000

The process of evaluating a system or component to determine whether the products of a given development phase satisfy the conditions imposed at the start of that phase.

The process of providing objective evidence that the system, software or hardware and their related products conform to requirements (e.g., for correctness, completeness, consistency, and accuracy) for all life cycle activities during each life cycle process (acquisition, supply, development, operation, and maintenance); satisfy standards, practices and conventions during life cycle processes; and successfully complete each life cycle activity and all the criteria for initiating succeeding life cycle activities. Verification of interim work products is essential for proper understanding and assessment of the life cycle phase product(s).

IEEE 1012 [IEE 12]

Verification aims to show that an activity was done correctly (doing it right), in accordance with its implementation plan and has not introduced defects in its output. It can be done on successive intermediate states of a product that is the outcome of an activity.

Validation

Confirmation, through the provision of objective evidence, that the requirements for a specific intended use or application have been fulfilled.

> Note 1 to entry: The objective evidence needed for a validation is the result of a test or other form of determination such as performing alternative calculations or reviewing documents.
>
> Note 2 to entry: The word "validated" is used to designate the corresponding status.
>
> Note 3 to entry: The use conditions for validation can be real or simulated.

ISO 9000

The process of evaluating a system or component during or at the end of the development process to determine whether it satisfies specified requirements.

The process of providing evidence that the system, software, or hardware and its associated products satisfy requirements allocated to it at the end of each life cycle activity, solve the right problem (e.g., correctly model physical laws, implement business rules, and use the proper system assumptions), and satisfy intended use and user needs.

IEEE 1012 [IEE 12]

Validation is composed of a series of activities which start, early in the development life cycle, with the validation of customer requirements. End-users or their representatives will also evaluate the behavior of the software product in the target environment, either real, simulated, or on paper.

Validation helps minimize the risk of developing the wrong items by ensuring that the requirements are adequate and complete. Subsequently, it will be ensured that these validated requirements are developed during the following phase, notably, the specifications. Validation also ensures that the software does not do what it should not do. This means that no unintended behavior should arise from it.

If quality assurance is the poor cousin of software development, validation has the same relationship with V&V. While verification practices, such as testing, have a very important place in academia and industry, we cannot say the same for validation techniques. Validation techniques are often absent or ignored by developers and the mandated development processes. Some organizations validate requirements early in a project. Unfortunately, they will carry out some validation only at the very end.

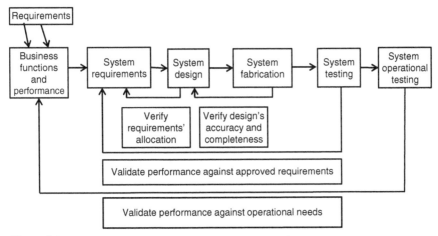

Figure 7.1   V&V activities in the software development life cycle.

Occasionally, we find validation practices embedded at different stages of the development cycle (as shown by Figure 7.1). The lines at the top of this figure indicate the development cycle phases where validation activities can be performed.

Some organizations explicitly have a phase called software validation in their software development process, and if they produce software that is integrated into a system, they will also explicitly have a system validation phase. Figure 7.2 illustrates this with a software life cycle named the V development life cycle process. It illustrates, along the center arrow, that system and software validation plans that will be executed during the validation phase originate from the system and software specification phases. These plans will be updated throughout the development phases

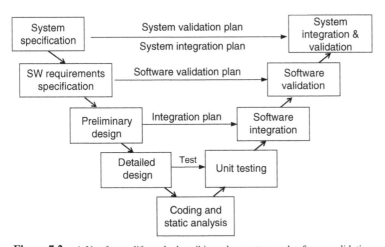

Figure 7.2   A V software life cycle describing when system and software validations are executed.

and will be used during the validation phases on the ascending line of activities shown on the right hand side of the figure.

One reason for preparing these plans early in the development life cycle is that validation activities may require special equipment or environments to perform the validation of the system that includes software. It is possible that a testing environment also needs to be established. For example, the validation of an air traffic control system that needs to operate in conditions where dozens of aircraft are in the air at the same time may require validation of the system near a busy airport. This would enable the validation of several functional and non-functional requirements of the system.

Figure 7.2 shows that system and software validation plans are developed during the descending part of the development cycle in the V-shaped diagram. These plans will be used during the subsequent phases of development to validate, among other things, the system requirements and software requirements against the needs and during the ascending part of the development cycle in the V-shaped diagram to validate the software during the validation phase. Since the software is a component of a system, it will be integrated with hardware or other software and subject to system validation activities.

Difference Between Verification and Validation

To design a graphical user interface where requirements ask that color indicators be shown to reflect the power level of certain connected devices, verification will check whether the necessary indicators are all present. The goal of the validation will be, among other things, to determine that the color indicators reflect the actual state of the power level of these devices. If the agreement was to display the red color to indicate that the devices need recharging and the displayed color is green, then the application does not do what it is supposed to do. It does something, but not correctly!

You can see that validation can only be performed after the verification activities are completed successfully. This avoids devoting effort too early in the process to validate a software product that would be incomplete or may contain too many errors.

Several of the V&V techniques are of a similar nature. For example, tests are used to perform V&V. Other examples of techniques include analysis, inspection, demonstration, or simulation. One should choose the most appropriate technique for the least cost.

After identifying risks and the required V&V techniques, the V&V activities need to be planned. In some projects, V&V activities are planned by a team belonging to different parts of an organization, for example, system engineers, software developers, supplier personnel, a risk manager, V&V or software quality assurance (SQA), software testers, a configuration manager, etc. The main objective of the V&V activity will be to develop a detailed V&V plan for the project.

7.2 BENEFITS AND COSTS OF V&V

As we have mentioned above, the goal of V&V is to build quality into the software early during its construction and not just try to fix this at the testing stage. Figure 7.3 shows an example of the software processes, in an American company, where defects are injected. The figure shows that a large percentage of defects, approximately 70%, are injected even before any line of code has been produced. It is therefore necessary that we include techniques in the software life cycle that will allow for the early detection and removal of these defects as close as possible to when they are created. Additionally, good detection techniques will greatly reduce the costly rework associated with corrections, which is an important cause of schedule delays.

Figure 7.4 describes the defect detection effectiveness in an American company [SEL 07]. These results originate from Northrop Grumman who collected data for 14 systems where 3418 defects where detected during 731 reviews. These systems contained between 25,000 and 500,000 lines of code and the corresponding teams ranged from 10 to 120 developers. This study shows that not only is it possible to detect errors, but also to eliminate them in the same phase where they were produced. For example, Figure 7.3, shows that 50% of the defects where injected in the requirements phase. Figure 7.4 also shows that 96% of these defects where eliminated in the same phase.

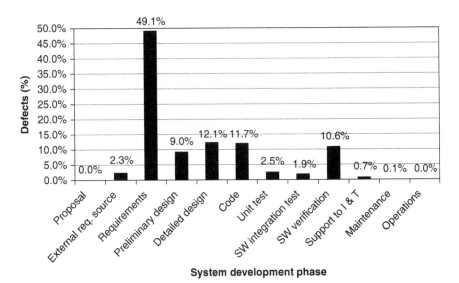

Figure 7.3   Example of software process phases where defects are injected [SEL 07].

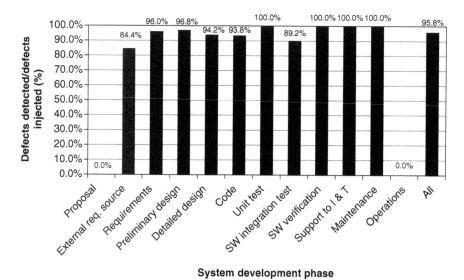

System development phase

Figure 7.4 Percentage of the defects detected by the development process phase [SEL 07].

What can be learned from this figure is that it is possible to estimate the percentage of injected defects and to detect and correct a high percentage of them such that they will not propagate from one phase to another. This process has been named the error containment process. It is therefore possible to describe, in the quality plan of the project, quantitative quality criteria concerning defect removal objectives at each phase of the project.

Why is it important to correct defects where they are injected? As shown in Chapter 2 (Figure 2.2), there is a compounding cost of a defect that is not corrected in the phase where it is injected. For example, a defect that arises during the assembly phase will cost three times more to fix than one that is corrected during the previous phase (during which we should had been able to find it). It will cost seven times more to fix the defect in the next phase (test and integration), 50 times more in the trial phase, 130 times more in the integration phase, and 100 times more when it is a failure for the client and has to be repaired during the operational phase of the product.

We are able to test only a small fraction of all the possible states of complex software based systems. For example, the software that is used for airplane collision avoidance systems, which is embedded on every modern airplane, includes approximately 1040 states. Our

> level of confidence about our ability to detect defects by only using test techniques is still very limited. We therefore have to use other methods and techniques to ensure the safe functioning of systems that include software.
>
> Leveson (2000) [LEV 00]

7.2.1 V&V and the Business Models

We recall here the main business models used by the software industry that were introduced in Chapter 1 [IBE 02]:

- Custom systems written on contract: The organization makes profits by selling tailored software development services for clients (e.g., Accenture, TATA, and Infosys).
- Custom software written in-house: The organization develops software to improve organizational efficiency (e.g., your current internal IT organization).
- Commercial software: The company makes profits by developing and selling software to other organizations (e.g., Oracle and SAP).
- Mass-market software: The company makes profits by developing and selling software to consumers (e.g., Microsoft and Adobe).
- Commercial and mass-market firmware: The company makes profits by selling software in embedded hardware and systems (e.g., digital cameras, automobile braking system, airplane engines).

These business models help us to understand the risks associated with each situation. V&V techniques can be used to detect defects and reduce these risks. The project manager, supported by SQA, will choose, budget and plan the adequate V&V practices for his project commensurate to the risks faced. The mass-market business model and embedded systems use these techniques extensively.

7.3 V&V STANDARDS AND PROCESS MODELS

The most important standards and process models that describe required processes and practices for V&V are presented next: the ISO 12207 [ISO 17], the IEEE 1012 [IEE 12], and the CMMI®. Some standards will go as far as to recommend, for critical software, that some programming languages be avoided. For example, a railway control software standard forbids that programmers use the "GoTo" programming instruction and that they remove "dead code" before the final delivery of the product. In the next section, the IEEE 1012 standard is explained, then other standards will be briefly covered.

7.3.1 IEEE 1012 V&V Standard

The IEEE 1012—*Standard for System and Software Verification and Validation* [IEE 12] is applicable to the acquisition, supply, development, operation and maintenance of systems, software, and hardware. This standard is applicable to all types of life cycles.

7.3.1.1 Scope of IEEE 1012

IEEE 1012 addresses all the life cycle processes of systems and software. It is applicable to all types of systems. In this standard, the V&V processes determine whether the products completed by a specific development activity meet the requirements of their intended use and the corresponding end-user needs. This assessment can include analysis, evaluation, reviews, inspections, and testing of the products and the development activity.

The verification process provides objective evidence that the system, software, or hardware and its associated products [IEE 12]:

- conform to requirements (e.g., for correctness, completeness, consistency, and accuracy) for all life cycle activities during each life cycle process (acquisition, supply, development, operation, and maintenance); refer to the quality characteristics of requirements listed in section 1.3.1 of Chapter 1;
- satisfy standards, practices, and conventions during life cycle processes;
- successfully complete each life cycle activity and satisfy all the criteria for initiating succeeding life cycle activities.

The validation process provides evidence that the system, software, or hardware and its associated products [IEE 12]:

- satisfy requirements allocated to it at the end of each life cycle activity;
- solve the right problem (e.g., correctly model physical laws, implement business rules, and use the proper system assumptions);
- satisfy intended use and user needs.

7.3.1.2 Purpose of IEEE 1012

The intention of this standard is to perform the following [IEE 12]:

- establish a common framework for all the V&V processes, activities and tasks in support of the system, software, and hardware life cycle processes;
- define the V&V tasks, required inputs, and required outputs in each life cycle process;
- identify the minimum V&V tasks corresponding to a four-level integrity scheme;
- define the content of the V&V Plan.

7.3.1.3 Field of Application

IEEE 1012 applies to all types of systems. When executing V&V for a system, software, or hardware element, it is important to pay special attention to the interactions with the system.

A system provides the capacity to satisfy a need or an objective by combining one or more of the following elements: processes, hardware, software, facilities, and human resources. These relationships require that the V&V processes address interactions with all of the system elements. Since software interconnects all the key elements of a digital system, the V&V processes also examine the interactions with every key component of the system to determine the impact of each element on the software. The V&V processes take the following system interactions into account [IEE 12]:

- environment: determines that the system correctly accounts for all conditions, natural phenomena, physical laws of nature, business rules, and physical properties and the full range of the system operating environment.

- operators/users: determines that the system communicates the proper status/condition of the system to the operator/user and correctly processes all operator/user inputs to produce the required results. For incorrect operator/user inputs, assure that the system is protected from entering into a dangerous or uncontrolled state. Validate that operator/user policies and procedures (e.g., security, interface protocols, data representations, and system assumptions) are consistently applied and used across each component interface.

- other software, hardware, and systems: determines that the software or hardware component interfaces correctly with other components in the system in accordance with requirements and that errors are not propagated between components of the system.

7.3.1.4 Expected Benefits of V&V

The expected benefits of V&V are [IEE 12]:

- facilitate early detection and correction of anomalies;
- enhance management insight into process and product risks;
- support the life cycle processes to assure conformance to program performance, schedule, and budget;
- provide an early assessment of performance;
- provide objective evidence of conformance to support a formal certification process;
- improve the products from the acquisition, supply, development, and maintenance processes;
- support process improvement activities.

7.3.2 Integrity Levels

IEEE 1012 uses integrity levels to identify V&V tasks that should be executed depending on the risk. High integrity level system and software require more emphasis on V&V processes as well as a more rigorous execution of the V&V tasks in the project.

Integrity Levels

A value representing project-unique characteristics (e.g., complexity, criticality, risk, safety level, security level, desired performance, and reliability) that defines the importance of the system, software, or hardware to the user.

IEEE 1012

Table 7.1 lists the IEEE 1012 definition of each of the four integrity levels and their expected consequences.

Table 7.2 presents an example of a four level integrity framework that takes into account the notion of risk. It is based on the possible consequences and risk mitigation.

Table 7.3 illustrates the risk-based framework using the four levels of integrity and their potential consequences described in Tables 7.1 and 7.2. Each cell of Table 7.3 attributes an integrity level on the basis of the potential consequence of a defect and its probability of occurring in an operating state that contributes to the failure. Some of the cells in this table reflect more than one integrity level. This is an indication that the final assignment of the integrity level by a project team can be selected to reflect the system requirements and the need for risk mitigation.

Table 7.1 Definition of Consequences [IEE 12]

Consequence	Definition
Catastrophic	Loss of human life, complete mission failure, loss of system security and safety, or extensive financial or social loss.
Critical	Major and permanent injury, partial loss of mission, major system damage, or major financial or social loss.
Marginal	Severe injury or illness, degradation of secondary mission, or some financial or social loss.
Negligible	Minor injury or illness, minor impact on system performance, or operator inconvenience.

Table 7.2 Integrity Levels and the Description of Consequences [IEE 12]

Software integrity level	Description
4	An error to a function or system feature that causes the following: – catastrophic consequences to the system with reasonable, probable, or occasional likelihood of occurrence of an operating state that contributes to the error; or – critical consequences with reasonable or probable likelihood of occurrence of an operating state that contributes to the error.
3	An error to a function or system feature that causes the following: – catastrophic consequences with occasional or infrequent likelihood of occurrence of an operating state that contributes to the error; or – critical consequences with probable or occasional likelihood of occurrence of an operating state that contributes to the error; or – marginal consequences with reasonable or probable likelihood of occurrence of an operating state that contributes to the error.
2	An error to a function or system feature that causes the following: – critical consequences with infrequent likelihood of occurrence of an operating state that contributes to the error; or – marginal consequences with probable or occasional likelihood of occurrence of an operating state that contributes to the error; or – negligible consequences with reasonable or probable likelihood of occurrence of an operating state that contributes to the error.
1	An error to a function or system feature that causes the following: – critical consequences with infrequent likelihood of occurrence of an operating state that contributes to the error; or – marginal consequences with occasional or infrequent occurrence of an operating state that contributes to the error; or – negligible consequences with probable, occasional, or infrequent likelihood of occurrence of an operating state that contributes to the error.

Table 7.3 Example of Probability Combinations of Integrity Levels and Consequences [IEE 12]

Error Consequence	Likelihood of occurrence of an operating state that contributes to the error (decreasing order of likelihood)			
	Reasonable	Probable	Occasional	Infrequent
Catastrophic	4	4	4 or 3	3
Critical	4	4 or 3	3	2 or 1
Marginal	3	3 or 2	2 or 1	1
Negligible	2	2 or 1	1	1

Tools that generate or translate source code (e.g., compilers, optimizers, code generators) are characterized by the same integrity level as the software they are used for. As a general rule, the integrity level assigned to a project should be the highest integrity level of any of the components of a system, even if there is only one critical component.

The integrity level assignation process should be consistent and reassessed throughout the project development life cycle. The rigor and intensity level of the V&V and documentation activities in the project should be commensurate to its integrity level. As the integrity level of a project lowers, the rigor and intensity level of V&V should also be diminished accordingly. For example, a risk analysis conducted for a project at integrity level 4 will be formally documented and will investigate failures at the module level, while risk analysis at integrity level 3 could assess only important failure scenarios and be documented informally during a design review process.

The four-level integrity framework is essentially used for the V&V practices recommended by IEEE 1012. The next section provides an example of V&V practices recommended for the software requirements activity.

7.3.3 Recommended V&V Activities for Software Requirements [IEE 12]

The recommended V&V activities for software requirements address functional and non-functional software requirements, interface requirements, system qualification requirements, security and safety, data definition, user documentation, installation, acceptance, operation, and ongoing maintenance of the software. The V&V test planning is initiated at the same time as V&V activities for software requirements and continues throughout many other V&V activities.

The objectives of the V&V activities for software requirements are to ensure that they are correct, complete, accurate, testable and consistent with the system software requirements. The V&V effort for software requirements, for any integrity level, shall perform:

– requirements evaluation;
– interface analysis;
– traceability analysis;
– criticality analysis;
– software qualification test plan V&V;
– software acceptance test plan V&V;
– hazard analysis;
– security analysis;
– risk analysis.

Qualification
Process of demonstrating whether an entity is capable of fulfilling specified requirements.
ISO 9000

The following table, presented in the IEEE 1012, indicates the minimum V&V tasks that must be executed at each integrity level. For example, concerning the traceability analysis task, the standard indicates an "X" when this task is recommended (e.g., for three integrity levels shown in Table 7.4). Alternatively, safety analysis is recommended for levels 3 and 4 only.

Table 7.4 Minimum V&V Tasks by Integrity Level

Minimum V&V tasks	Integrity level			
	1	2	3	4
Traceability analysis		X	X	X
Security analysis			X	X

Source: Adapted from IEEE (2012) [IEE 12].

Table 7.5 describes the V&V tasks recommended for the traceability analysis of software requirements.

7.4 V&V ACCORDING TO ISO/IEC/IEEE 12207

The ISO 12207 [ISO 17] standard also presents the requirements for V&V processes. We will not describe all the details here but provide a high level view of the V&V processes, their purpose, and outcomes.

Table 7.5   Description of the Traceability Task [IEE 12]

Requirements for V&V (Process: Development)		
V&V tasks	Required inputs	Required outputs
Traceability analysis Trace the software requirements (SRS and IRS) to the system requirements (concept documentation) and the system requirements to the software requirements. Analyze identified relationships for correctness, consistency, completeness, and accuracy. The task criteria are as follows: – Correctness Validate that the relationships between each software requirement and its system requirement are correct. – Consistency Verify that the relationships between the software and system requirements are specified to a consistent level of detail. – Completeness ○ Verify that every software requirement is traceable to a system requirement with sufficient detail to show conformance to the system requirement. ○ Verify that all system requirements related to software are traceable to software requirements. – Accuracy Validate that the system performance and operating characteristics are accurately specified by the traced software requirements.	Concept documentation (system requirements) Software requirements specifications (SRS) Interface requirements specifications (IRS)	Task report(s)— Traceability analysis Anomaly report(s)

7.4.1 Verification Process

The purpose of the verification process is to provide objective evidence that a system or system element fulfills its specified requirements and characteristics.

The verification process identifies the anomalies (errors, defects, or faults) in any information item (e.g., system/software requirements or architecture description), implemented system elements, or life cycle processes using appropriate methods, techniques, standards, or rules. This process provides the necessary information to determine resolution of identified anomalies.

As a result of the successful implementation of the verification process [ISO 17]:

- constraints of verification that influence the requirements, architecture, or design are identified;
- any enabling systems or services needed for verification are available;
- the system or system element is verified;
- data providing information for corrective actions are reported;
- objective evidence that the realized system fulfills the requirements, architecture, and design is provided;
- verification results and anomalies are identified;
- traceability of the verified system elements is established.

7.4.2 Validation Process

The purpose of the validation process is to provide objective evidence that the system, when in use, fulfills its business or mission objectives and stakeholder requirements, achieving its intended use in its intended operational environment.

The objective of validating a system or system element is to acquire confidence in its ability to achieve its intended mission, or use, under specific operational conditions. Validation should be approved by the stakeholders of the project. This process provides the necessary information so that identified anomalies can be resolved by the appropriate technical process where the anomaly was created.

As a result of the successful implementation of the validation process [ISO 17]:

- validation criteria for stakeholder requirements are defined;
- the availability of services required by stakeholders is confirmed;
- constraints of validation that influence the requirements, architecture, or design are identified;
- the system or system element is validated;
- any enabling systems or services needed for validation are available;

– validation results and anomalies are identified;

– objective evidence that the realized system or system element satisfies stakeholder needs is provided;

– traceability of the validated system elements is established.

7.5 V&V ACCORDING TO THE CMMI MODEL

Another perspective of V&V can be seen in process models like the CMMI. The staged representation of the CMMI for Development [SEI 10a] has two process areas, at maturity level 3, dedicated to V&V. Preparation for verification is the first step suggested by the CMMI. It consists of selecting the life cycle phase outputs and the methods chosen for each product, in order to prepare the verification activity, or environment, depending on the specific needs of the project. It also suggests that verification success criteria and an iterative procedure be put in place, in parallel to the product design activities.

The purpose of verification is to ensure that selected work products meet their specified requirements. The verification process area includes the following specific goals (SG) and specific practices (SP) [SEI 10a]:

SG 1 Prepare for verification

SP 1.1 Select work products for verification,
SP 1.2 Prepare the verification environment,
SP 1.3 Establish verification procedures criteria;

SG 2 Perform peer reviews

SP 2.1 Prepare for peer reviews,
SP 2.2 Conduct peer reviews,
SP 2.3 Analyze peer review data;

SG 3 Verify selected work products

SP 3.1 Perform verification,
SP 3.2 Analyze verification results.

CMMI-DEV recommends inspections and walk-throughs for peer reviews as they have been described in a previous chapter.

The purpose of validation is to demonstrate that a product or product component fulfills its intended use when placed in its intended environment. The validation process area includes the following SG and SP [SEI 10a]:

SG 1 Prepare for validation

SP 1.1 Select products for validation,
SP 1.2 Establish the validation environment,
SP 1.3 Establish validation procedures and criteria;

SG 2 Validate product or product components

SP 2.1 Perform validation,
SP 2.2 Analyze validation results.

Validation can be applied to all aspects of the product within its target operational environment: operation, training and maintenance and support. The validation should be executed in a real operational environment with actual data volumes.

Example of Recommended CMMI Validation Methods

– discussions with end-users, perhaps in the context of a formal review;

– prototype demonstrations;

– functional demonstrations (for example, system, hardware units, software, service documentation and user interfaces);

– pilot use of training materials;

– tests of products and product components by end-users and other relevant stakeholders;

– incremental delivery of working and potentially acceptable product;

– analyses of product and product components (e.g., simulations, modeling, user analyses).

Software Engineering Institute (2010) [SEI 10a]

V&V activities are often executed together and can use the same environment. End-users are usually invited to conduct the validation activities.

7.6 ISO/IEC 29110 AND V&V

The ISO 29110 standard for very small entities has already been introduced. Elements of ISO 12207 V&V processes have been used to develop ISO 29110 standards and guides. This section shows how these very small organizations can conduct V&V using one of the four recommended profiles: the ISO 29110 Basic profile. This profile describes two processes: a project management (PM) process and a software implementation (SI) process.

One of the seven objectives of a PM process is to prepare a project plan describing the activities and tasks for the development of a software for a specific customer. Required tasks and resources are sized and estimated early on. In this plan, V&V tasks are described and are reviewed between the development team and the customer, and then approved.

One of the objectives is that the V&V tasks, for each identified work product, be done according to stated exit criteria to ensure the coherence between the outputs and the inputs of each development task. Defects are identified and corrected and the quality records stored in the V&V report.

Table 7.6 lists the V&V tasks. The table shows the role of the person executing the task, a brief description of the task, the input and output, and their states (in brackets). The following acronyms are used for roles: TL for technical lead, AN for analyst, PR for programmer, CUS for customer, and DES for designer. Table 7.6 is limited to describing only the first task in detail and then lists only the subsequent task names.

ISO 29110 Basic profile imposes a minimum number of V&V tasks to ensure that the end product will meet the requirements and needs of the customer even with a small budget.

ISO 29110 also suggests that a verification result file be updated in order to record the V&V activity results. Table 7.7 shows an example of the proposed format for this important project quality record.

7.7 INDEPENDENT V&V

IV&V are V&V activities conducted by an independent organization. This can be used to supplement internal V&V and is often used for very critical software: medical devices, metro and railway control, and airplane navigation systems.

Independent Verification and Validation (IV&V)
V&V performed by an organization that is technically, managerially, and financially independent of the development organization.

IEEE 1012 [IEE 12]

Technical independence requires that the V&V effort use personnel who are not involved in the development of the system or its elements. Managerial independence requires that the responsibility for the IV&V effort be vested in an organization separate from the development and program management organizations. Financial independence requires that control of the IV&V budget be vested in an organization independent of the development organization.

Table 7.6 V&V Task List of the Implementation Process of ISO 29110 [ISO 11e]

Role	Task list	Input work products	Output work products
AN TL	SI.2.3 Verify and obtain approval of the Requirements specification. Verify the correctness and testability of the requirements specification and its consistency with the product description. Additionally, review that requirements are complete, unambiguous and not contradictory. The results found are documented in a verification results and corrections are made until the document is approved by AN. If significant changes were needed, initiate a change request.	Requirements specifications Project plan	Verification results Requirements specification [verified] Change request [initiated]
CUS AN	SI.2.4 Validate and obtain approval of the Requirements Specification Validate that requirements specification satisfies needs and agreed upon expectations, including the user interface usability. The results found are documented in a validation results and corrections are made until the document is approved by the CUS.	Requirement Specifications [verified]	Validation results Requirement specifications [validated]
AN DES	SI.3.4 Verify and obtain approval of the Software Design. Verify correctness of software design documentation, its feasibility, and consistency with their requirement specification. Verify that the traceability record contains the adequate relationships between requirements and the software design elements. The results found are documented in a verification results. Results and corrections are made until the document is approved by DES. If significant changes are needed, initiate a change request.	Software design Traceability record Requirements specifications [validated, baselined]	Verification results Software design [verified] Traceability record [verified] Change request [initiated]

(continued)

Table 7.6 (*Continued*)

Role	Task list	Input work products	Output work products
DES AN	SI.3.6 Verify and obtain approval of the Test Cases and Test Procedures. Verify consistency among requirements specification, software design and test cases and test procedures. The results found are documented in a verification results and corrections are made until the document is approved by AN.	Test cases and test procedures Requirements specification [validated, baselined] Software design [verified, baselined]	Verification results Test cases and test procedures [verified]
PR DES	SI.5.8 Verify and obtain approval of the *Product Operation Guide. Verify consistency of the product operation guide with the software. The results found are documented in a verification results and corrections are made until the document is approved by DES. *(Optional)	*Product operation guide Software [tested]	Verification results *Product operation guide [verified]
AN CUS	SI.5.10 Verify and obtain approval of the *Software User Documentation. *(Optional)	*Software user documentation Software [tested]	Verification results *Software user documentation [verified]
DES TL	SI.6.4 Verify and obtain approval of the Maintenance Documentation. Verify consistency of Maintenance Documentation with Software Configuration. The results found are documented in a Verification Results and corrections are made until the document is approved by TL.	Maintenance documentation Software configuration	Verification results Maintenance documentation [verified]

Table 7.7 Example of a Verification Result File [ISO 11e]

Name	Description
Verification results	Documents the verification execution. It may include the record of: – participants – date – place – duration – verification check-list – passed items of verification – failed items of verification – pending items of verification – defects identified during verification

7.7.1 IV&V Advantages with Regards to SQA

SQA and V&V are the main organizational processes, that is, the "watchdogs," put in place to ensure process, product, and service quality. Since software development is under pressure to deliver, there is a need to counter balance the situation so that quality is not forgotten. Internal politics can interfere with these processes and this is why IV&V can be useful.

Given that SQA is part of the development organizational process, this function sometimes has very little influence when there are schedule and cost pressures. The IV&V process is like an external watchdog representing the client's interests and not those of the developers.

Figure 7.5 describes the relationships between customer, supplier, and IV&V.

7.8 TRACEABILITY

Software traceability is a simple V&V technique that ensures that all the user requirements have been:

- documented in specifications;
- developed and documented in the design document;
- implemented in the source code;
- tested;
- delivered.

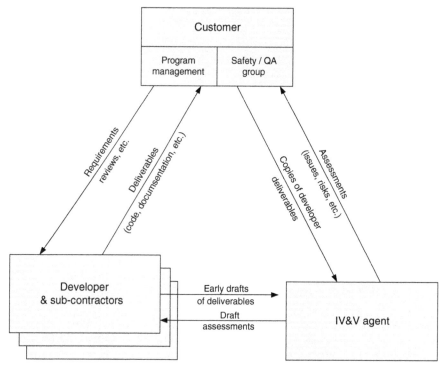

Figure 7.5   Relationship between IV&V, supplier, and customer [EAS 96].

Traceability facilitates the development of test plans and test cases. It ensures that the resulting tests have covered all the approved requirements. With traceability, we focus on detecting the following situations: a need without a specification, a specification without a design element, or a design element without source code or tests.

Traceability
Ability to trace the history, application or location of an object.
　　Note 1 to entry: When considering a product or a service, traceability can relate to:

　　– the origin of materials and parts;
　　– the processing history;
　　– the distribution and location of the product or service after delivery.

ISO 9000

The degree to which a relationship can be established between two or more products of the development process, especially products having a predecessor–successor or master–subordinate relationship to one another.

ISO 24765 [ISO 17a]

A discernable association among two or more logical entities such as requirements, system elements, verifications, or tasks.

CMMI

Bidirectional Traceability

An association among two or more logical entities that is discernable in either direction (i.e., to and from an entity).

CMMI

Requirements Traceability

A discernable association between requirements and related requirements, implementations, and verifications.

CMMI

7.8.1 Traceability Matrix

A software traceability matrix is a simple tool that can be developed to facilitate traceability. This matrix is completed at each phase of the development life cycle. But in order for the matrix information to be useful, it requires user requirements that have been well defined, documented, and reviewed. During the project, requirements will evolve (e.g., requirements will be added, deleted, and modified). The organization must use process management to ensure the matrix is kept up to date or it will become useless. Traceability of requirements is explained by the CMMI-DEV in two separate process areas: (1) requirements development and (2) requirements management. You can read more about traceability by referring to this source.

Traceability Matrix

A matrix that records the relationship between two or more products of the development process.

Example: a matrix that records the relationship between the requirements and the design of a given software component.

ISO 24765 [ISO 17a]

For small projects that have only 20 requirements, it is easy to develop such a matrix. For large projects, specialized tools like **IBM Rational DOORS** are available to support this functionality.

Traceability Matrix

Here is a list of attributes necessary to complete the requirements traceability:

– a unique identifier for each requirement;

– a link to a document explaining the requirement (e.g., operational concept);

– a descriptive text of the requirement;

– for derived requirements, a link to the parent requirement;

– a forward link, in the development process towards the architecture or the design;

– an explanation of the requirement verification method (e.g., review, test, demonstration);

– a link to the test plan, test scenario and test result;

– the last verification result date;

– the name of the specialist responsible for the quality of this requirement.

INCOSE Handbook [INC 15]

Table 7.8 presents an example of a basic traceability matrix with only four columns: (1) requirements; (2) source code; (3) tests; and (4) test success indicator.

To illustrate the importance of traceability, the failure of the "Mars Polar Lander" mission landing on Mars in 1999 is explained in the following text box. The NASA failure report pointed to the premature shutdown of the propulsion engine 40 meters above the surface of Mars [JPL 00].

Table 7.8   Example of a Simple Traceability Matrix

Requirement	Code	Test	Test success indicator
Ex 001	CODE 001	Test 001	Pass
		Test 002	Pass
		Test 003	Fail
Ex 001	CODE 002	Test 004	… …
		Test 005	… …
Ex 002	CODE 003	Test 006	… …
		Test 007	… …
		Test 008	… …
Ex 003	CODE 004	Test 010	… ..
		Test 011	… ..

Mars Polar Lander Failure Analysis

The engines of the Polar Lander have to be stopped automatically on landing. Sensors, on each of the three legs of the lander, send signals when touching the ground, and the computer shuts down the descent engines immediately.

System engineers had written the requirements that are described in the table below. This table shows the system requirements on the left and the software requirements on the right. We can see that the last part of system requirement 1 states that a precaution be taken not to read the sensors during the deployment of the legs, at 457 meters (1500 feet) from the planet, as sensors can wrongly signal a landing during this process.

Note that the last system requirement on the left is not traced to a software requirement. The corresponding software check was not expressed as a requirement so as not to take into account the signals at that time. The requirement that is missing would allow developers to add <u>one</u> line of code to reflect the transient signal produced by the touchdown sensors when deploying the legs.

System requirements		Flight software requirements	
ID number	Description	ID number	Description
1	Touchdown sensors shall be sampled 100 times per second.	3.7.2.2.4.1.a.	The lander flight software shall cyclically check the state of each of the three touchdown sensors (during entry, descent, and landing).
	The sampling process shall be initiated prior to lander entry to keep processor demand constant.	3.7.2.2.4.1.b.	The lander flight software shall be able to cyclically check the touchdown event state with or without touchdown event generation enabled.
	However, the use of the touchdown sensor data shall not begin until 12 meters above the surface. (Note: The altitude was later changed from 12 meters to 40 meters above the surface.)		

This is an example of the importance of tracing software requirements to design and code. This missing line of code in the flight control software was never designed, programmed and tested. According to the NASA report, this is most probably what caused the Polar Lander to shut down its descent engine too soon hitting Mars at 22 metres per second instead of 2.4 meters per second.

[JPL 00]

7.8.2 Implementing Traceability

The first step is to document the traceability process indicating "who does what." We will also assign the task of documenting and updating the content of the matrix for the project. Then, the matrix can be created, as illustrated in this chapter, using each requirement identification number. When other components pertaining to the requirements are produced, like design, code, or tests, they are added to the matrix. This is done until all tests are successful.

Use of Traceability

– Certification
  o Certification in critical applications requires a high level of safety (e.g., commercial airplanes). In order to demonstrate that all the requirements have been implemented and verified, traceability facilitates certification.

– Impact analysis during development and maintenance
  o Helps in quickly finding the interrelated elements of a system that could require a modification. Without traceability, an inexperienced programmer could be unaware of the ripple effects of a change.

– Project management
  o Brings a higher and more precise state of the project as blank spaces in the matrix point to work products or deliverables that have not yet been created or finalized.

– Follow-up of development by the customer
  o Facilitates the monitoring of the progress of the project by looking at the supplier traceability matrix.
  o Helps in better understanding the impact of a change request on the project as the customer will see the potential impacts (e.g., effort, schedule, and cost) on the matrix.

– Reduced losses and delays
  o Prevents the development of an unneeded component or forgetting the development of one component.

o Helps to check that all the components have been tested and all the documents have been verified for accuracy before their delivery to the customer.
  – Reuse
o Facilitates reuse of documented and tested software components by clearly identifying the requirements, their design, tests, and other documentation.
  – Risk reduction
o The documentation of links between artifacts reduces the risk associated with loss of key personnel (e.g., architect).
  – Reengineering
o Facilitates the identification of all the functions developed, requirements, architecture, code components, and tests.
o When a requirements document is not available, you will need to read the source code to find its backward traceability. With the traceability matrix already available, reengineering the system is made easier.

Adapted from Wiegers (2013) [WIE 13]

Once the development team has accepted this new practice, then additional information can be added to the traceability matrix. For example, on the left of Table 7.8 we could add a column to paste the original text of the needs of the customer. Finally, at the far right we could add what technique was used to verify the requirement, that is, a test (T), a demonstration (D), a simulation (S), an analysis (A), or an inspection (I).

7.9 VALIDATION PHASE OF SOFTWARE DEVELOPMENT

In some organizations, validation activities have been regrouped into a single development life cycle phase. It is often located at the end of the process. The objective of this last phase is to prove that the software meets the initial requirements, for example, that the right product was developed. The software is tested by the end-users to ensure it is fit for use in a real environment. The validation plan scenarios and test cases are developed and baselined during the integration and test phase.

Figure 7.6 presents a validation process using the Entry-Task-Verification-eXit (ETVX) notation presented earlier in this book. In certain situations, the validation phase will be split into many steps [CEG 90]:

  – testing in the presence of the customer or its representative;
  – installation of the software in the operational environment;
  – user acceptance testing, where the software is either accepted as is or accepted providing defects are corrected or not accepted. In the case where the software is accepted providing defects are corrected, the errors detected during the tests

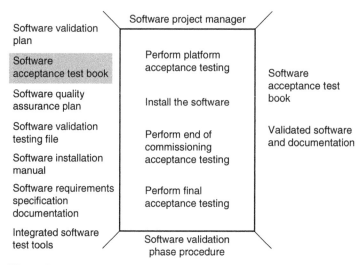

Figure 7.6   Validation representation of a process using the ETVX process notation [CEG 90] (© 1990 - ALSTOM Transport SA).

must be corrected and the software tested again before the customer accepts this software;

– end-user trial testing: use, in production, of the software in trial mode;

– warranty period, where the system is delivered and used, defects are corrected and change requests are processed;

– software final acceptance.

The validation phase is very important for the organization. Indeed, the success of this phase will lead to the transfer of the software to the client and, more importantly, to payment to the supplier when a contract is involved. For the developer it is often followed by a final project review where lessons learned are compiled to be used for process improvement.

The end of the validation also leads to the use, in production, of the software and the start of the support phase. The transition to maintenance is also an important phase of the life cycle. Even if there are still minor defects, they will be addressed during the maintenance phase.

During the validation phase, a series of tests are performed. It is not uncommon for anomalies to be detected and that minor changes are required. In addition, corrections or changes must be made, testing the corrected components as well as regression testing must be performed, and the configuration management process must be used to ensure that the changes are reflected in all of the documentation. At this time, the traceability matrix is used to ensure that all documents in the process have been corrected.

Validation can also lead to product qualification or even external certification in certain domains. For example, the Food and Drug Administration (FDA) requires a pre-market submission to the FDA before the release of the software in some situations.

7.9.1 Validation Plan

A software validation plan, written by the project manager, lists the organization and resources required to validate software. It should be approved during the software specification review and it describes:

- the validation activities planned as well as the roles, responsibilities, and resources assigned;
- the grouping of the test iterations, steps, and objectives.

To develop this plan, the project manager can use the following source documents: contractual documents, project plan, specifications document, system validation plan (if applicable), software quality plan, and the organization template for the validation plan and the validation plan checklist. Figure 7.7 describes a typical table of contents for the validation plan.

The many roles and responsibilities of the individuals involved in creating this plan can be summarized as follows (adapted from [CEG 90]):

- the project manager:
  - o write the validation plan;
  - o get the approval of the client during a review that takes place at the end of the software specification phase;
  - o update the plan as required during the subsequent phases;
  - o supervise the execution of the validation plan;
- the tester:
  - o execute the validation plan;
  - o organize and lead test iterations;
  - o produce the test iteration report;
  - o raise defect reports and agree on defect severity;
- test execution support personnel:
  - o prepare and configure the test environment;
  - o get the testing documents from configuration management;
  - o execute test procedures;
  - o find defects;
  - o correct defects;
  - o correct any documentation impacted by the correction of defects;
- customer:
  - o approve and sign the software validation plan;
  - o approve and sign the test iteration minutes;
  - o approve the defect correction list;

Title page (document title, project name, customer name, etc.)

Page listing the evolution of the plan (versions and changes)

Summary

1. Introduction
    1.1 Objectives
    1.2 Description of the software to validate
    1.3 Validation steps (e.g., installation, qualification)
    1.4 Reference documents

2. Organisation of the validation activities
    2.1 Activity "name of the activity"
        2.1.1 Definition of the activity
        2.1.2 Schedule of the activity
        2.1.3 Results of the activity

3 Organisation of test iterations
    3.1 Participants
    3.2 Agenda of test iterations
    3.3 Defect report process and defect severity scheme
    3.4 Defect iteration reporting and decision

4. Validation resources
    4.1 Tools
    4.2 Environment

5. Roles and responsibilities
    5.1 Approval of the plan
    5.2 Customer or his representative
    5.3 SQA
    5.4 SCM
    5.5 Test personnel
    5.6 Support personnel

6. Approval of the validation plan
    6.1 Signature of customer or representative
    6.2 Signature of project manager or management

Attachment 1: Validation terminology guide

Figure 7.7   Typical table of contents of a validation plan [CEG 90].

    – SQA personnel:
        o review the software validation plan and provide comments;
        o verify that the right versions of documents are used;
        o assist with testing iterations;
        o assist the project team during the lessons learned review;

– configuration management personnel:
  o provide the latest approved versions of documents required for tests;
  o assist the testing team when an error is found or with a minor modification request;
  o prepare deliverables identified in the contract and project plan;
  o archive project artifacts according to guidelines.

The validation plan does not necessarily need to be a document of its own. The information presented here can also be a section within the SQA plan or of the project plan depending on the size of the project.

7.10 TESTS

Tests are central to the V&V of a software. There are four major categories of tests: development tests, qualification, acceptance, and operational tests. The following text box provides their definitions.

Test

An activity in which a system or component is executed, under specified conditions, the results are observed or recorded, and an evaluation is made of some aspect of the system or component.

ISO 24765 [ISO 17a]

Development Testing

Formal or informal testing conducted during the development of a system or component, usually in the development environment by the developer.

ISO 24765 [ISO 17a]

Acceptance Testing

Testing conducted to determine whether a system satisfies its acceptance criteria and to enable the customer to determine whether to accept the system.

IEEE 829 [IEE 08a]

Qualification Testing

Testing, conducted by the developer and witnessed by the acquirer (as appropriate), to demonstrate that a software product meets its specifications and is ready for use in its target environment or integration with its containing system.

ISO 12207 [ISO 17]

Operational Testing

Testing conducted to evaluate a system or component in its operational environment.

IEEE 829 [IEE 08a]

7.11 CHECKLISTS

A checklist is a tool that facilitates the verification of a software product and its documentation. It contains a list of criteria and questions to verify the quality of a process, product, or service. It also ensures the consistency and completeness of the execution of tasks.

An example of a checklist is one that helps detect and classify a defect (e.g., an oversight, a contradiction, or an omission). A checklist can also be used to ensure that a list of tasks to be accomplished was completed, like a *"to do list."* Elements of a checklist are specific to the document, activity, or process. For example, a verification checklist to review a plan is different than a code review checklist. In this section, the following topics are presented:

– how to develop a checklist;

– how to use a checklist;

– how to improve and manage a checklist.

We also provide examples of different types of checklists. Following is the description of an anecdote about the creation of the first checklist.

History of Checklists—October 30, 1935, Dayton Military Airport, Ohio, USA

On the day that the last phase of the evaluation of three new aircraft models for the US defense was made, one of these aircraft, a Boeing model 299 took off. It began to climb gently, then suddenly dropped out of the sky. The aircraft crashed and exploded into flames. The accident investigation revealed that the cause was a pilot error. The pilot was unfamiliar with the new aircraft and had forgotten to remove the elevator lock before take-off. Once airborne, he realized what was happening and tried to manipulate the lock, but it was too late.

After this accident, a group of pilots sat down together to find a way to ensure that everything was done to prepare for flight and that nothing was forgotten. What resulted is what is called a "pilot's checklist." Four checklists were developed: a list for takeoff, a list once in flight, one before landing, and one after landing.

The Boeing 299 aircraft was too complex for anyone to memorize everything. With checklists and training, the twelve units purchased by the Defense department flew more than 1.8 million miles without any serious accident. The US Defense Department accepted the model 299, and eventually ordered more than 12,000 more. This aircraft model, renamed the B-17, was widely used during World War II.

The pilot checklists became so popular that other checklists were developed for crew members and other aircrafts.

Schamel [SCH 11]

Checklists for Astronauts

Astronauts are among the best trained professionals. They spend several years learning the operation of complex equipment and how to perform a variety of activities like flying a spaceship. Despite all the training, astronauts also require checklists. The photograph below shows a checklist sewn onto the sleeve of a glove used by astronaut Buzz Aldrin during the Apollo 11 mission.

http://www.internationalspacearchives.com/assets/
http://community.internationalspacearchives.com/blog/2009/06/19/apollo-11-highlight-videos/

7.11.1 How to Develop a Checklist

There are two popular approaches used in the development of a checklist. The first is to use an existing list, such as the ones available in this book or those found on the Internet, and adapt them to your needs. The second approach is to develop a checklist from a list of errors, omissions, and problems that were already noted during document reviews and lessons learned reviews. We will see how to improve these lists below.

According to Gilb, checklists are developed according to some rules [GIL 93]:

- a checklist must be derived, among others things, from process rules or from a standard;
- a checklist should include a reference to the rule it is inspired from and that it is interpreting;
- a checklist should not exceed one page because it is difficult to memorize and effectively use a list containing more than twenty items to be checked;
- a keyword should describe each item in the list, this facilitates its retention;
- a checklist must include a version number and the date of the last update;
- the checklist items can be stated using a sentence structure that responds in the affirmative if the condition is satisfied. For example, regarding the clarity of a requirement: "the requirement is clear" and not "the requirement is not ambiguous";
- a checklist can contain a classification, for example, the severity of defects: major or minor;
- a checklist should not contain all possible questions or details, as a concise list should focus on key issues and steps that need to be executed sequentially;
- a checklist should be kept updated to reflect the experience gained by the organization and its developers.

During a course on peer reviews given by Professor Laporte in a large Swedish company, a software engineer proudly presented the checklist for code reviews he had developed. This list included more than 250 items!

He was asked to present some elements of the checklist to other course participants. Participants came to the conclusion that a large number of items on this list could be detected by a source code formatting tool. A large number of elements could also be detected by setting compiler options. Students also identified a list of items that addressed minor issues. Following this discussion, the checklist was reduced to one page of significant problems that a developer should avoid and could detect.

Lastly, a checklist should be included in the training of the individual user. It does not replace the knowledge required to perform the tasks listed. Table 7.9 describes a checklist used to classify defects.

Figure 7.8 shows an example of a checklist to verify software requirements, hence the abbreviation used for this list is REQ. Note, that for each item of the

Table 7.9 Example of Defect Classification Scheme [CHI 02]

Defect class number	Defect type	Description
10	Documentation	Comments, messages
20	Syntax	Spelling, punctuation, instruction format
30	Build, package	Change management, library, version management
40	Assignment	Declaration, name duplication, scope, limits
50	Interface	Procedure call, input/output (I/O), user format
60	Validation checks	Error messages, inadequate validation
70	Data	Structure, content
80	Function	Logic, pointers, loops, recursion, calculations, function call defect
90	System	Configuration, timing, memory
100	Environment	Design, compilation, test, other system support problems

checklist, a keyword has been added. Keywords greatly facilitate the memorization of the items of the checklist.

- REQ 1 (Testable) – All the requirements must be objectively verifiable.
- REQ 2 (Traceable) – All the requirements must be traceable to a system specification or a contract clause or to the proposal.
- REQ 3 (Unique) – Requirements should be stated only once.
- REQ 4 (Elementary) – Requirements should be broken into their most elementary form.
- REQ 5 (High level) – Requirements should be stated in terms of final needs that have to be fulfilled and not perceived means (solutions).
- REQ 6 (Quality) – Their quality attributes are defined.
- REQ 7 (Hardware) – Its hardware is completely defined (if necessary).
- REQ 8 (Solid) – Requirements are a solid base for the design.

Figure 7.8 A software requirements checklist [GIL 93].

7.11.2 How to Use a Checklist

We present two ways to use a checklist. The first way is to review a document while keeping in mind all the elements of the checklist. The second way is to review the entire document using only one element of the checklist at a time. This second approach is carried out as follows:

– Use the first item in the checklist to review the document in full. When finished, check off that item on the checklist and move to the next;

Table 7.10   Example of the Use of a Checklist to Verify Component # 1

Name	Description	1	2	3	4
Initialization	• Variables and initialization values:	√			
	• When the program starts				
Interfaces	• At the start of each loop	√			
	• Internal interface (procedure call)				
	• Input/Output (e.g., display, printout, communication)				
	• User (e.g., format, content)				
Pointers	• Initialization of pointers to NULL	√			

– Continue the review using the second element of the checklist and check off that item on the checklist when done;

– Continue reviewing the document until all the items on the checklist are checked off;

– During the review, note defects or errors with the document;

– After completing the review of the entire document, correct all defects listed;

– After completing the correction of defects, print the updated version of the document and check all the corrections to ensure none are forgotten;

– If there were many important corrections, review the entire document again.

Table 7.10 provides an example of the use of a checklist designed for code review using this approach. The columns on the right are used during the review of a section of the document. For example, for program source code, consider the first item on the checklist, that is, review the initialization step of the program. After checking this item, check off this box and then move on to the next item on the checklist.

Typical exit or completion criterion with this approach is to ensure that all elements of the checklist have been checked.

For both approaches, unless the document to be reviewed is very short (i.e., less than one page), it is suggested not to review a document on screen, but use a hard copy in order to highlight the identified defects. During a review, a paper copy makes it easier to navigate from one page to another of the document and facilitates the identification of omissions and contradictions that may occur in large documents.

7.11.3 How to Improve and Manage a Checklist

Every professional, whether due to training, experience or writing style, makes mistakes. We must update checklists periodically as we learn from our mistakes.

The disadvantage of using a checklist is that the reviewer will focus his attention only on the items that are on the list. This may leave defects in software that are not listed on the checklist. It is therefore important to update the checklist based on results obtained and not just to follow it blindly.

7.12 V&V TECHNIQUES

Tools and techniques can be used to help perform V&V activities. Using these tools is highly dependent on the integrity level of the applications, the maturity of the product, the corporate culture, and the type of development, modeling, and simulation paradigm of individual projects.

The degree to which verification activities can be automated directly influences the overall efficiency of V&V efforts. As there is no formal process for selecting tools, it is important to select the right tool. Ideally, modeling and simulation tools used during the design and the development phases should be integrated with the verification tools. Moreover, validation does not permit a detailed match with modeling and simulation processes.

The market for verification tools is large. It is easy to find a list of at least a hundred vendors on the Internet today. These tools fall into the following two categories:

– generic tools supporting data results from validation:
  o database management systems;
  o data manipulation tools;
  o data modeling tools;

– formal methods:
  o formal language;
  o mechanized reasoning tools (automated theorem proofs);
  o model verification (checker) tool.

7.12.1 Introduction to V&V Techniques

Wallace et al. [WAL 96] wrote an excellent technical report presenting the different V&V techniques and it is still current today. First, we present three types of V&V techniques, then we briefly describe these techniques, and finally, we propose techniques for each of the development life cycle phases.

The V&V tasks are composed of three types of techniques: static analysis, dynamic analysis, and formal analysis [WAL 96]:

– Static analysis techniques are those that directly analyze the content and structure of a product without executing the software. Reviews, inspections, audits, and data flow analysis are examples of static analysis techniques;

– Dynamic analysis techniques involve the execution or simulation of a developed product looking for errors/defects by analyzing the outputs received following the entry of inputs. For these techniques, the output values or expected ranges of values must be known. Black box testing is the most widely used and well known dynamic V&V technique;

– Formal analysis techniques use mathematics to analyze the algorithms executed in a product. Sometimes, the software requirements may be written in a formal specification language (e.g., VDM, Z), which can be verified using a formal analysis technique.

7.12.2 Some V&V Techniques

7.12.2.1 Algorithms Analysis Technique [WAL 96]

The algorithms analysis technique examines the logic and accuracy of the configuration of a software by the transcription of the algorithms in a structured language or format. The analysis involves re-deriving equations or evaluating whether specific numerical techniques apply. It checks that the algorithms are correct, appropriate, stable, and that they meet the accuracy requirements, timing, and sizing. The algorithms analysis technique examines, among other things, the accuracy of equations and numerical techniques, the effects of rounding and truncations.

7.12.2.2 Interface Analysis Technique [WAL 96]

Interface analysis is a technique used to demonstrate that program interfaces do not contain errors that can lead to failures. The types of interfaces that are analyzed include external interfaces to the software, internal interfaces between components, interfaces with hardware between software and system, between software and hardware, and between the software and a database.

7.12.2.3 Prototyping Technique [WAL 96]

Prototyping demonstrates the likely results of the implementation of software requirements, especially the user interfaces. The review of a prototype can help identify incomplete or incorrect software requirements and can also reveal whether the requirements will not result in undesirable system behavior. For large systems, prototyping can prevent inappropriate designs and development which can be a costly waste.

Prototyping

A hardware and software development technique in which a preliminary version of part or all of the hardware or software is developed to permit user feedback, determine feasibility, or investigate timing or other issues in support of the development process.

ISO 24765 [ISO 17a]

Prototype
A preliminary type, form, or instance of a system that serves as a model for later stages or for the final, complete version of the system.
    Note: A prototype is used to get feedback from users for improving and specifying a complex human interface, for feasibility studies, or for identifying requirements.
ISO 24765 [ISO 17a]
A method of obtaining early feedback on requirements by providing a working model of the expected product before actually building it.
PMBOK® Guide

7.12.2.4 Simulation Technique [WAL 96]

Simulation is a technique used to evaluate the interactions between large complex systems composed of hardware, software, and users. The simulation uses an "executable" model to examine the behavior of the software. Simulation can be used to test the operator's procedures and to isolate installation problems.

7.13 V&V PLAN

The V&V plan essentially answers the following questions: what do we verify and/or validate? How and when, and by whom will the V&V activities be performed and what level of resources will be required?

IEEE 1012 specifies that the V&V effort starts with the production of a plan that addresses the following list of elements. If there is an irrelevant item that the project should not cover in its plan, it is better to state that "This section is not applicable to this project" instead of removing the item from the plan. This allows the SQA to clearly see that the item was not forgotten by the project team. Of course, additional topics may be added. If elements of the plan are already documented in other documents, the plan should refer to it instead of repeating it. The plan must be maintained throughout the software life cycle. The V&V plan proposed by IEEE 1012 includes the following (without listing the system and hardware V&V elements) [IEE 12]:

1. Purpose
2. Referenced documents
3. Definitions
4. V&V overview
    4.1 Organization
    4.2 Master schedule
    4.3 Integrity level scheme
    4.4 Resources summary

4.5 Responsibilities

4.6 Tools, techniques, and methods

5. V&V processes

   5.1 Common V&V processes, activities, and tasks

   5.2 System V&V processes, activities, and tasks

   5.3 Software V&V processes, activities, and tasks

      5.3.1 Software concept

      5.3.2 Software requirements

      5.3.3 Software design

      5.3.4 Software construction

      5.3.5 Software integration test

      5.3.6 Software qualification test

      5.3.7 Software acceptance test

      5.3.8 Software installation and checkout (Transition)

      5.3.9 Software operation

      5.3.10 Software maintenance

      5.3.11 Software disposal

   5.4 Hardware V&V processes, activities, and tasks

6. V&V reporting requirements

   6.1 Task reports

   6.2 Anomaly reports

   6.3 V&V final report

   6.4 Special studies reports (optional)

   6.5 Other reports (optional)

7. V&V administrative requirements

   7.1 Anomaly resolution and reporting

   7.2 Task iteration policy

   7.3 Deviation policy

   7.4 Control procedures

   7.5 Standards, practices, and conventions

8. V&V test documentation requirements

7.14 LIMITATIONS OF V&V

No technique can prevent all errors or defects. Regarding V&V, we note the following limitations [SCH 00]:

– Impracticability of testing all the data: for most programs, it is virtually impossible to try to review the program with all possible inputs, due to the multitude of possible combinations;

- Impracticability of testing all the branch conditions: for most programs, it is impractical to try to test all the possible execution paths of a software. This is also due to the multitude of possible combinations;
- Impracticability of obtaining absolute proof: there is no absolute proof of correctness of a software based system unless formal specifications can prove it to be correct and accurately reflect user expectations.

It is not uncommon for test plans to be designed by the developer of the system and then approved by the V&V staff. This practice is far from ideal to guarantee a high level of quality. Although the V&V role should not be part of the development team, sometimes the developer becomes the evaluator of his own software. It is therefore important that the V&V role consist of people who have good knowledge and experience of systems in order to provide sound evaluations of the quality of the resulting product.

7.15 V&V IN THE SQA PLAN

The IEEE 730 standard discusses V&V and starts with a statement that SQA activities need to be coordinated with the verification, validation, review, audit, and other life cycle processes needed to ensure the conformity and quality of the final product. There is no need to duplicate efforts here. The standard asks the project team to ensure that the V&V concerns have been well explained in the V&V or SQA plan.

For verification activities, the standard lists the following questions that the project team members should ask themselves [IEE 14]:

- has verification between the system requirements and the system architecture been performed?
- have verification criteria for software items been developed that ensure compliance with the software requirements allocated to the items?
- has an effective validation strategy been developed and implemented?
- have appropriate criteria for validation of all required work products been identified?
- have verification criteria been defined for all software units against their requirements?
- has verification of the software units compared with the requirements and the design been accomplished?
- have adequate criteria for verification of all required software work products been identified?
- have required verification activities been performed adequately?
- have results of the verification activities been made available to the customer and other involved parties?

For the validation of the software, the standard recommends that the tools that will be used for validation be chosen and assessed based on product risk and that the project team evaluate whether these tools need validation. If they are validated, they are to keep the records of this validation. It also asks the team to answer each of the following questions [IEE 14]:

- have all tools that require validation been validated before using them?
- has an effective validation strategy been developed and implemented?
- have appropriate criteria for validation of all required work products been identified?
- have required validation activities been performed adequately?
- have problems been identified, recorded, and resolved?
- has evidence been provided that the software work products as developed are suitable for their intended use?
- have results of the validation activities been provided to the customer and other involved parties?

Of particular interest in the SQA plan, the project team will pay special attention to the acceptance process and how to classify defects until an exit criterion is met. It is important to clarify the final testing process stage of a project as it is the last line of defense before going to production.

7.16 SUCCESS FACTORS

The execution of the V&V practices can be helped or slowed down depending on a number of organizational factors. The next text box lists some of these factors.

Factors that Foster Software Quality
1) A documented process that includes V&V tasks is available.
2) V&V tasks are planned early in the project and executed throughout.
3) Developers are trained on V&V.

Factors that may Adversely Affect Software Quality
1) Management forces a rapid and inadequate completion of the work.
2) Individuals do not feel valued.
3) Training is inadequate.
4) Principles of software quality are missing.

5) The view of the final product does not correspond to the client's vision.

6) No improvement process.

7) Life cycle processes are not formalized.

8) Quality is not the first priority.

9) Absent or superficial peer reviews.

7.17 FURTHER READING

SCHULMEYER G. G. (DIR.) *Handbook of Software Quality Assurance*, 4th edition. Artech House, Norwood, MA, 2008.

WIEGERS K. *Software Requirements*, 3rd edition. Microsoft Press, Redmond, WA, 2013.

7.18 EXERCISES

7.1 List the key activities of a procedure to verify requirements.

7.2 Classify the V&V techniques listed in the following table according to three categories [WAL 96]:

a) Static analysis: analysis of the structure and form of a product without executing it;

b) Dynamic analysis: executing or simulating a developed product with the objective of detecting defects by analyzing its outputs based on input scenarios;

c) Formal analysis: use of mathematical equations and techniques to rigorously analyze algorithms used in a product.

Technique	Static analysis	Dynamic analysis	Formal analysis
Algorithm analysis			
Boundary value analysis			
Code reading			
Coverage analysis			
Control flow analysis			
Database analysis			
Data flow analysis			
Decision (truth) tables			
Desk-checking			
Error seeding			
Software fault tree analysis or Software failure mode			

Technique	Static analysis	Dynamic analysis	Formal analysis
Finite state machines			
Functional testing			
Inspections			
Interface analysis			
Interface testing			
Performance testing			
Petri-nets			
Prototyping			
Regresion analysis and testing			
Reviews			
Simulation			
Sizing and timing analysis			
Software failure mode, effects, and criticality analysis			
Stress testing			
Structural testing			
Symbolic execution			
Test certification			
Walk-throughs			

7.3 Provide examples of selection criteria for an IV&V service supplier.

7.4 Your manager asks you to develop a job description for the position of V&V engineer for critical software products. List the qualifications/experience, accountability, and responsibilities typically required for that position.

7.5 For some critical systems, a standard imposes that the developer demonstrates to the client that there is no dead code in the final product. Explain why this requirement is imposed by the customer?

7.6 List three V&V techniques for each of the development life cycle phases.

Chapter 8

Software Configuration Management

After completing this chapter, you will be able to:

- understand the software configuration management activities as they are recommended by ISO 12207, IEEE 828, and the CMMI®;
- understand how the change control process is used;
- learn about code control and its branching strategies;
- see how configuration management is possible in a very small project or organization;
- list what is included in a configuration management plan;
- understand what is recommended by the IEEE 730 standard for the project software quality assurance plan.

8.1 INTRODUCTION

In many industries, the result of a production process is a product that you can see, touch, and measure. In software, code is the most important deliverable and it is an intangible product for most people. To give it the most visibility, it is necessary to document it and communicate its characteristics at each step of its development. For the same reasons, during its life cycle, it will be possible to review it, improve it, and expand the documents supporting it. This progression originates from change requests, omissions, defects, and problems encountered when developing the product. Also, when software must reside on a processor that receives data and controls a process, changes to the hardware can result in change requests to the software. As long as

Software Quality Assurance, First Edition. Claude Y. Laporte and Alain April.
© 2018 the IEEE Computer Society, Inc. Published 2018 by John Wiley & Sons, Inc.

the software supports a business process of the organization, modifications will need to be made to keep it current with the evolution of business rules and technology.

All these documentation activities and constant changes can become costly. In large projects, these activities represent a percentage of the effort that cannot be ignored. This is why it is important to be familiar with these activities, thus optimizing this effort. In this chapter, all the different activities of software configuration management (SCM) are explained. This topic is one of the knowledge areas that a software engineer should master and where software quality assurance (SQA) is actively involved to ensure project teams understand its importance and are guided given that it is an area that is often audited.

We have already discussed the importance of creating, updating, and managing quality records in the project file to keep track of all the changes that happen over time. Today, this information can be found in documents, emails, wikis, ticket systems, and team chats and contain a wealth of knowledge for the organization. Some say it is strategic to be able to harness this knowledge. Be it for knowledge transfer, project management, laws and regulations, or the obligation to conform to standards, organizations must be able to manage all this information coherently.

8.2 SOFTWARE CONFIGURATION MANAGEMENT

The configuration of a system [BUC 96] is composed of many perspectives: functional from software, firmware and/or physical hardware, as normally described in the accompanying documentation of the product.

System

Combination of interacting elements organized to achieve one or more stated purposes

Note 1 to entry: A system is sometimes considered as a product or as the services it provides.

Note 2 to entry: In practice, the interpretation of its meaning is frequently clarified by the use of an associative noun, e.g., aircraft system. Alternatively, the word "system" is substituted simply by a context-dependent synonym, e.g., aircraft, though this potentially obscures a system principles perspective.

Note 3 to entry: A complete system includes all of the associated equipment, facilities, material, computer programs, firmware, technical documentation, services and personnel required for operations and support to the degree necessary for self-sufficient use in its intended environment.

ISO 15288 [ISO 15]

In this section, we present SCM processes as recommended by ISO 12207, IEEE 828, and the CMMI.

Configuration Management
Discipline applying technical and administrative direction and surveillance to:

– identify and document the functional and physical characteristics of a configuration item,
– control changes to those characteristics,
– record and report change processing and implementation status, and
– verify compliance with specified requirements.

IEEE 828 [IEE 12b]

Essentially, SCM describes answers to the following questions [STS 05]:

– who can make changes?
– what changes were made?
– what were the impacts of a change?
– when did this change occur?
– why were these changes made?
– what version is currently in which environment?
– what branching approach are we using for this project?

8.3 BENEFITS OF GOOD CONFIGURATION MANAGEMENT

Good configuration management (CM) will bring about the following benefits [STS 05]:

– reduce confusion, organize, and better manage software items;
– organize the required activities that ensure the integrity of the many software products;
– ensure traceable and current configuration of products;
– optimize the cost of development, maintenance, and after-sales support;
– facilitate the validation of the software with respect to its requirements;

- provide stable development, maintenance, testing, and production environments;
- improve quality and compliance to software engineering standards;
- reduce rework costs.

8.3.1 CM According to ISO 12207

According to the ISO 12207 standard, the purpose of CM is to [ISO 17]: manage and control system elements and configurations over the life cycle. CM also manages consistency between a product and its associated configuration definition. SCM is part of SQA practices. SQA must give assurance that processes and products used in the development life cycle meet their requirements by planning, using and executing activities that ensure that quality is integrated in the final product during its production. The SCM activities help SQA meet its goals and the goals of the project. It is recommended that it be used by every software project, maintenance activity as well as for its infrastructure.

ISO 12207 defines the requirements of CM as an essential process for software developers. We will not describe the detailed requirements here but provide only a high-level description and focus on the expected results of using this process.

As a result of the successful implementation of the CM process [ISO 17]:

- items requiring CM are identified and managed;
- configuration baselines are established;
- changes to items under CM are controlled;
- configuration status information is available;
- required configuration audits are completed;
- system releases and deliveries are controlled and approved.

The six activities of the ISO 12207 CM process are [ISO 17]:

- plan CM;
- perform configuration identification;
- perform configuration change management;
- perform release control;
- perform configuration status accounting;
- perform configuration evaluation.

The six CM activities of ISO 12207 are composed of 19 tasks that are not described here. The last activity, however, includes audit configuration tasks that are described later in this chapter.

CM in ISO 12207 is also applicable to maintenance, hardware and base software. The maintainer will implement or use the CM process for managing modifications to the existing system. The scope also recommends that the configuration of the infrastructure should be planned and documented as well: the infrastructure (e.g., software tools) shall be maintained, monitored, and modified as necessary. As part of maintaining the infrastructure, the extent to which the infrastructure is under CM shall be defined.

8.3.2 CM According to IEEE 828

IEEE 828 is the IEEE Standard for Configuration Management in Systems and Software Engineering [IEE 12b] and provides the requirements and the details concerning the SCM process. IEEE 828 also supports the ISO 12207 standard. It states the minimum acceptable requirements for CM in both systems and software. Consequently, it applies to all class/type of systems or software.

IEEE 828 describes the CM activities to be performed at what point of the life cycle and describes planning and the required resources. The standard details the items that should be included in a CM plan and are aligned with ISO 12207 (for software engineering), ISO 15288 (for systems engineering), and ISO 15939 (measurement standard presented in a later chapter). According to IEEE 828, the purpose of CM is to [IEE 12b]:

- identify and document the functional and physical characteristics of a product, component, output or of a service;
- control any changes to these characteristics;
- record and report each change and its implementation status;
- support the audit of the products, results, services, or components to verify conformance to requirements.

IEEE 828 also specifies [IEE 12b] that CM establishes and protects the integrity of a product or product component throughout its lifespan, from determination of the intended users' needs and definition of product requirements through the processes of development, testing, and delivery of the product, as well as during its installation, operation, maintenance, and eventual retirement. In so doing, CM processes interface with all other processes involved in the product's life.

8.3.3 CM According to the CMMI

The CMMI-DEV has a process area named "configuration management". The purpose of this process area is to [SEI 10a]: establish and maintain the integrity of work products using configuration identification, configuration control, configuration status accounting, and configuration audits. This process area describes the following activities be conducted [SEI 10a]:

- identifying the configuration of selected work products that compose baselines at given points in time;
- controlling changes to configuration items (CI);
- building or providing specifications to build work products from the CM system;
- maintaining the integrity of baselines;
- providing accurate status and current configuration data to developers, end-users, and customers.

These CMMI activities apply to both systems and software engineering. The following text box describes the specific goals (SG) and specific practices (SP) of this CMMI process.

 Configuration Management According to the CMMI [SEI 10a]

Specific Objectives and Practices

SG 1 Establish baselines

- SP 1.1 Identify configuration items
- SP 1.2 Establish a configuration management system
- SP 1.3 Create or release baselines

SG 2 Track and control changes

- SP 2.1 Track change requests
- SP 2.2 Control configuration items

SG 3 Establish integrity

- SP 3.1 Establish configuration management records
- SP 3.2 Perform configuration audits

Figure 8.1 shows a graphical representation of the CMMI recommendations for CM.

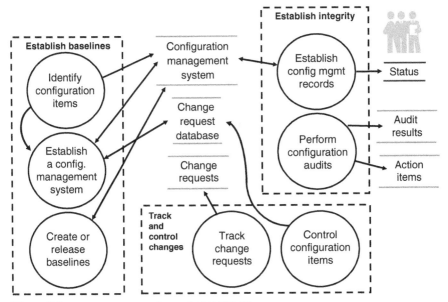

Figure 8.1 CM activities according to CMMI [SEI 00].

8.4 SCM ACTIVITIES

8.4.1 Organizational Context of SCM

In large organizations, the responsibility for SCM is often assigned to a separate department. In small organizations, the SCM tasks can be shared with developers performing software development tasks. Software is often developed as part of a system including hardware, software, and documentation. SCM activities included with project team responsibilities will need to align with the infrastructure CM (e.g., hardware and software). In this chapter, we will only focus on SCM.

SCM constraints that can make it difficult to meet these obligations originate from many situations. Published policies and procedures could impose or influence the use of SCM in projects. In the case of acquisition of software, an agreement or a contract may contain specific SCM clauses.

When software products planned to be developed may present safety issues, external regulating authorities and standards will likely impose the use of SCM.

Examples from the medical devices sector were described in Chapter 2 (Therac-25), however, similar problems can be happening right now all over the world in other domains such as avionics, nuclear energy, automobiles and banking, just to name a few.

8.4.2   Developing a SCM Plan

A SCM plan is typically developed during the planning of a project. It should be ready and approved by the time the project reaches the end of the software specification stage. During planning, the following should be considered: a list of SCM-related activities, the work and role assignments, the required resources and schedule, the choice and installation of tools, and the identification of supplier responsibilities.

The planning outputs will be placed in a separate SCM plan or in a section of the project planning documentation for agile projects and will be typically accessible for SQA review/audit. IEEE 828 [IEE 12b] defines six SCM information categories to be placed in the SCM plan (adapted from [IEE 12b]):

1) Introduction to the plan (the purpose, scope, and terminology);
2) SCM management (organization, responsibilities, authorities, applicable policies, guidelines, and procedures);
3) SCM activities (configuration identification, configuration control, and other activities);
4) SCM schedule (identification of the SCM activities in the project schedule);
5) SCM resources (tools, servers, and human resources);
6) SCM maintenance and updates.

IEEE 828 describes, in a normative annex, the content of a SCM plan. Following is a typical table of contents for a SCM plan:

– introduction;
– purpose;
– scope;
– relationship/dependencies with the organization and other projects;
– references (e.g., policies, guidelines, procedures);
– criteria for identifying which software elements will be placed under SCM;
– description of the configuration to be managed;
– development and testing configurations;
– delivery configuration;
– configuration identifiers and assignment;
– versions and fixes numbering rules;
– source code branching strategy;
– marking rules;
– repository content, location;

- library management project rules;
- architecture;
- procedures for the creation of the library;
- recording of elements in the project library;
- access rules;
- backups;
- archiving;
- project library control;
- planned baselines and states.

If the software is developed and maintained for many years, as for the controls for railway and planes, it is necessary to add:

- the source of the modifications;
- a formal modification procedure;
- the tools that were used to develop the software;
- the verification records.

Organizational units that have to be involved in the SCM process have to be clearly identified. Table 8.1 illustrates the assignment roles for the execution of SCM of a complex project involving many players.

Depending on the size of the project, the tasks listed could be managed by the project manager, whereas for very small projects, one or more different people could be directly involved.

Updates to the plan are approved when the need arises during development. Depending on the importance of the project, SQA can be done to ensure the plan is executed correctly. In more involved projects, SCM compliance audits can be called to ensure adherence to plans.

It can be a bit overwhelming to be assigned the development of a SCM plan for the first time. Most organizations have examples and templates that can be used to kick start this process. It is common to reuse the majority of sections of existing SCM plans in an organization. Therefore, SQA should ensure that the reference documents are of high quality so it can be reused in the future. Table 8.1 shows an example of SCM task allocations for a project.

8.4.3 Identification of CI to be Controlled

This activity aims at identifying which software elements will need to be controlled during the development and maintenance life cycle of the project to avoid creating elements without proper identification or links with other elements already identified.

Table 8.1 Example of SCM Task Allocations [CEG 90a] (© 1990, ALSTOM Transport SA)

Task	SCM manager	Project manager	Participants	SQA	Implementation manager
Collect and centralize change requests	Produce				
Manage change request process	Lead and request	Participate		Participate	Participate if system is modified
Manage access rights	Produce	Participate			
Name components			Produce		
Identify relationships	Participate	Produce			Participate
Mark items			Produce		
Store items	Keep items active and request archiving				
Organize libraries	Produce				
Control changes to the libraries	Produce (form)	Produce (content)			
Control completeness and coherence of the libraries	Produce (form)	Produce (content)			
Provide copies	Produce				
Report on configuration status	Produce				

Configuration Item (CI)

Aggregation of work products that is designated for configuration management and treated as a single entity in the configuration management process.

ISO 24765 [ISO 17a]

Configuration Item Identification

An element of configuration management, consisting of selecting the configuration items for a system and recording their functional and physical characteristics in technical documentation.

ISO 24765 [ISO 17a]

The identification of CI requires: (1) an already established classification reference nomenclature for items and (2) a predefined list of planned baselines identified uniquely throughout the project schedule.

Baseline

Specification or product that has been formally reviewed and agreed upon, that thereafter serves as the basis for further development, and that can be changed only through formal change control procedures.

ISO 24765 [ISO 17a]

The definition of a baseline implies that a document or software product was approved following a formal review. It is then available to the customer or ready for another step in the development process. Baselines for the production of new versions are also used.

The different project artifacts are typically located in a document or on the project server in folders (see an example in Figure 8.2). This folder structure is often mandated by the SQA or the project office as good practice for all projects. We will see in section 8.8 that this structure can be repeated in all the branches of the CM repository. It is recommended that developers do not change this hierarchy so that it is consistent during the project and can also be compared with other projects.

8.4.3.1 Identification of CI

Each CI should be assigned a unique identifier that helps with recognizing the evolution of the item over time and can be recognized by everyone involved in a project.

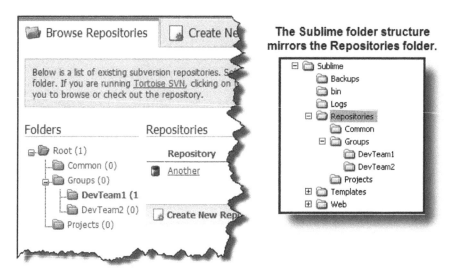

Figure 8.2 Example of a directory structure to control configuration with the Subversion CM tool.

An identifier is composed of a fixed part and a variable part. The fixed part is the name and the variable part reflects its evolution. When an evolution of the item does not add value, we call it a revision. This is the case when an item has been corrected or improved but no functionality, interface or operational constraints have been altered.

How to Number Software Versions?

Software versioning is the process of affecting a name and a unique version to the initial software and its documentation and to subsequent releases of a software of a specific baseline. Typically, within the same category of changes (majors or minors) the numbers will be incremented gradually. There are all sorts of approaches to versioning. In this book, we will use the indicator sequence approach. For example, when version 3.4 of a software goes through a small change or a correction, it will progress to version 3.4.1. But if a major functionality change occurs, the numbering will go to version 3.5. The use of the zero, for example version 0.5, is used before initial release, that is, when software is still in alpha or beta testing.

To indicate that an evolution of a configuration item affects descending compatibility but ensures it still works with previous versions of other CI, we use a new release version. This is the case when functionalities or existing interfaces were not modified but new interfaces and functionalities were added.

Version Control
Establishment and maintenance of baselines and the identification and control of changes to baselines that make it possible to return to the previous baseline.

ISO 24765 [ISO 17a]

For example, for a radar display software-based system, we could use the following naming convention: Identification of the contract—System identification—Identification name of the software component—evolution indicator (version/revision). For version 1 and revision 2 of the architecture document of the radar display document, we would obtain: C001-Rad-Aff-DA-01-02.

8.4.3.2 Configuration Item Marking and Labeling

Marking identifies a configuration item such as a software or a requirement specification document.

A software item could be identified with the help of the following information (e.g., its metadata): the software component identification information, its author, its creation date, its functional description, its history of modifications including a date, the modification reason as well as the author and finally, a summary of its design. For documents, this metadata could be saved using the following structure: a presentation page, a section explaining the evolution of the document, the main text, and possibly attachments.

The presentation page could contain the following information:

– the name and address of the organization;
– the document author name;
– the document creation date (yyyy-mm-dd);
– the document version;
– the security level (e.g., confidential, limited circulation, secret).

The evolution section is updated when the document is modified and includes the following information:

– the version indicator;
– the date of change: yyyy-mm-dd;
– the section or page that was modified;
– a brief description of the modification.

The evolution section is often formatted as a table as shown in Table 8.2.

Table 8.2 Example of Table Describing the Evolution of a Document

Revision	Date (yyyy-mm-dd)	Page or section modified	Author	Description of the modification
00			C. Laporte	First version
1.1	2006-03-01	P 5	A. April	Added section 2.1
1.2	2007-02-13	Section 3.1	A. Abran	Corrected a reference

8.4.3.3 Selecting CI

A software configuration item (SCI) is an aggregation of software components identified for CM and treated as a whole (e.g., one entity) by the SCM process. In addition to the source code, there are a number of software work products, such as plans, requirements, specifications, and design documents, that can be controlled by SCM. The project manager and his team must decide which ones to select to be controlled by the SCM process.

The SCI selection is always some form of compromise between enough visibility to provide tight control on an element versus not managing it at all. It is not possible to tightly manage all the elements of a software project at a reasonable cost. Once an item is chosen, it will be subject to formal review and acceptance by a number of individuals. Some reviews are not too formal, as we have seen in the review chapter. Formal reviews, on the other hand, require that a process be followed, minutes be issued and that defects be tracked, corrected and verified. This is why choosing the right amount of CI is an important task. The following text box lists criteria to help with this decision.

Criteria to Help Choose SCIs
- Number of changes anticipated
- Complexity and dimensions
- Criticality
- Security
- Impact on development schedule
- Impact on implementation schedule
- Commercial items that are not modified
- Items provided by the customer
- Maintenance performed by different suppliers
- Involves more than one supplier

– Location—when components are developed at many sites
– Multiple usage—when a component is used by many systems

8.5 BASELINES

A software development life cycle model defines each of the development processes uniquely and places them in a pipeline. Each step is defined as a coherent group of activities and is characterized by a dominant activity. Normally, to ensure its quality, a review is performed at the end of each stage or iteration.

A baseline is a set of CI that have been carefully pre-selected and are fixed along with specific milestones throughout the project. A baseline can only be changed through a change control procedure. A given baseline, plus all the changes subsequently approved to it, represents its current configuration. This is important because we want our employees to work on the right version of things and know where they are. The baselines for key project milestones commonly referred to by team members and listed in a typical CM plan are:

– specification;
– design;
– construction;
– integration;
– validation;
– delivery.

Figure 8.3 shows an example of a product development cycle that includes hardware and software. The dotted lines indicate some of the milestones. For example, the system requirements analysis phase produces the software and hardware requirements document. A review of this document, called "System Requirement Review," is then performed with the customer. Once this document is approved, it is placed in the functional baseline of the project. Then, the document software requirements are reviewed by the development team. It can then be presented to the customer for approval at a formal review called the "Software Specification Review." Once this document is approved, it is stored in the repository called "allocated baseline." Other software reviews that could occur are: a review of the preliminary design (Preliminary Design Review), a detailed design review (Critical Design Review), and a review to ensure that software components are ready for testing (Test Readiness Review).

Once the CI are tested and corrections are made, they are incorporated into the hardware. Lastly, the system is production tested and is then validated with the client.

Figure 8.4 presents elements of the specification phase of a project using the Entry-Task-Verification-eXit (ETVX) process notation [RAD 85] described in an

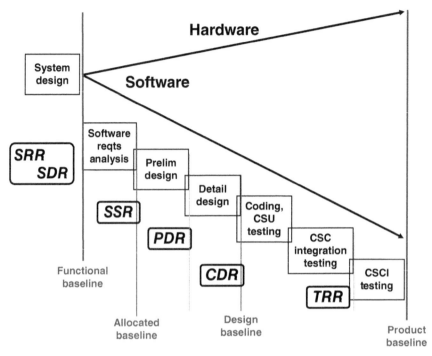

Figure 8.3   Example of a project life cycle where planned baselines are indicated.

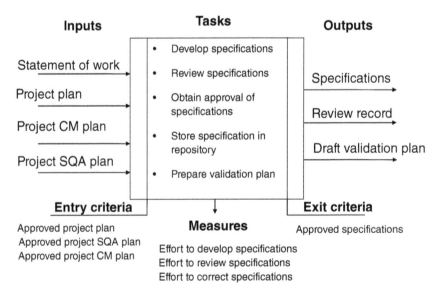

Figure 8.4   Description of the specifications steps using ETVX notation.

earlier chapter. Notice that a CM plan is part of the inputs. Other CI will also be added to the baseline throughout the software life cycle.

Following the incorporation of a SCI into the SCM repository, any further changes to the SCI will need to be approved as described in the SCMP. Following this approval process, the item will be incorporated to its proper location in the project repository.

Delivery
Release of a system or component to its customer or intended user.

Release
A delivered version of an application that may include all or part of an application.
Collection of new and/or changed configuration items that are tested and introduced into the live environment together.

Version
An initial release or re-release of a computer software configuration item, associated with a complete compilation or recompilation of the computer software configuration item.
An initial release or complete re-release of a document, as opposed to a revision resulting from issuing change pages to a previous release.

ISO 24765 [ISO 17a]

8.6 SOFTWARE REPOSITORY AND ITS BRANCHES

There are many options for SCM repository tools, often called version control software or software repository. They have been created to facilitate working as a group on software development, maintenance, and production, where code and documentation need to be shared and updated.

Software Repository
A software library providing permanent, archival storage for software and related documentation.

ISO 24765 [ISO 17a]

Table 8.3   Examples of Functions Provided by the SCM Repository

Functions of a software repository
Supports multiple levels of SCM (managers, developers, SQA, systems engineering, etc.);
Enables the storage and retrieval of CI;
Enables the sharing and transfer of CI between groups and developers;
Enables the storage and recovery of archive versions of CI;
Provides verification of status, of the presence of all the elements, and allows for the integration of changes into new baselines;
Ensures the correct build and correct versions of products;
Provides storage, update, and retrieval of SCM records;
Supports the production of SCM lists and reports;
Supports backward and forward traceability of the requirements throughout the life cycle;
Provides safe storage and restricted access of CI so that they cannot be changed without authorization.

Source: Adapted from [CEG 90a].

The software repository is central to development, maintenance, and release management activities during a project. Table 8.3 lists some of the functions of a software repository. Several types of libraries may be employed. For example, one of following three types of libraries can be used:

- Private library: used by the developer to make or modify CI, for development or unit testing activities;

- Project library: accessible to all the members of a project, it contains the elements likely to be used by the team members. This library is the official source of all information about the project. Its access is normally controlled;

- Public library: is often where the library elements common to several projects, such as common tools and reusable components, are located.

When configuring the SCM repository, a number of decisions must be made by the librarian. It might be a bit confusing (see Figure 8.5) the first time you do this but after using it for a time, this becomes second nature. One key item of the SCM plan is the proposal of a branching strategy that will be appropriate for the project. Nearly every software repository tool has some form of branching support. Branching means you diverge from the main line of development (also called the trunk) and continue to do work without affecting them. This is called isolating yourself. In many tools, this is a somewhat expensive process often requiring you to create a new copy of your source code directory, which can take a long time for large projects. You will also find many other resources online. The following text summarizes the basic strategies for typical branches used in software development projects.

Figure 8.5   Choosing a bad branching strategy.
Source: Used under license from Shutterstock.com.

Microsoft has published guidance on branching https://www.visualstudio.com/en-us/articles/branching-strategies-with-tfvc

Trunk, Branches, Commit, and Synchronize

Tag: A symbolic name assigned to a specific release or a branch.

Trunk: The software's main line of development; the main starting point of most branches.

Conflict: A change in one version of a file that cannot be reconciled with the version of the file to which it is applied. Note: can occur when versions from different branches are merged or when two committers work concurrently on the same file.

Branch:
1) A computer program construct in which one of two or more alternative sets of program statements is selected for execution.
2) A point in a computer program at which one of two or more alternative sets of program statements is selected for execution.

Note: Every branch is identified by a tag. Often, a branch identifies the file versions that have been or will be released as a product release. May denote unbundling of arrow

> meaning, i.e., the separation of object types from an object type set. Also refers to an arrow segment into which a root arrow segment has been divided.
>
> *Commit*: To integrate the changes made to a developer's private view of the source code into a branch accessible through the version control system's repository.
>
> *Development Branch*: A branch where active product development takes place.
>
> *Stable Branch*: A branch where stability-disrupting changes are discouraged.
>
> *Synchronize*:
> 1) To pull the changes made in a parent branch into its (evolving) child (for example, feature) branch.
> 2) To update a view with the current version of the files in its corresponding branch.
>
> ISO 24765 [ISO 17a]

Before getting into the description of the different branching strategies, we draw your attention to the fact that each branch incurs a certain management cost. It is therefore important to choose a strategy that will have a minimal impact on your specific project. Adding a new branch has the potential of additional costs that will be incurred later, in the integration and testing phases. With this in mind, here are just some of the many popular branching strategies (e.g., patterns) used in software projects:

- simple: two to three developers working together on a project;
- typical: four or more developers working on a project requiring several versions;
- advanced: five or more developers working on the same project;
- functional: five or more developers working on specific functions, resulting in several versions.

Abhishek Luv's Podcast: SDP Show #4: Source control and basics of Git & GitHub
http://www.softwaredevelopmentpodcast.in/

The elements of these strategies are iterative, so it is possible to start with a simple strategy and gradually move toward an advanced strategy. Again, trying to minimize the complexity of your project in order to facilitate its development is the key to success with SCM.

The moment there is more than one person working on a software, there is a need to isolate them so they can work independently. The main branch must always serve as a point of departure and return for a branch. That is to say, you create branches to isolate yourself from the main branch and then you need to bring these changes back to the main branch. Avoid staying away for too long or creating branches from branches, because this pattern is not a preferred approach and can create confusion. Typically, only a backup branch uses this approach with little impact.

8.6.1 A Simple Branching Strategy

This strategy is appropriate for simple cases, for the development of a website for example, and requires the creation of two types of branches (see Figure 8.6):

– the main branch;
– the versions branch.

This simple branching strategy proposes that the main branch is used by the developers of the software. It implies that the main branch must remain stable at all times as we are always in it. Once the software is quite advanced and the team wants to deliver a version to its client, a version branch is created (e.g., 1.0.1) that will contain a complete version of all artifacts from the first production version. On the main branch, the team can continue working and improving the software. Over time, the development team will be ready to deliver a subsequent release and create another release branch (e.g., 1.1.3) and this branch will become the new production version for the client. The previous branch can be kept as historical or may be archived given that the customer has been provided a new version.

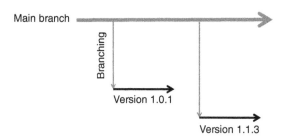

Figure 8.6   The simplest branching strategy of versions by branch.

This strategy is appropriate for the following conditions:

- a very small team who works in the same location in order to communicate easily;
- corrections to production software require a fusion to the main branch. These must be closely controlled because they are made directly within the development branch;
- new version shipped from the main branch includes all previous changes from this branch.

8.6.2 A Typical Branching Strategy

This strategy responds to the needs of more than 80% of the SCM requirements for software projects. It requires the implementation of three branches (see Figure 8.7):

- a main branch;
- a development branch;
- a production branch.

In this strategy, no one works directly in the main branch. It is only used for the integration of components from the team members. This strategy will force team members to work in the development branch. It is in this branch that most of the activity and changes will take place. The development team must control its content. As a general principle, components in a branch should always be able to compile without error. The main branch will receive an intermediary version from time to time that marks progress and can be used for demonstrations as well. Its rate of change will depend on the number of merges coming from development (typically every few days, but at least once a week). Finally, the version branch will also be quite stable as it contains the production version. This branch can be used to fix production problems

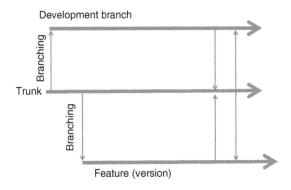

Figure 8.7 A typical strategy of development and production branches.

but if it is changed, then someone must merge these changes to the development branch to ensure it is kept in sync. This strategy can be applied in the following situations:

– delivery of a unique major version;

– delivery of a major version at fixed intervals;

– every new version delivered, coming from the version branch, includes all the changes from the previous versions.

Using the GIT CM tool, here is a simple example that describes the use of two branches simultaneously. Suppose that the development team has created a file "mytext.txt" that is ready to be reviewed by the maintainers. The development team sends the new file onto the remote server GIT branch "master".

    % git add mytext.txt
    % git commit –m "Create file mytext.txt"
    % git push origin master

The development team, wanting to make changes without affecting the maintenance team, create a branch to isolate their future changes.

    % git checkout –b "Development"
    % git push origin Development

Meanwhile, the maintainers review the file created by the developers by downloading the GIT server file.

    % git pull origin master

Once they have completed their changes, they update the file on the server:

    % git commit -a -m "Correction done by maintainers"
    % git push origin master

During this time, the developers have made several changes and are ready to re-validate the file. Here is the state of the GIT server at that time.

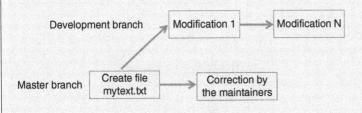

To re-validate their changes, developers must update their changes in the "master" branch. The first step is to get the changes to the branch "master" server.

    % git checkout master
    % git pull origin master

Then developers merge the two branches.

    %git merge Development –no-ff

GIT notifies developers that automatic merging is not possible because changes were made to the files. Developers must change the files that are causing the conflict and resubmit the changes to the server.

    % vim mytext.txt
    % git commit –a –m "Fusion of the Development branch"
    % git push origin master

Here is the result after the merge of the two branches.

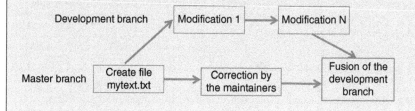

8.7 CONFIGURATION CONTROL

Configuration control (also named change control) is concerned with managing changes during the project. Change control identifies and documents the relative importance of proposed or needed changes to the product, when they will be deployed and who will approve the change (except urgent fixes that will follow a fast path process).

Configuration Control
An element of configuration management consisting of the evaluation, coordination, approval or disapproval, and implementation of changes to configuration items after formal establishment of their configuration identification.

CMMI [SEI 10a]

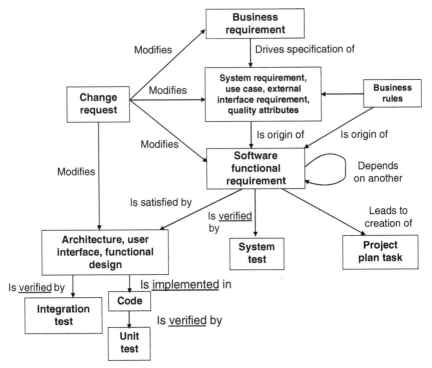

Figure 8.8   Impact of a change request on a software [WIE 13].

Changes to the software can have several origins:

– problem reports (PR), or trouble report (TR), from the development team, maintenance team, or customer;

– modification/evolution requests (MR) communicated as a result of an environment change or the need for new features;

– requests from maintenance and infrastructure (preventive and perfective) issued to improve the maintainability of the software.

In some organizations, these types of changes require that an engineering change request be raised in a centralized system (e.g., often referred to as a change report or ticket). Figure 8.8 describes the impact of a change request involving software and its intermediate products.

8.7.1 Requests, Evaluation, and Approval of Changes

A change request management process, as shown in Figure 8.9, describes the typical sequence for processing a change request (called a ticket in some environments):

- the customer (internal or external) or a team member submits a request;
- a maintainer evaluates the priority, impact and cost;
- the Change Control Committee (CCC) or Configuration Control Board (CCB) reviews and approves the change;
- the change is made to the software and associated documentation;
- the change is tested and approved by the end-user;
- the change is moved to production or installed in a system and verified afterward;
- the change request or ticket is closed and then later archived.

Note that the request will change state at each stage of this process (see Figure 8.9).

During the verification activity, developers and maintainers will also typically perform a series of regression tests to ensure that a change has no impact on other characteristics. Some automation servers, like Jenkins (https://jenkins.io/) or locally developed test robots are also used to perform accelerated regression testing.

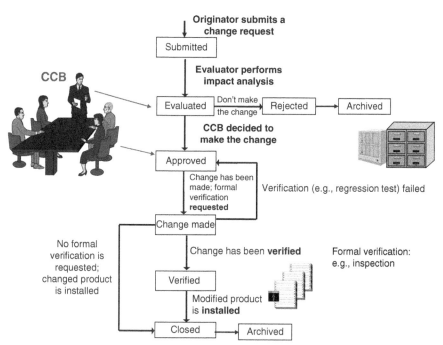

Figure 8.9 Change request change management workflow.
Source: Adapted from Wiegers (2013) [WIE 13].

Change Management
Judicious use of means to effect a change, or a proposed change, to a product or service.
CMMI [SEI 10a]

Regarding critical software, it is common to add evaluation tasks to assess the level of risk for each of the proposed changes. Individuals who can conduct this type of risk evaluation will need to be invited to contribute to the impact analysis of this change request.

The notion of traceability was introduced in an earlier chapter. Traceability is also used in SCM to facilitate the impact analysis of a change request.

8.7.2 Configuration Control Board

Typically, the authority that can accept or reject a proposed change to a software is named a configuration control board (CCB) or a change management office. In smaller projects, this authority may be directly delegated to the project manager or a maintainer. There may be several levels of change authority depending on your organizational processes, the criticality of the software involved, the nature of change (e.g., impact on budget or schedule), or the current step of the life cycle of the project.

Configuration Control Board
A group of people responsible for evaluating and approving or disapproving proposed changes to configuration items, and for ensuring implementation of approved changes.
ISO 24765 [ISO 17a]

The composition of a CCB with respect to a particular project changes on a case by case basis depending on the criticality of the change. The project specialists, according to project size and criticality, may be asked to give their opinion at CCB meetings. In addition to the project manager, a SCM representative and even a SQA representative may be present to verify that the change process and CM plan are followed.

The main tasks of the CCB are: take a go/no-go decision, assign a priority, assign the change to a future version, answer request issuers, issue access rights to libraries and write a change order with decisions taken. The change request includes the following information: identification of the affected software, the list of items in a baseline to modify, SQA tasks to ensure quality checks, if any.

8.7.3 Request for Waivers

The constraints imposed during the development activities may lead to situations where some constraints may need to be relaxed to ensure the success of the project (e.g., a process is not adequate to meet project needs, a request for a non-compliance). A waiver may then be raised on project plans or obligations (e.g., agreement or contract) of the approved life cycle processes. A waiver is an authorization to depart from an obligation. When approved, it also allows the use of that item when completed although it does not meet all of its requirements.

Waiver
A written authorization to accept a configuration item or other designated item which, during production or after having been submitted for inspection, is found to depart from specified requirements, but is nevertheless considered suitable for use as is or after rework by an approved method.

ISO 24765 [ISO 17a]

A procedure for submitting and approving waivers should be described in the organizational processes.

8.7.4 Change Management Policy

A policy sets out general principles that have to be followed and are reflected in the processes and procedures of an organization. The elements of a change management policy are:

- the original text describing a change request will not change or will not be deleted from the organizational repository;
- each changed requirement will be traceable to an approved change request;

- all changes to requirements should follow the SCM process. If a change request is not documented, it will not be considered;
- a CCB is created for each project;
- the contents of the change request library must be available to all project team members;
- no design or development/maintenance work can be initiated on unapproved changes, with the exception of exploration or feasibility studies required by the CCB.

8.8 CONFIGURATION STATUS ACCOUNTING

The status accounting of CI represents all the recording and reporting activities required for managing the SCM.

Configuration Status Accounting
An element of configuration management consisting of the recording and reporting of information needed to manage a configuration effectively.
This information includes a list of the approved configuration, the status of proposed changes to the configuration, and the implementation status of approved changes.

CMMI [SEI 10a]

8.8.1 Information Concerning the Status of CI

Information concerning the state of configuration elements must be identified, collected, and maintained. The following information should be available to managers and developers: the approved configuration identification, as well as the identification and current status of changes, deviations and waiver approvals. SCM items are recorded in sufficient detail so that previous versions can be recovered when needed. Questions answered by the status of CI:

- what is the status of item X?
- was change request X approved or rejected by the CCB?
- what has changed in this new version of this item?
- how many defects were detected last month and how many are corrected?
- what is the cause of this change request?

An Example of Status Control Information using Microsoft Team Foundation System 2010 [GHE 09]

Imagine I have a main branch and one isolation branch: Branch 1 features originate from the main branch. Now let us assume that a member makes some changes to the items of Branch 1. These changes are merged (backward fusion) to the main branch and the

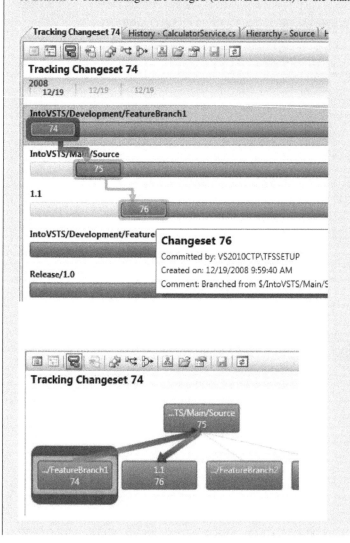

team named this new release branch (version 1.1) in a new version branch (which is connected to the main branch). Using the annotate function, we can see the status details of that change, such as that line X has been changed by person Y on date and time Z. The question that now arises is: what detail do we want to see or report for a given branch? The answer is, of course, you want to see which branch we originated from to make these changes.

The Tooltip function of TFS 2010 allows you to find all the activities that took place on a branch. Here is what we can see from the version 1.1 branch. The first part of the figure shows a chronological view where you can follow the history of the merge of change number 74 (named Changeset 74) and the second part of the figure shows a hierarchical view where you can follow the hierarchical dependencies between the branches.

8.8.2 Configuration Item Status Reporting

Reported information can be used by various organizational units and the project team. The CMMI noted that this report typically includes [SEI 10a]:

- the meeting minutes of the change management board;
- the summary and status of change requests;
- the summary and status of problem reports (including corrective measures);
- the summary of changes to software repositories;
- the revision history of CI;
- the state of software repositories;
- the results of audits of the software repositories.

Figure 8.10 shows the output of a tool that reports on CI.

8.9 SOFTWARE CONFIGURATION AUDIT

As presented already, the IEEE 1028 standard defines audits as an independent examination of a software product, software process, or set of software processes performed by a third party to assess compliance with specifications, standards, contractual agreements, or other criteria [IEE 08b].

A SCM audit could be called for a software project to assess how CIs satisfy the functional and physical characteristics needed and as well as to assess how the SCM plan was implemented in the project. Two types of formal audits are typically performed: a functional configuration audit (FCA) and the physical configuration audit (PCA). Details about these audits can be found in the informative annex J of IEEE 828 titled "Examples of how configuration auditing is applied." We will now present an overview of these two audits.

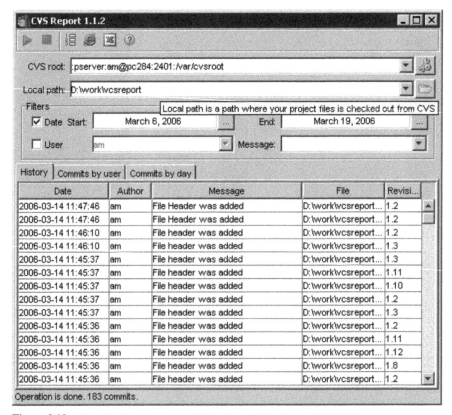

Figure 8.10 Example of a status report with the Commit Monitor tool [COL 10].

Configuration Audit

An audit conducted to verify that a configuration item or a collection of configuration items that make up a baseline conforms to a specified standard or requirement.

CMMI

Physical Configuration Audit (PCA)

An audit conducted to verify that a configuration item, as built, conforms to the technical documentation that defines it.

Note: For software, the purpose of the software physical configuration audit is to ensure that the design and reference documentation is consistent with the as-built software product.

IEEE 828 [IEE 12b]

Functional Configuration Audit (FCA)

An audit conducted to verify that the development of a configuration item has been completed satisfactorily, that the item has achieved the performance and functional characteristics specified in the functional or allocated configuration identification, and that it is operational and support documents are complete and satisfactory.

ISO 24765 [ISO 17a]

8.9.1 Functional Configuration Audit

The objective of the FCA is to provide an independent evaluation of software products to verify that the actual functionality and performance of each configuration item is within specifications. FCA should be concerned not only with the functionality, but also with the non-functional requirements. FCA typically includes the following [KAS 00]:

- an audit of test documentation and test results;
- audit reports of V&V to ensure their accuracy;
- a review of all approved changes to ensure they have been properly incorporated and verified;
- a review of updated documents to ensure their accuracy;
- a sampling of minutes of design meetings to ensure that all findings have been incorporated;
- a sampling of performance and other non-functional test results for completeness.

FCA of the outputs of the design process verifies the following [IEE 12b]:

- traceability between the design items and their sources (requirements);
- every requirement is linked to at least one design element;
- every design element is linked to at least one requirement that justifies it.

8.9.2 Physical Configuration Audit

The objective of the PCA is to provide an independent assessment of CI to confirm that each element that makes up the software as delivered is present and traceable to specifications [KAS 00]. This audit verifies that the software and documentation are correct, consistent and ready for delivery. The available documentation typically includes: installation manuals, operation manuals, maintenance manuals, and version description documents.

A PCA will typically include the following elements [KAS 00]:

– an audit of specifications to ensure completeness;
– a review of the problem reporting and change management process;
– a comparison between the architectural design and components to ensure consistency;
– a code review to assess compliance to coding standards;
– audit documentation to ensure the completeness and compliance with the format and with functional descriptions. These manuals include user manuals, programmer's manual, and operator's manuals.

PCAs of the outputs of the requirements process in the life cycle verify the following [IEE 12b]:

– requirements assets have been placed under configuration control;
– requirements assets have been properly labeled in accordance with the CMP;
– an inventory of requirements assets exists and correctly reflects the attributes of each CI;
– there is evidence of the use of change control procedures for each of the changes made to previous baselines, if any (for example, in a previous iteration).

8.9.3 Audits Performed During a Project

These audits are required during the phases of design and development (before the PCA and FCA) to verify the consistency of the design during its evolution. These audits are performed [KAS 00]:

– to verify that the interfaces between hardware and software comply with the design;
– to verify that the code is fully tested and that business requirements are met;
– to check whether the product design throughout the development meets the functional requirements;
– to check whether the code adheres to the detailed design.

"The cost of rework adds greatly to the cost of developing software and can be reduced by implementing an effective SCM program. The principles behind the modern cost-of-quality concept were derived for manufacturing applications and can be found in the works of J. M. Juran."

Kasse and Mcquaid (2000) [KAS 00]

8.10 IMPLEMENTING SCM IN VERY SMALL ENTITIES WITH ISO/IEC 29110

As described in the management and engineering guide of the Basic profile of the ISO 29110 standard for a small organization, a simple approach to CM is to perform the following tasks [ISO 11e]:

- identify components;
- describe the standards that will be used during a typical project;
- formalize the reviews before a configuration item is deposited in the repository;
- establish a simple change control process;
- establish the library and control access;
- manage change requests;
- occasionally check that the backups to the repository were correctly performed.

ISO 29110 recommends that a repository be set up to store work products and their versions. Table 8.4 describes one task of the software process associated with CM.

Table 8.4   A CM task of ISO 29110 [ISO 11e]

Role	Task list	Input work products	Output work products
TL	SI.3.8 Incorporate the Software Design, and Traceability Record to the Software Configuration as part of the baseline. Incorporate the Test Cases, and Test Procedures to the project repository.	Software Design [verified] Test Cases, and Test Procedures [verified] Traceability Record [verified]	Software Configuration - Software Design [verified, baselined] - Test Cases, and Test Procedures [verified] - Traceability Record [verified, baselined]

Table 8.5 shows an example of the suggested content of a change request as proposed by the ISO 29110 management guide. The column on the right shows the source of the request, that is, who asked for the change. In this example, a change request could be raised by the customer, by the development team or the project manager. Note that the states of the request are also listed at the bottom of the description column.

Table 8.5 A Change Request as Recommended by ISO 29110 [ISO 11e]

Name	Description	Source
Change request	Identifies a software, or documentation problem or desired improvement, and requests modifications. It may have the following characteristics:	Customer Project manager Software implementation
	– identifies purpose of change;	
	– identifies request status;	
	– identifies requester contact information;	
	– impacted system(s);	
	– impact to operations of existing system(s) defined;	
	– impact to associated documentation defined;	
	– criticality of the request, date needed.	
	The applicable statuses are: initiated, evaluated, and accepted.	

8.11 SCM AND THE SQAP

The IEEE 730 standard [IEE 14] defines the requirements for SQA activities during the project. As we said earlier, each project needs to have one SQAP. Here are the questions that the project team must answer before project planning is approved:

- has an appropriate and effective SCM strategy, including change control processes, been developed?
- has the completeness and consistency of CI been ensured?
- has consistency and traceability been planned between CI?
- has the storage, handling, and delivery of CI been controlled?

The SQAP should describe the SCM activities that are planned, how they should be performed, and who is responsible for SCM tasks. This plan should also define the methods and tools used to conserve, store, secure and control software records and artifacts, and all their versions, created during all phases of the software life cycle.

During project execution, the team should ask the following questions related to the evaluation of their adherence to the SQAP as well as the tracking of the status of the CIs:

- has an appropriate and effective SCM strategy, including change control processes, been implemented?
- have items generated by the process or project been identified, defined, and baselined?
- are the requirements allocated to the system elements and their interfaces traceable to the customer's requirements baseline?
- have modifications and releases of the items been controlled?
- has documentation been maintained in accordance with defined criteria?
- have modifications and releases been made available to affected parties?
- has the status of the items and modifications been recorded and reported?
- has the completeness and consistency of CI been ensured?
- has the storage, handling, and delivery of CI been controlled?

Concerning the tracking of change requests and problem reports, the following questions should be addressed:

- has an appropriate and effective problem management strategy been developed?
- are problems recorded, identified, and classified?
- are problems analyzed and assessed to identify an acceptable solution(s)?
- are problem resolutions implemented?
- are problems tracked to closure?
- is the status of all problems reported and known?

If a project is required by its customer to have a stand-alone CM plan, then information listed in IEEE 730 could be used to develop a CM plan separate from the SQAP. IEEE 828 could be used to develop the CM plan. In such a case, the SQAP would have to refer to the CM plan regarding CM issues.

8.12 SUCCESS FACTORS

Following are some factors related to an organization's CM practices that can promote or discourage the development of quality software.

Factors that Foster Software Quality

1) Organizational culture and management support for SCM.

2) Presence of a SCM vision, mission, and policies.

3) Adequate resource allocation and SQA support.

4) Early CM planning and effective communications.

5) Stable toolset and competent CM practitioners.

6) SCM certification and training available.

Factors that may Adversely Affect Software Quality

1) No management support for SCM.

2) No SCM training or certification available.

3) Rigid and complex SCM process.

4) Lack of required human resources and necessary budgets for SCM execution during the project.

5) Uncoordinated communications with the CCB.

Here are some excuses that you may hear when you want to implement SCM in your project [SPM 10]:

- SCM applies only to code;
- SCM applies only to documents;
- SCM is not required because we use the latest technology and agile processes;
- It's not such a big project;
- It slows down our technical staff when it's time to make a quick change during testing;
- We made the change without bothering to submit a change request because we want a satisfied customer and this is only internal paperwork;
- We do not need a separate CM system because it is automatically done by the IDE tool we use for development;
- Changing documentation is a thing of the past-we change the source code directly;
- We do not have the SCM on many of our external interfaces because they are the responsibility of other agencies;
- SCM is a practice used by the department of defense (DoD) and we do not make software for the DoD;

– We do not put software architecture and detailed design under the constraint of formal change management during development because it limits our flexibility and productivity;

– We do not put the operating software, custom software, libraries and compilers under SCM because its available online.

8.13 FURTHER READING

CASAVECCHIA D. Reality configuration management, *Crosstalk*, November 2002.

DJEZZAR L. *Gestion de configuration: maitrisez vos changements logiciels*, Dunod, Paris, 2003.

JOHANSSEN HASS A. Finding CM in CMMI, *Crosstalk*, July 2005.

LEISHMAN T. and COOK D. But I Only Changed One Line of Code!, *Crosstalk*, 2003.

PHILLIPS D. Go Configure, *Understanding the principles and promise of configuration management*, STQE, vol. 4, Issue 3, May–June 2002.

RINKO-GAY W. Preparing to choose a CM Tool, *STQE Magazine*, July–August 2002.

WIEGERS K. Creating a Software Engineering Culture, *Control Change Before It Controls You*. Dorset House, New York, 1996, Chapter 15.

8.14 EXERCISES

8.1 Present a definition and the expected results of using SCM.

8.2 What are the issues/problems addressed by SCM?

8.3 What are the six SCM information categories that need to be included in a software configuration management plan (SCMP)?

8.4 Why do we need to place our tools, software and the libraries used for developing a software under configuration management?

8.5 How do we identify the CI to be controlled in a project?

8.6 List the potential tasks to be executed when changing an existing requirement in an ongoing project. Using this list, create a checklist of items that are impacted when there is a change to an existing requirement during a project.

8.7 Extend the checklist developed in question 6 by adding the software items affected by the proposed requirement change.

8.8 Describe the typical branching strategy used to manage the software configuration of source code during a software development project?

8.9 Develop a process map that explains how to process a change in your software project.

8.10 Your project includes firmware. Are the SCM concepts that apply to hardware and software independently also applicable to the firmware portion of your project?

8.11 Your manager asks you to perform an impact assessment for a major change planned in your portion of the existing software. He also sends you a copy of the new requirements to be added to the software. Develop a list of SCM factors you need to consider in order to provide a better estimate of this impact (e.g., effort, schedule, etc.).

8.12 Regarding the criteria for selecting the software CI: what are the consequences (e.g., risks) of the lack of choosing all the necessary CI for your project?

8.13 According to the CMMI-DEV model, what is suggested in a SCM report?

8.14 Write a checklist for a FCA.

8.15 Write a checklist for an audit of physical configuration.

Chapter 9

Policies, Processes, and Procedures

After completing this chapter, you will be able to:

- understand the documentation system of an organization;
- understand the role of a policy in an organization;
- understand the importance of software process documentation;
- understand how to document processes and procedures;
- understand a few popular notations such as ETVX, IDEF, and BPMN;
- understand the requirements of the ISO 12207 standard and the CMMI® model concerning the documentation of processes and procedures;
- understand the process notation of the ISO/IEC 29110 management and engineering guides;
- understand the personal improvement process;
- understand the relationship between policies, processes, procedures, and the software quality assurance plan.

9.1 INTRODUCTION

An article of the online encyclopedia Wikipedia, summarizes critics' comments about software development as follows: "In traditional engineering, there is a clear consensus about how things should be built, how they should respect engineering standards and what risks must be taken into account. If an engineer does not comply with his code of practices and something breaks down, he can be sued. There is no such consensus yet, in software engineering, where everyone is promoting his own methods and tools, claiming their advantages in productivity, which, in general are not supported by any objective and scientific evidence whatsoever."

Software Quality Assurance, First Edition. Claude Y. Laporte and Alain April.
© 2018 the IEEE Computer Society, Inc. Published 2018 by John Wiley & Sons, Inc.

"Future large systems will no longer be delivered late, with cost overruns and poor quality, and they will probably not be delivered at all."

Humphrey et al. (2007) [HUM 07]

In this chapter, we present approaches that address some of the issues with this criticism. We will use, as in previous chapters, known standards such as the ISO standards and the CMMI® model, and present how to use and adapt these standards to address the problems and needs of the organization. The SWEBOK Guide (see Figure 9.1) includes a knowledge area that describes the importance of processes in software engineering and the existing knowledge that should be used.

We will not discuss process assessment or measurement here as these topics are discussed in detail in other chapters of this book.

We know that a quality management system includes a number of documents. Figure 9.2 shows an example of a pyramid model to classify the many types of documents found in an organization. Quality objectives and organizational policies are normally at the top of the pyramid. At the second level, we find processes used at

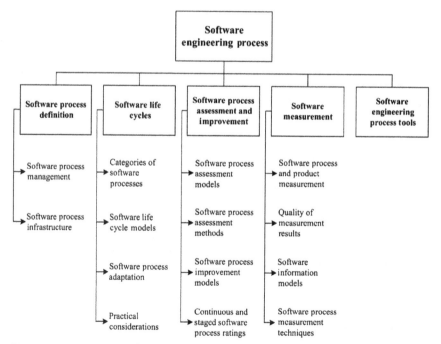

Figure 9.1 The SWEBOK® Guide software engineering process knowledge area [SWE 14].

Figure 9.2   Example of a quality system documentation model.

all levels of the organization. At the third level, we find more detailed procedures, checklists and templates/examples to help ensure efficient daily operations. Finally, at the base of this documentation pyramid, we find the quality records that are the accumulated evidence, or proof, resulting from the execution of both processes and procedures in accordance with the organization's policies and quality objectives.

In this book, we propose an organizational process approach which entails the daily use of a quality management system consisting of organizational processes that are identified, formalized, interact with each other, and are managed and improved. In this chapter, we describe how to develop, document, and improve policies, processes, and procedures to ensure the effectiveness and efficiency of the organization. Remember that software quality assurance (SQA) plays a key role in defining, and particularly in improving, the organizational processes in many organizations, since SQA has the mission of auditing these processes.

Efficiency
Relationship between the result achieved and the resources used.
Effectiveness
Extent to which planned activities are realized and planned results are achieved.

ISO 9000

Two key reference documents will be used in this chapter: the ISO 9000 standard and the CMMI-DEV model. The ISO 9001 standard [ISO 15] is useful with respect to how a quality management system is implemented. As illustrated in the following text box, having previously identified its business objectives, the organization must, among other things, develop its quality objectives, policy, and organizational processes.

Quality Management Systems Approach of ISO 9000

An approach to developing and implementing a quality management system consists of several steps including the following:

– determining the needs and expectations of customers and other interested parties;
– establishing the quality policy and quality objectives of the organization;
– determining the processes and responsibilities necessary to attain the quality objectives;
– determining and providing the resources necessary to attain the quality objectives;
– establishing methods to measure the effectiveness and efficiency of each process;
– applying these measures to determine the effectiveness and efficiency of each process;
– determining means of preventing nonconformities and eliminating their causes;
– establishing and applying a process for continual improvement of the quality management system.

Such an approach is also applicable to maintaining and improving an existing quality management system.

An organization that adopts the above approach creates confidence in the capability of its processes and the quality of its products, and provides a basis for continual improvement. This can lead to increased satisfaction of customers and other interested parties and to the success of the organization.

ISO 12207 [ISO 17] also asks the software organization to develop its policies and processes. The business perspective is comprised in its portfolio management process. Note that this standard also requires that the software development life cycle models used, such as waterfall, iterative, or agile be formalized. We briefly discussed this in previous chapters. We will address this in more detail in the chapter entitled "Risk management" where the criteria for selecting the appropriate life cycle for a project is associated with its criticality and perceived risks.

Life Cycle Model Management Process of ISO 12207

Purpose
The purpose of the Life Cycle Model Management Process is to define, maintain, and assure availability of policies, life cycle processes, life cycle models, and procedures for use by the organization with respect to the scope of this International Standard.

This process provides life cycle policies, processes, and procedures that are consistent with the organization's objectives, that are defined, adapted, improved, and maintained to support individual project needs within the context of the organization, and that are capable of being applied using effective, proven methods and tools.

Outcomes

As a result of the successful implementation of the Life Cycle Model Management Process:

– organizational policies and procedures for the management and deployment of life cycle models and processes are established;

– responsibility, accountability, and authority within the life cycle policies, processes, models, and procedures are defined;

– life cycle processes, models, and procedures for use by the organization are assessed;

– prioritized process, models, and procedure improvements are implemented.

The CMMI-DEV model [SEI 10a] contains several process areas that include topics about software development and process improvement. In this chapter, the following processes are presented:

– organizational process definition;

– organizational process focus;

– organizational performance management;

– organizational process performance

– causal analysis and resolution;

– organizational training.

"It is always faster to do the job right the first time."

Watts S. Humphrey

Only the first two process areas will be described in this chapter. The reader should refer to the staged representation of the CMMI-DEV for descriptions of the other process areas:

Organizational process definition [SEI 10a]:

– The purpose of the organizational process definition is to establish and maintain a usable set of organizational process assets, work environment standards, and rules and guidelines for teams.

– This process area has one specific goal (SG) "Establish Organizational Process Assets" and the following seven specific practices (SP):

- o SP 1.1 Establish Standard Processes
- o SP 1.2 Establish Life Cycle Model Descriptions
- o SP 1.3 Establish Tailoring Criteria and Guidelines
- o SP 1.4 Establish the Organization's Measurement Repository
- o SP 1.5 Establish the Organization's Process Asset Library
- o SP 1.6 Establish Work Environment Standards
- o SP 1.7 Establish Rules and Guidelines for Teams

Organizational process focus [SEI 10a]:

- – The purpose of organizational process focus is to plan, implement, and deploy organizational process improvements based on a thorough understanding of current strengths and weaknesses of the organization's processes and process assets.
- – This process area has the following three SG and nine SP:
  - o SG 1 Determine Process Improvement Opportunities
    - □ SP 1.1 Establish Organizational Process Needs
    - □ SP 1.2 Appraise the Organization's Processes
    - □ SP 1.3 Identify the Organization's Process Improvements
  - o SG 2 Plan and Implement Process Actions
    - □ SP 2.1 Establish Process Action Plans
    - □ SP 2.2 Implement Process Action Plans
  - o SG 3 Deploy Organizational Process Assets and Incorporate Experiences
    - □ SP 3.1 Deploy Organizational Process Assets
    - □ SP 3.2 Deploy Standard Processes
    - □ SP 3.3 Monitor the Implementation
    - □ SP 3.4 Incorporate Experiences into Organizational Process Assets

The CMMI model pays particular attention to the artifacts and the means that are deployed to support the process. The following text box describes some of them.

Process Asset
Anything the organization considers useful in attaining the goals of a process area.
Organization's Process Asset Library
A library of information used to store and make process assets available that are useful to those who are defining, implementing, and managing processes in the organization.
    This library contains process assets that include process related documentation such as policies, defined processes, checklists, lessons learned documents, templates, standards, procedures, plans, and training materials.

CMMI

The CMMI model proposes that the process assets are regrouped into an organizational process repository. This repository can be digital, such as an organizational intranet or wiki. In this way, all employees can have access to the latest versions of the process information. In addition to the items listed in the last text box, this intranet may contain examples of already developed/validated documents originating from completed projects. Given that an organization operates in a particular industry and business domain, software projects could share/reuse some artifacts that are similar across projects, such as plans, configuration management, and quality assurance processes. The project manager who is starting a new project would be advised to consult the organizational process repository to identify reusable and customizable artifacts before creating his own. This could not only save effort by reusing items, but also allows him to benefit from the experience of others.

9.1.1 Standards, the Cost of Quality, and Business Models

The cost of quality and business models concepts were presented earlier. The cost of quality, policies, processes, and procedures are classified as preventive cost elements, which are costs incurred by an organization to prevent the occurrence of errors in various development or maintenance processes. Detection costs are the verification and appraisal costs of software products or services during the various life cycle phases of a software development project. Detection costs also represent the additional cost of controlling these internal standards (e.g., the cost of their maintenance and management). Appraisal costs represent the associated audit cost invested to determine compliance as well as the costs associated with the certification/attestation/compliance to a standard such as ISO 9001 or a process model like the CMMI for development.

Table 9.1 presents the different preventive, detection or appraisal costs. Development costs, training, and implementation of policies, processes, and procedures are considered preventive costs.

Policies, processes, and procedures are commonly used in the following business models: custom systems written on contract and commercial and mass-market firmware. In these business models, policies, processes and procedures are used to control development and minimize errors and risks. With respect to the custom systems written on contract business model, it is the customer who decides whether or not to impose his processes and procedures on the supplier.

In the following sections, we describe the policies, processes, and procedures of the quality system documentation pyramid for the software organization.

9.2 POLICIES

Essentially, organizational policies aim at showing, publicly and officially, how the software organization intends to meet its business objectives. They are typically

Table 9.1   Preventive and Detection (or Appraisal) Costs Related to Policies, Processes, and Procedures

Major category	Sub-category	Definition	Typical cost element
Preventive costs	Establish the foundation of the quality management system.	Efforts to define quality, quality objectives and thresholds, and quality standards. Quality compromise analysis.	Definition of acceptance test success criteria and of quality standards.
	Interventions oriented toward projects and processes.	Efforts to prevent defects or to improve the quality of the processes.	Training, process improvement, measurement and analysis.
Detection or appraisal costs	Discover the state of the product.	Discover the non-conformity level.	Tests, software quality assurance. Inspections, reviews.
	Ensure that quality objectives are met.	Quality control mechanism.	Product quality audits, new versions delivery decision criteria.

Source: Adapted from Krasner (1998) [KRA 98].

developed and approved by upper management and, once deployed, help in guiding projects and product development decisions as well as the behavior of personnel. The term "organizational directive" is also used as a synonym to the term "organizational policy."

Policy
Clear and measurable statements of preferred direction and behavior to condition the decisions made within an organization (ISO/IEC 38500).
Organizational Policy
A guiding principle typically established by senior management that is adopted by an organization to influence and determine decisions (CMMI).

Quality Policy

Overall intentions and direction of an organization related to quality as formally expressed by top management.

> Note 1: Generally the quality policy is consistent with the overall policy of the organization, can be aligned with the organization's vision and mission and provides a framework for the setting of quality objectives.
>
> Note 2: Quality management principles presented in this International Standard can form a basis for the establishment of a quality policy (ISO 9000).

In order for policies to be effective, management should:

- ensure that the commitment is clearly communicated to all levels of the organization;
- initiate, manage, and monitor the implementation of policies;
- require that a strong business case be presented before accepting any deviations to the policy;
- show their support for the policies by providing adequate resources (e.g., budget, competent personnel and appropriate tools) for their implementation, monitoring, evaluation, and improvement;
- provide adequate training to support the understanding of the daily execution of the policies.

In the CMMI, one of the generic practices, the GP 2.1, requires that the organization establish and maintain an organizational policy for planning and performing a process. The organization may write a policy that covers all the software process areas or one different directive for each CMMI process area. The following text box describes what is intended and asked for in the CMMI model.

GP 2.1 Establish an Organizational Policy [SEI 10a]

Establish and maintain an organizational policy for planning and performing the process.

The purpose of this generic practice is to define the organizational expectations for the process and make these expectations visible to those members of the organization who are affected. In general, senior management is responsible for establishing and communicating guiding principles, direction, and expectations for the organization.

Not all direction from senior management will bear the label "policy." The existence of appropriate organizational direction is the expectation of this generic practice, regardless of what it is called or how it is imparted.

For example, for the "project planning" process area, the CMMI asks that the organization establish an organizational policy for their "project planning" activities, such as: "This Directive establishes organizational expectations concerning the estimates for internal and external commitments and for developing the project management plans" [SEI 10a].

Process Owner

Person (or team) responsible for defining and maintaining a process.

    Note: At the organizational level, the process owner is the person (or team) responsible for the description of a standard process; at the project level, the process owner is the person (or team) responsible for the description of the defined process. A process may therefore have multiple owners at different levels of responsibility.

ISO 24765 [ISO 17a]

The following text box provides an actual policy example and describes the policy objectives, scope, target processes, and the responsibilities of key stakeholders of an organization of about 35 software engineers. This company uses the concept of process owner.

Acme Corporation – Software Policy
Introduction

It is a policy at Acme to use a number of software engineering processes to achieve its project quality objectives, estimated costs and schedule. As a reference framework for the development of software engineering processes, Acme uses the Software Engineering Institutes' Capability Maturity Model.

    Software based systems are strategic to the Acme business market. Software is a competitive differential factor for our products. The basic principle of this policy is that personnel must use predictable software engineering processes to design, develop and maintain software products to ensure they are reliable, scalable and portable.

    The policy applies to the following software products or for the purchase of software products that require a capital investment:

– real-time software;

– software that supports software development (e.g., compilers, editors, operating systems, etc.);

– scientific software (e.g., modeling, simulation, test, etc.);

– embedded software;

– artificial intelligence software.

Objectives

Once the system requirements are defined, the software engineering process (SEP) defines the reference framework to be applied to research, design, develop, maintain or acquire the software products. In turn, the SEP provides a management tool for Acme to ensure that the personnel, products and processes are supported by a set of methods and tools capable of meeting the performance objectives set by the company.

Scope of the Policy

The software engineering process applies to the following activities:

– engineering: for software used during engineering design, coding and debugging – this also includes reverse engineering (refactoring) activities carried out to maintain or improve existing systems and activities to improve or correct software development defects;

– integration and testing: for software used for activities concerning the integration and testing of our products or services;

– acquisitions: for software used to support the selection and acquisition activities of software packages (e.g., commercial-off-the-shelf software) and software developed by an external supplier.

The process can be adapted to meet specific project requirements and constraints. However, minimum requirements apply and the adaptation and associated risks must be approved by the process owner. Conflicts concerning the adaptation of the software process, between the process owner and a project manager, will be resolved by the vice president responsible for the project.

The software engineering process is continuously improved to meet the following objectives:

– ensure the quality of products and services that meet the customer requirements;

– ensure technical compliance of delivered products and services;

– minimize negative impacts on schedule and cost.

9.3 PROCESSES

At the second level of the pyramid illustrated in Figure 9.2, we find processes. The processes are the implementation of policies that guide the development of products and services. A process describes what must be done to produce the expected results. We will describe procedures in the next section. Procedures are very detailed and describe how to do a job, step by step.

Process

Set of interrelated or interacting activities that use inputs to deliver an intended result.

Note 1: Whether the "intended result" of a process is called output, product or service depends on the context of the reference.

Note 2: Inputs to a process are generally the outputs of other processes and outputs of a process are generally the inputs to other processes.

Note 3: Two or more interrelated and interacting processes in series can also be referred to as a process.

Note 4: Processes in an organization are generally planned and carried out under controlled conditions to add value.

Note 5: A process where the conformity of the resulting output cannot be readily or economically validated is frequently referred to as a "special process."

ISO 9000

Process Description

Documented expression of a set of activities performed to achieve a given purpose.

Note: A process description provides an operational definition of the major components of a process. The description specifies, in a complete, precise, and verifiable manner, the requirements, design, behavior, or other characteristics of a process. It also may include procedures for determining whether these provisions have been satisfied. Process descriptions can be found at the activity, project, or organizational level.

ISO 24765 [ISO 17a]

Product

Output of an organization that can be produced without any transaction taking place between the organization and the customer.

Note 1: Production of a product is achieved without any transaction necessarily taking place between provider and customer, but can often involve this service element upon its delivery to the customer.

Note 2: The dominant element of a product is that it is generally tangible.

Note 3: Hardware is tangible and its amount is a countable characteristic (e.g., tyres). Processed materials are tangible and their amount is a continuous characteristic (e.g., fuel and soft drinks). Hardware and processed materials are often referred to as goods. Software consists of information regardless of delivery medium (e.g., computer programme, mobile phone app, instruction manual, dictionary content, musical composition copyright, driver's license).

ISO 9000

A defined (i.e., formalized) and documented process provides the organization with [SEI 10a]:

- a defined framework for planning, monitoring, and managing the work;
- a guideline to do the job correctly and completely describing a series of sequential steps to do the work;
- a basis for measuring the work and monitoring progress against targets, and allows for refining the process in future iterations;
- a basis for refining the process;
- a planning and quality management tool for the products being developed;
- procedures to be used to coordinate work in order to produce a common product;
- a mechanism that allows team members to support each other throughout the project.

Acme Corporation – Software Policy
To achieve these objectives, the following process should be used:

- software development process;
- software maintenance process;
- planning and project monitoring processes;
- configuration management process;
- software quality assurance process;
- management process of outsourcing software;
- software reverse engineering process.

Responsibilities
Process Owner
The manager of the software engineering department is the owner of the software engineering process. He is responsible for defining, implementing and maintaining a software engineering organization that includes processes, people and tools needed to develop software products at the lowest cost and fastest time to market. He is also responsible for informing project managers of the risks associated with their project plans and approving the software related sections. In this role, he is the leader of the software process improvement group. As the leader, he is responsible for approving the process changes to ensure that the objectives of Acme are met.

The Software Process Improvement Group

The software process improvement group is responsible for defining, maintaining and improving the software assets, verifying and validating the effectiveness and efficiency of the software engineering process by continually measuring their performance. The group is also responsible for managing the software sections of the process improvement plan.

Project Team Members

Project team members must use the software engineering processes as approved in the project plan. They are responsible for informing the project manager and the process owner of all risks related to their use or the needed improvements they believe should be done to the software engineering processes.

ISO/IEC TR 24774 - Systems and Software Engineering — Life Cycle Management — Guidelines for Process Description [ISO 10a]

Intended Audience

The editors, working group members, reviewers and other participants in the development of process standards and technical reports.

This technical report describes the following process elements:

– the title is a descriptive heading for a process;

– the purpose describes the goal of performing the process;

– the outcomes express the observable results expected from the successful performance of the process;

– the activities are a list of actions that may be used to achieve the outcomes. Each activity may be further elaborated as a grouping of related lower level actions;

– activities are sets of cohesive tasks of a process;

– the tasks are specific actions that may be performed to achieve an activity. Multiple related tasks are often grouped within an activity;

– the information items are separately identifiable bodies of information produced and stored for human use during a system or software life cycle.

When documenting a process, we must think of those who will use it. This statement may seem obvious, but often processes are documented for the wrong audience. What are audiences? They are, for example, developers who have a wealth of knowledge and experience. Another audience can be expert developers that require very little documented processes (see Figure 9.3). Experts are like experienced pilots.

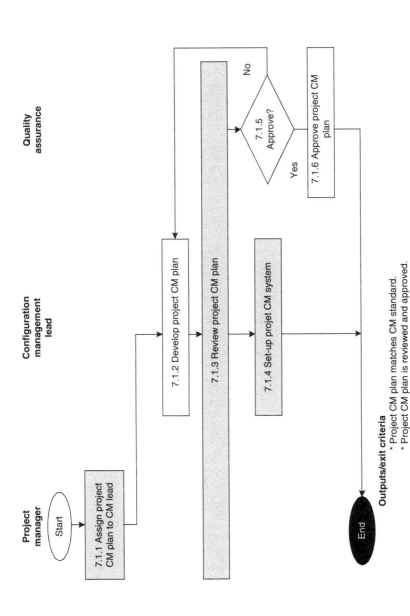

Project manager

Start

7.1.1 Assign project CM plan to CM lead

Configuration management lead

7.1.2 Develop project CM plan

7.1.3 Review project CM plan

7.1.4 Set-up projet CM system

End

Quality assurance

7.1.5 Approve?

No

Yes

7.1.6 Approve project CM plan

Outputs/exit criteria
* Project CM plan matches CM standard.
* Project CM plan is reviewed and approved.
* CCB and CM lead are identified.
* CM system is set up according to the set up CM system procedure.

Figure 9.3 Processes documented for an expert.
Source: Adapted from Olson (2006) [OLS 06].

Table 9.2 Description of Process Guidance for Intermediate Software Process Users [OLS 06]

Process step ID	Role	Process step description
7.1.1	Project manager	**Assign the responsibility for configuration management planning** At the appropriate time, typically during the project planning phase, the project manager assigns the configuration management (CM) plan development work. **Advice** An individual, responsible for CM should have experience in the installation and execution of CM systems. His supervisor should also have been trained in CM.
7.1.2	Configuration Management manager	**Develop the CM plan** The configuration management manager develops the CM plan using organizational standards and CM guidelines. The organizational process references and standards must be used. **Advice** CM Plan and templates, examples and guidelines can be found on the organizational process repository that is accessible on the Intranet of the software division.

Once behind the controls, they commonly only use checklists. They do not have to dig through detailed manuals to find out what to do next.

Other audiences can be junior programmers and new employees of the organization. They require more detailed processes. Documentation may include a tutorial or contain explanatory material. Between these types of audiences are intermediate users that still cannot be referred to as expert process users but do not need as much information as beginners. For this last audience, we could add to Figure 9.3 the more detailed description shown in Table 9.2.

A well-thought out, ready and usable process includes the following elements: a definition of the process, inputs required for its execution, the impacted agents, resources needed (e.g., people, equipment, time, and budget) and its exit criteria. A process defines precisely what to do by listing the tasks, in sufficient detail, to guide its user during its execution. Processes should provide sufficient details for team members and individuals in order for them to develop detailed project plans and then be able to execute the planned process to guide and monitor their work [SEI 09]. The following sections describe process notations used to graphically represent processes.

When reviewing a process, you should be able to answer the following questions [OLS 94]:

- Does this process refer to at least one of the organization's policies?
- Why is this process executed?
- Who performs this process (e.g., its roles.)?
- What software products are used by this process?
- What tools are used?
- What software products are produced by this process?
- When does the process begin?
- When does this process end?
- What happens to the products developed by this process?
- How is this process implemented (e.g., what are the procedures)?
- Where is the process implemented?
- What is the typical effort needed to execute this process?
- What other resources have been used to execute this process?
- Is the process terminology used understandable in your environment?
- Can the performance of this process be measured?

9.4 PROCEDURES

Procedures support process execution. They define and clarify each step of the process. These procedures can take many forms: they may be descriptions of actions to take, templates of documents and forms that include instructions on how to use them, or even checklists that can be referred to or completed.

Procedure
Specified way to carry out an activity or a process.
    Note 1: Procedures can be documented or not.

ISO 9000 [ISO 15b]

Ordered series of steps that specify how to perform a task.

ISO/IEC 26514 [ISO 08]

> **Template**
> A partially complete document in a predefined format that provides a defined structure for collecting, organizing, and presenting information and data.
>
> PMBOK® Guide

When a procedure has been developed, the checklist, as described in the next text box, can be used to verify that it contains all the required elements of a good procedure.

During the review of a procedure, you should be able to answer yes to the following questions [OLS 94]:

– Does it support at least one process?
– Are the procedural steps described executed in the right order?
– Is each step of the procedure clearly and correctly described?
– Is the terminology used objective?
– Is the terminology used understandable in your environment?
– Does this procedure make sense in your environment?
– Can it be applied?
– Can you measure the successful use of the procedure?

9.5 ORGANIZATIONAL STANDARDS

ISO 9001 defines the work environment as a set of conditions under which daily work is performed. Organizational standards and processes define expectations and acceptable performance by providing clear definitions that guide daily activities, the collection and use of data, and coding standards for example. These standards are described so that they can be applied uniformly in all development projects, maintenance and operational activities. These standards also enable a developer or programmer to work the same way from one project to another. In this way, they do not have to learn a new programming guide for a particular project, increasing the productivity of the development team who has to develop the source code, review and test it. Table 9.3 shows some elements of a C++ program guide used at NASA.

These types of conventions facilitate the understanding of documents and can reduce maintenance time. A new employee will be more quickly productive once he

Table 9.3   Example of Coding Standard Elements [NAS 04]

Naming convention	Comments
Heading conventions	Language instruction recommended use (functions, variables, constant, pointers, operators, etc.)
Readme guidelines	Empty lines, spacing and identification conventions
Comments conventions (e.g., comments should be next to the instructions to be commented.)	Table and figure conventions
Exception handling conventions	Decision branching conventions
Use of uppercase and lowercase convention	

has learned the local conventions, rather than disturb everyone to ask them how work should be done. It is also possible to use tools to automatically check compliance with organizational conventions. In this case, code peer reviews will also be easier as the guidelines are published and should be used by all programmers. Reviewers can focus on detecting more significant problems and defects and will not spend a significant amount of time on coding conventions.

The organizational standards also define, among other things, software approved for project use, developers that are approved to have a copy of certain software, and describe how that software can be obtained. Similarly, the standards can define the hardware characteristics that development projects can use, such as number of platforms, type of computers, and approved peripherals that are fully supported by the acquisition department, operation, and maintenance of the organization.

9.6 GRAPHICAL REPRESENTATION OF PROCESSES AND PROCEDURES

The best and most effective documentation is a concise graphical representation of a document, a process, or a procedure. It may include, in addition to tasks to be executed, its inputs and outputs, input and output criteria, measures, roles, audit activities, tools, checklists, templates, and examples. There are many techniques available to help document processes and procedures. In this section, we present some of the graphical notations that are commonly used. A few of these are very simple to use and do not require any tool, while others are supported by specialized software. Before diving into the documentation representation on your intranet, the important question is: why should we graphically represent and document processes and procedures?

Here are some of the reasons to document processes and procedures graphically (adapted from [OLS 94], [SEI 10a]):

- allows for easier process improvement—it is difficult to improve a process when it is in the mind of developers;
- improves productivity—documented processes, which include the use of best practices, should help to improve overall organizational productivity;
- facilitates individual knowledge capture—if experienced people leave an organization, a part of their knowledge will have been documented and will remain;
- reduces defects—documented processes help reduce and prevent errors;
- saves time and money—users of documented processes can reduce development time and reduce rework;
- allows for measurement—when a process is documented, it is possible to measure its characteristics, such as the effort made to implement it and the size and number of defects it generates;
- facilitates training—documents describing the processes and procedures can be used to train new employees; and
- facilitates audits—preparing an audit will be faster when the organization can prepare the documentation requested by the auditor to demonstrate compliance to a standard for a project.

Here are some basic principles to guide the development of the process and procedure documentation (adapted from [OLS 94], [SEI 10a]):

- We must keep the organization's business objectives and the elements that make these objectives difficult to attain in mind—in some cases, a process documentation project becomes the ultimate goal. The processes must be used to achieve the business objectives;
- The repository of assets should contain the description of the development life cycle of the organization;
- We must use only the relevant information for each type of document (e.g., the information concerning training should only be in the training materials and policies should only contain information that does not change frequently). By locating the relevant information only where it belongs in the documentation reference model, developers will always know where to look for the information;
- It is necessary to manage changes and improvements—once defined, policies should not change frequently. A process will probably not need to change if only one step in one of its procedures should be amended;
- Use organizational standards and conventions to document processes and procedures (process descriptions and procedures should be consistent to allow for

their efficient use, and processes must answer the following questions: why, what, who, when, where, or be described according to the convention used, such as the Entry-Task-Verification-eXit (ETVX) graphical notation);

– Use drawings or mathematical formulas and complete the description of processes and procedures using text;

– When possible, use the templates as vehicles to communicate the information which is reflected in a procedure—we then avoid creating procedures that will have to be maintained;

– Add a checklist to the document template that will be used by the author to ensure that the document meets the organization's standards and can also be used for quality assurance during internal audits;

– Use a label to identify each document—developers want to find information quickly;

– Separate the process into cohesive chunks—it will be easier to understand a process that is broken down into logical parts, or sub-processes, instead of a complex process on several pages.

Processes should also be documented according to the level of expertise of its users. For example, you can use three levels of expertise: expert, intermediate, and beginner (adapted from [OLS 94], [SEI 10a]):

– for expert—the documentation includes just enough information for use by people who have implemented the process many times and just need a reminder, as these people use the process often;

– for intermediate—the intermediate level uses the expert level documentation, but the objective of the activity is added as well as some tips or applicable lessons learned;

– for beginner—the beginner level is aimed at people that are new to the process, that is, they have not yet used the process, and are in need of more detailed guidelines and training. The beginner level uses the documentation of the intermediate level to which is added training material. Beginners should feel free to use the training materials until they are familiar with the process.

"If you cannot describe what you are doing as a process, you do not know what you do."
W. Edwards Deming

9.6.1 Some Pitfalls to Avoid

Even if documentation provides advantages, we must pay attention to some documentation difficulties. Here are some pitfalls to avoid:

- large and complex documentation—when documentation becomes too big and complex, it becomes difficult to find the relevant information; additionally, the expert user does not want to rummage through a mass of documents to find simple information. Documentation that is too voluminous can quickly become shelf-ware.
- paper documentation—today's developers require documents that are available electronically to facilitate information retrieval;
- textual only documentation—we often say that a picture is worth a thousand words. This quote reflects well the situation with process and procedure documentation. In addition, a large number of users prefer graphical representations to textual ones. However, they will prefer a textual description when graphics cannot convey the same detailed information;
- process documentation developed by an external consultant to the organization—sometimes developers reject the process description made by someone who does not understand their organization. In some cases, a consultant cannot reflect the culture, the terminology, and the existing processes of the organization;
- documentation that mixes different types of documentation (e.g., policy and processes);
- complex notation—do not use a modeling notation that is too complex as users risk spending a lot of time trying to understand the notation that describes the process.

In a government agency, a contract was awarded to an external consultant to develop processes, procedures and other documents. This organization had generous budgets for the development of its processes. Since consultants were paid according to the amount of documents produced, the resulting repository of assets contained several thousand pages. After years of availability, it was found that several documents had never been updated and others had never been used. The organization had therefore paid considerable sums to develop unused and often unusable documents.

9.6.2 Process Mapping

Process mapping has been done for many years. Flow charts were the initial graphical representations and notation used for process documentation. These charts represented the following object types: decisions, input devices, output devices, and data storage media. Descriptive information as well as decisions where represented directly into the lozenges and the connection between objects was made using arrows. The ability to simply and explicitly describe the flow and the decisions made this notation very popular. Examples of graphical objects used by this notation are illustrated in Figure 9.4.

Although we still find diagrams using flow charts, is it less popular today. The next section introduces the ETVX notation.

9.6.3 ETVX Process Notation

The ETVX notation was used in the 1980s by IBM [RAD 85]. Given its simplicity, it has been adopted by many organizations such as NASA and the SEI. Figure 9.5 describes the concept by illustrating how the ETVX notation works.

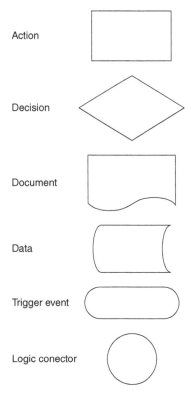

Action

Decision

Document

Data

Trigger event

Logic conector

Figure 9.4   Control flow graphical notation objects.

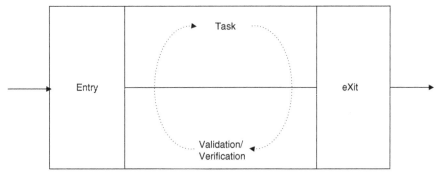

Figure 9.5 ETVX notation concept [RAD 85].

This notation includes the following components:

- Inputs
  - Assets (e.g., document) received from outside the process are required for its execution. It is often the case that not all inputs are mandatory when executing the process the first time. When other inputs become available, additional iterations may be performed.
- Tasks
  - Actions to be carried out to achieve the goal of the process and create the required outputs.
- Validation/Verification
  - A mechanism to ensure that the process tasks were carried out as required and that the deliverables meet the quality required.
- Outputs
  - Assets produced following the execution of the process and that will be used outside of this process.

Some organizations have added elements, such as those that follow, to the initial ETVX notation to better document processes (see Figure 9.6):

- Title (Process title)
- Entry criteria
  - Measurable conditions that must be satisfied before the process tasks can be executed.
- Exit criteria, also called a completion criteria
  - Measurable conditions that must be satisfied before you can exit this process.
- Measures taken during the execution of this process

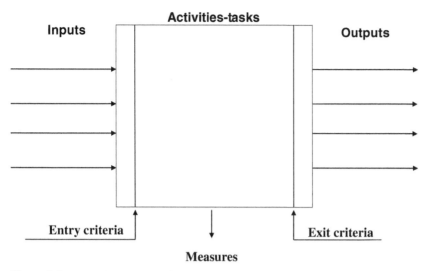

Figure 9.6   Illustration of the modified ETVX notation [LAP 97].

o The measurements are used to monitor the progress of activities, and can be used by other projects to better estimate the effort required to execute this process. Measures such as effort, for example, can be captured starting at the time the entry criterion is satisfied and stopped when the exit criterion is met. Measurement units should be specified (e.g., person-hour for effort).

– Tools
  o A list of tools/software required to execute the process.
– Risks
  o A section that lists the risks encountered when this process was used in the past.
– References
  o A section that lists reference documents that can be used to explain and support the process, its inputs, and its execution.

It is also possible to specify additional exit criteria as follows:

– Each artifact generated by the process is consistent with the organizational policies, standards and procedures;
– Each artifact produced by the process was verified, approved and stored in the organizational process repository.

Procedure: <Name of process/procedure>   Phase: <Name of phase where the
                                                procedure is used>

Process/procedure owner: <owner of this process/procedure>

Description: a brief description, background and purpose (value) of the process/procedure

Entry criteria: Exit criteria:
• <entry criteria>   • <exit criteria>

Inputs: Outputs:
• <work products as input>   • <work products as output>

Roles:
• <list of all the actors and their responsibilities>

Reference(s)
• <Document required to use this procedure>

Assets:
• <Tools; methodologies; references; guidelines; checklists; other procedures>

Tasks:
• <Itemized list of tasks (summarized) which need to be accomplished to satisfy this
  process/procedure (using an active verb and a noun)>

Measures:
• <Measures captured during execution of process/procedure>

Figure 9.7   Template of a textual process description using the ETVX notation.

Process ETVX diagram

Inputs	Entry criteria	Major tasks	Exit criteria	Outputs
Input #1	Entry criteria #1	1. *Task* 2. *Task*	Exit criteria #1	Output #1
AND	**AND**	3. *Task*	**AND**	**AND**
Input #2	Entry criteria #2		Exit criteria #2	Output #2
AND	**AND**		**OR**	**OR**
Input #3	Entry criteria #3		Exit criteria #3	Output #1
OR	**OR**		**OR**	**OR**
Input #4	Entry criteria #4	1. *Task* 2. *Task*	Exit criteria #4	Output #3
	AND			**AND**
	Entry criteria #5			Output #4
		Verification and validation		

Figure 9.8   Example of a NASA process graphically represented using the ETVX notation [NAS 04].

ISD ETVX* Diagram

Number: 580-TM-011-01
Effective Date: August 1, 2004
Expiration Date: August 1, 2009

Approved By: (signature)
Name: Joe Hennessy
Title: Chief, ISD

Responsible Office: 580/Information Systems Division (ISD)	**Asset Type:** Template
Title: ETVX Diagram	**PAL Number:** 3.5.2.2

*GUIDANCE: This template can be used to define an ETVX Diagram.
The above header must be changed for new ETVX diagram. Changes include:
Name (ISD ETVX* Diagram) should be "ISD [process title] ETVX* Diagram"
Also change Number, Dates, Asset Type, Title, and PAL Number*

Process ETVX Diagram

Inputs	Entry Criteria	Major Tasks	Exit Criteria	Outputs
Input #1	Entry Criteria #1	1. Task 2. Task	Exit Criteria #1	Output #1
AND	**AND**	3. Task	**AND**	**AND**
Input #2	Entry Criteria #2		Exit Criteria #2	Output #2
AND	**AND**		**OR**	**OR**
Input #3	Entry Criteria #3		Exit Criteria #3	Output #1
OR	**OR**		**OR**	**OR**
Input #4	Entry Criteria #4	1. Task 2. Task	Exit Criteria #4	Output #3
	AND			**AND**
	Entry Criteria #5			Output #4
		Verification and Validation		

Formatting Conventions	*GUIDANCE: Description of the formatting for the above template. Delete this section for final version.*	
	ETVX Labels	Arial or Helvetica 10-point bold font
	Text formatting	Arial or Helvetica 9-point font
	AND/OR Booleans	Arial or Helvetica 10-point bold font
	Dashed Line	Separates Usage Scenarios.

Figure 9.9   A NASA template used to explain the ETVX notation [NAS 04].

Complexity of ETVX Diagram	*GUIDANCE: In situations where the Boolean expressions for Inputs and Entry Criteria or for Outputs and Exit Criteria are too complex to be easily and clearly labeled, make a separate ETVX Diagram for each Usage Scenario.* *Delete this section for final version.*

Definitions	*GUIDANCE: Description of the parts of the ETVX diagram.* *Delete this section for final version.*	
	Usage Scenario	A set of inputs and entry criteria that define unique conditions for execution of a process.
	Entry Criteria	Conditions that must be satisfied in order to start the process.
	Exit Criteria	Conditions that must be satisfied in order to exit the process.
	Inputs	Items received from outside the process that are needed for performance of the process.
	Outputs	Items produced as a product of the process for use outside of this process.
	Tasks	Activities, which taken together, will perform the work required by the process.
	Validation	Steps to determine whether the product(s) fulfills its specific intended use.
	Verification	Steps to determine whether the product(s) of a task fulfills the requirements or conditions imposed on them in the previous tasks.

Figure 9.9   (*Continued*)

It is also possible to use textual notation to describe the ETVX notation. Figure 9.7 provides this template. This format, as well as the one presented in the above figure, could be used by more mature developers.

Professor Laporte has used the ETVX notation for software process documentation and project management (PM) at Rheinmetal Corporation and Bombardier Transport [LAP 97]. He added title and activities to the notation. A rule was added to ensure that both of these additions include an action verb and a noun (e.g., "estimate the size of the product"). In addition, each activity needed to be assigned a unique number to facilitate its reference in sequence. NASA also uses the ETVX notation to document its processes and procedures. Figure 9.8 shows a NASA process that is described using the ETVX notation.

NASA has also documented how to complete the ETVX table shown in Figure 9.8. Figure 9.9 presents the result.

Figure 9.10 shows an ETVX representation of a configuration management process. It is not mandatory to include both entry and exit criteria.

GUIDANCE for Development History (Below): Description of major changes to the ETVX diagram under development and the author performing the change. Delete this section for final version.

Development History	Version	Date	Description of Development Changes
	0.1	Feb 16, 2004	Created initial version of the User's View Template PGArnold
	0.2	Mar 16, 2004	Added alternative formats for data tables. Dropped appendix, which was not necessary. Added standards for asset numbering & formatting in a new appendix. Added changes agreed to at March 16 ISD team meeting. PGArnold
	0.3	Mar 24, 2004	Incorporated comments from reviewers at March 23 ISD team meeting. Dropped table format for most sections. Added bulleted list format for most sections. PGArnold
	0.4	April 2, 2004	Incorporated comments from reviewers at March 30 ISD team meeting. Added in QMS Records and table for Training. PGArnold
	0.5	April 14, 2004	Incorporated comments from reviewers at April 2 ISD team meeting. This included addition of mapping tables to various entry scenarios, addition of more tables for sections recommended, and rewording of some sections to delete mandatory language. PGArnold
	0.6	April 19, 2004	Incorporated comments from reviewers. PGArnold
	0.7	April 21, 2004	Major rewrite to add comments from reviews, including dropping some tables, reformatting of many sections, & addition of comments for section headers. PGArnold
	0.8	June 21, 2004	Changes to address CCB and Sally Godfrey's comments. PGArnold
	0.9	June 23, 2004	Final changes to address ISD process team review PGArnold
	0.91	July 2, 2004	Minor changes to improve formatting and problems associated with grayed backgrounds. PGArnold
	0.92	July 30, 2004	Minor changes. PGArnold

GUIDANCE for Change History (Below): Description of improvements to the approved ETVX diagram, the Change Request responsible, and the author performing the change.

Change History	Version	Date	Description of Improvements
	1.0	TBD	Initial approved version by CCB

Figure 9.9   *(Continued)*

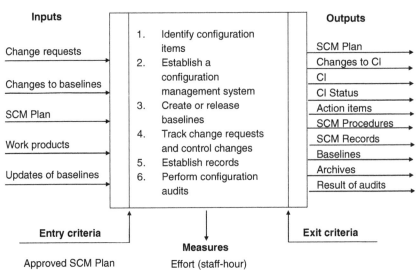

Figure 9.10   A configuration management process graphically represented using the ETVX notation ETVX [LAP 97].

Bombardier Transport Software Development Process [LAP 12]

The Bombardier software engineering process (BSEP) describes a disciplined approach to the assignment of activities and responsibilities in a software development team. Its purpose is to ensure the production of high quality software that meets the user requirements within the allocated budget and time.

The BSEP was developed using in-house knowledge (e.g., development process, history of approved practices) based on the Software Engineering Institute CMM model, international standards such as ISO 12207 and ISO 9001, the Guide to the Project Management Body of Knowledge (PMBOK® Guide) and IBM's RUP framework.

An overview of the BSEP process is shown in Figure 9.11. Two dimensions are represented: first, the dynamic aspect of the process is expressed in terms of phases, iterations, milestones and baselines; second, the static aspect of the process is expressed in terms of ISO 12207 processes and activities.

Three key process elements are represented by the roles, activities and artefacts:

– Roles: A role defines the behavior and responsibilities of a person or group of people working as a team, in this context, in a software engineering organization. The role and associated responsibilities define how the work will be performed as well as their author. A project member can play different roles during the project.

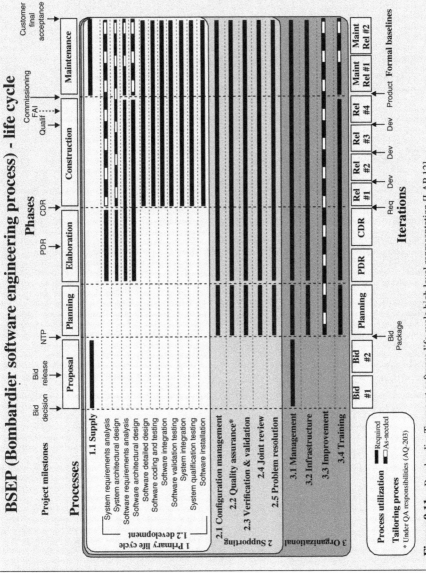

Figure 9.11 Bombardier Transport software life cycle high level representation [LAP 12].

– Activities: Roles have activities that define the work they are to perform. Activities are processes or functions to perform, both intellectually and physically, in order to achieve a given objective. An activity is a unit of work that a person, having the responsibility described by the role, may be asked to perform. An activity also refers to any work done by managers and technical staff to complete the project activities. An activity is used as a planning and progress monitoring element.

– Artefacts: Activities have artefacts as input or output. An artefact is a result of the process execution (e.g., a work product). Individuals that play a role use artefacts to carry out activities, and to produce some during the execution of activities. Artefacts can be internal or external to the project and take various forms:
  o a model, such as a use case model or a design model;
  o a document, such as a project plan, a requirements document or SRS (Software Requirements Specifications) document.
  o code.

9.6.4 IDEF Notation

During the 1970s, an integrated computer aided manufacturing program (ICAM) for the US aviation industry sought to increase productivity by the increased use of information technology. The ICAM program developed a series of graphical modeling notations known as ICAM Definitions (IDEF). The IDEF0 notation is derived from an existing graphical modeling notation known as the structured analysis and design technique (SADT). IDEF0 was developed by the original authors of SADT [IEE 98].

Figure 9.12 describes the IDEF0 notation. Arrows entering the left side of the box are inputs. Inputs are transformed by the function performed in the rectangle (which represents an activity to be performed) to produce outputs. The arrows entering the top of the box are controls. Controls specify the conditions required for the function to produce correct results. The arrows, located in the lower part of the rectangle and pointing up, are mechanisms that identify means that support the delivery of the function (e.g., tools). The arrows pointing down enable information sharing between

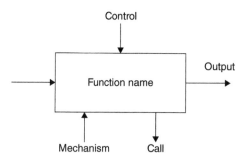

Figure 9.12   Notation IDEF0 [IEE 98].

models or between parts of the same model. Rectangles can be interconnected to form a process.

In the following example, we illustrate the description of a process at three levels: at the process level, at the step level and at the ETVX level.

Engineering and the Integration of the Software Engineering Process, Systems Engineering, and Project Management [LAP 97]

The project tracking and planning process is described here to illustrate the activities to be done. At the highest level of detail, we can see three processes (see Figure 9.13): a planning process during the proposal phase, a project planning process following the award of a contract and the tracking process for the project.

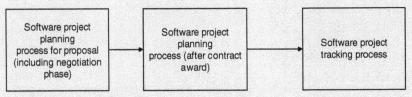

Figure 9.13   Three processes of the project planning phase.

The following figure shows the seven steps of the second phase of software project planning (SPP).

The following figure shows an ETVX representation of step SPP-120—Prepare estimates and project schedule.

The proposal phase considers the original vision of a potential product and transforms it into a business case, where if it is to be subcontracted development, the requirements of the project are analyzed: first, its size, cost and timeline is estimated, then a risk analysis is performed. In both cases, the main result of this phase amounts to determining whether the project is acceptable or not. Given that during the contract negotiation phase, it is possible that certain requirements (e.g., execution schedule, software requirements) have been modified, it is necessary that the planning phase, after contract award, reviews the plans submitted during the bidding phase. During the third phase, we collect and analyze project data to further adjust the initial plans.

The second level of planning and monitoring activity details occurs during the proposal phase and is shown in Figure 9.14. As shown, each step is numbered (e.g., SPP-120). In addition, each step name uses a verb and a noun. The steps can be connected together as required by the project. It is the responsibility of the project team to create a relevant process. Although the steps are represented as a sequence, feedback loops to earlier stages is permitted. Feedback loops are not shown so as not to clutter the example.

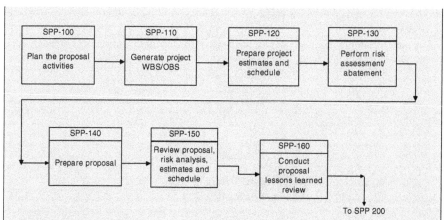

Figure 9.14   Planning phase.

Figure 9.15 shows the third level of detail. This figure shows an ETVX representation of step SPP-120. As the ETVX may not provide all the necessary information to execute a particular step, it can be supplemented by a text in the developer's manual (e.g., an estimation procedure).

Development of a Systems Engineering Process

The Generic Systems Engineering Process (GSEP) document describes, using the IDEF notation, the management, technical operations, and the documents produced by each activity. The main management activities (see Figure 9.16) are: understanding the

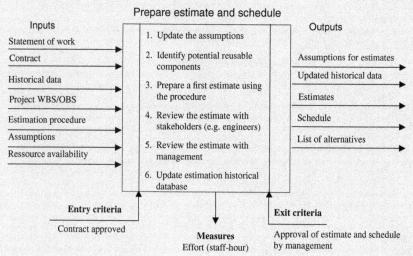

Figure 9.15   ETVX representation of step 120.

context, risk analysis, planning a development increment, tracking an increment, and developing the system. The main technical activities are: analyze needs, define requirements, define the functional architecture, synthesize allocated architecture, evaluate alternatives, validate and verify the solution, and manage the repository. Like the software process, each major activity is broken down into a number of smaller activities that are described individually using the ETVX notation.

Integration of the Software Engineering Process and Systems Engineering Process

We used a document entitled "Integrated Systems and Software Engineering Process (ISSEP)" as an integration structure. The ISSEP describes activities at three different levels: the system level, the configuration item level (CI) and the component level. The activities at the system level are: managing system development, design and verify the system, integrate and test the system. At the CI level, activities are: manage the development of the CI, design and verify the CI, develop components, and integrate and test the CI.

Configuration items can be decomposed into one or more components. Component level activities are: build components, develop test cases units, and perform unit testing and analysis. It is at the component level that software is coded and that the product is manufactured. Figure 9.16 shows the links between system engineering processes, software engineering processes, subsystems engineering processes, and the link with the manufacturing process.

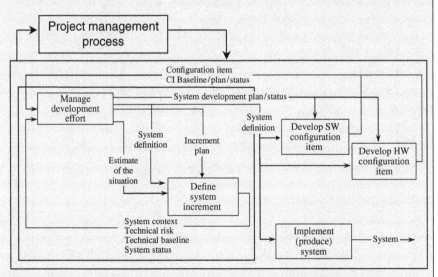

Figure 9.16   Integration of software engineering to the system engineering process [LAP 97].

9.6.5 BPMN Notation

The Business Process Modeling Notation (BPMN) is a standard of the Object Management Group (OMG) [OMG 11]. This notation was originally developed by the Business Process Management Initiative, and in 2009, the second version of BPMN was published. BPMN defines a set of graphical objects used in the description of the process. This notation is based on four families of items:

- flow objects
- connection objects
- activity corridors
- artifacts

Each family contains objects in object categories. The following sections introduce you to the concepts of objects used in the BPMN notation.

9.6.5.1 Event Objects

Generally, a BPMN graph contains the following three types of objects: (a) events; (b) activities; and (c) connectors.

9.6.5.1.1 Events   An event graphically represents all events that can take place, and that are likely to trigger one or more activities. The BPMN notation authors have designed this object in order to provide an accurate description of the process. Experts will notice that each new version of BPMN enriches the event library of the notation. Generally, there are three types of popular events as illustrated in Table 9.4.

There are many more events available. Figure 9.17 illustrates the types of events used by this classification.

9.6.5.1.2 Activities   As its name implies, an activity means work concerning an organization's business transactions. BPMN distinguishes between two types of activities:

- Task: an indivisible action.
- Sub-process: an action that includes or regroups a number of tasks.

Table 9.5 shows the symbols used to illustrate these two types of activities.

Table 9.4   BPMN Event Types

Event type	Symbol	Description
Start	◯	Event that starts a process
Intermediate	◎	Event that occurs during a process
End	⬤	Event that ends a process

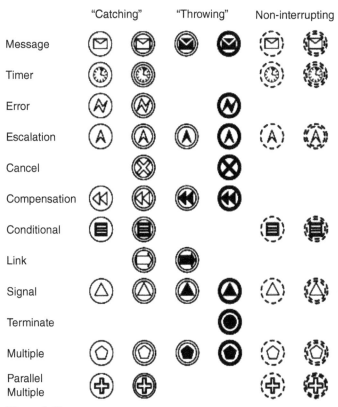

"Catching"   "Throwing"   Non-interrupting

Message

Timer

Error

Escalation

Cancel

Compensation

Conditional

Link

Signal

Terminate

Multiple

Parallel
Multiple

Figure 9.17 BPMN list of event types [DEC 08].

9.6.5.1.3 Connectors The last of the three objects is the connector. Also known as the gateway, this type of object illustrates the convergence of decision points as well as the divergence of a process activity. Connectors are represented by an empty diamond in the case of a conventional decision. Otherwise, an additional notation is used to indicate more complex cases, such as the disconnected

Table 9.5 BPMN Activity Types

Activity Type	Symbol
Task	
Sub-process	

Table 9.6 BPMN Connexion Objects

Connexion objects	Symbol
Sequence flow	———————————▶
Association	----------------------------▷
Message flow	o--------------▷

node/union node as well as the fusion node. The next section describes the connection objects.

9.6.5.2 Connection Objects

These objects serve as connectors between objects presented in the previous section. Table 9.6 shows the difference between the three types of connection objects used by the BPMN notation.

The concepts for structuring business process activities are discussed next.

9.6.5.3 Swim Lanes

BPMN notation uses the concept of the swim lanes of activities to organize and structure the processes. Table 9.7 provides an explanation of the two types of swim lanes proposed by the notation.

The following section introduces the way in which the BPMN notation allows for additional information to be included in the process model.

Table 9.7 The Types of BPMN Swim Lanes

Type of lane	Description	Graphical representation
Swim lane	A lane is used to represent the activities of a specific role	
Pool	A pool is usually used to represent a process in an organization. A lane, on the other hand, represents an activity of a department within that organization. By using pool and lane, you can identify how a process is done and which department performs that activity.	

Table 9.8   BPMN Artifacts

Artifact	Description	Symbol
Annotation	Object used for comments	Text Annotation Allows a Modeler to provide additional Information
Group	Object used to regroup tasks	
Data object	Object used to represent data needed (and produced) during the execution of a task	Name [State]

9.6.5.4 BPMN Artifact

Artifacts are additional objects that provide more detail to ensure a more complete understanding of a process. Table 9.8 shows the three types of BPMN artifacts, their descriptions and associated symbols.

9.6.5.5 BPMN Modeling Levels

The modeling of typical business processes is achieved using several logical levels, depending on the methodology used and the target clientele. BPMN allows for the production of high-level conceptual diagrams as well as very detailed levels.

Using BPMN notation can be done in three levels [SIL 09]:

1) Descriptive level: The goal of this first modeling level is to describe the business process at a general and conceptual level. It is meant to represent the overall flow of the process (sometimes called a meta process).

2) Analytical level: Analysis of the descriptive level diagrams does not allow for the evaluation of its quality or performance. This requires the production of more detailed diagrams that describe all the possible branching and interaction scenarios of a process. This is, indeed, the reason for the analytical level. This second level of modeling, mainly used by architects and business analysts, aims to describe the details of a process in an accurate manner.

3) Executable level: This level is dedicated specifically to software and system developers. It is used to produce executable process models (i.e., that reflect the business processes). When a process is represented at this level, it can be executed by many popular BPM commercial solutions/tools.

These three levels of modeling are aligned with the concept of model-driven architecture (MDA). MDA software implementation (SI) of a process can be used at the beginning of a project to produce a model independent of the implementation (CIM), which can be transformed into a platform-independent model (PIM model), that in turn can be transformed into a platform specific model (PSM model). Note that this classification is not part of the specification of the BPMN notation. Figure 9.18 shows an example of BPMN representation.

In conclusion, we have seen that BPMN is a notation that can be used for representing business processes. This description can be done at different levels of detail, based on a rich library of graphical objects.

Davies [DAV 06] conducted a study to determine the current most popular graphical representation models in industry. He surveyed the market to see what tools and notations are actually used by practitioners. He questioned 312 members of the Australian Computer Society (ACS). The survey listed 24 different modeling tools that were pre-selected by studying their popularity in the literature. Respondents were practitioners in the field of IT with only 15% declaring themselves as end-users or management. Davies research reveals that 61% of respondents use a version of Microsoft Visio as a modeling tool with no specific notation in mind.

The Design of a Process Diagram using Swim Lanes

– identify the roles of participants in the process;

– name roles using a generic approach, for example, do not give the name of a person playing that role but rather assign a name that can be used from one project to another without having to modify it;

– draw one lane for each identified role. These lanes can be traced horizontally or vertically. Horizontal lanes are more in line with the notion of steps that occur over time on a horizontal axis;

– add, at the bottom of the diagram, a lane to describe the deliverables associated with tasks and roles;

– identify the task and the role that initiates the process;

– identify other tasks and place them in the appropriate lane. It is possible that a task is carried out jointly by two or more roles. If it is, draw a box around these tasks. This box will overlap two or more lanes. If a task requires the participation of two non-adjacent roles, it is then possible to draw a box and for each role add a dotted box to show that these two roles are involved in this task;

– complete the diagram by adding a textual description of the following information:

  o a pointer to detailed procedures;

  o additional information for roles;

  o a pointer to the templates, checklists and tools to use.

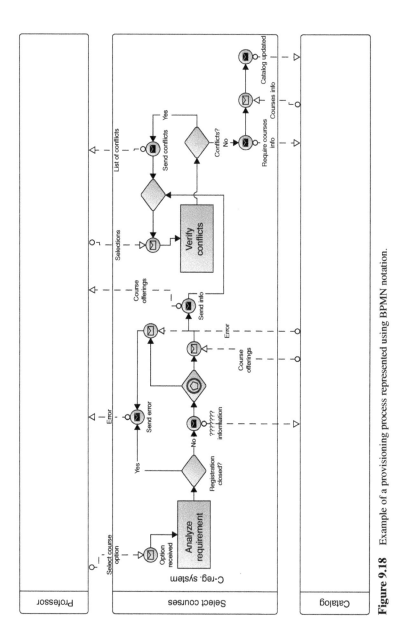

Figure 9.18 Example of a provisioning process represented using BPMN notation.

9.7 PROCESS NOTATION OF ISO/IEC 29110

As mentioned in Chapter 4, ISO/IEC 29110 describes processes, objectives, activities and tasks for Very Small Entities (VSEs) developing software or systems. A VSE is an entity (enterprise, organization, department or project) having up to 25 people. ISO/IEC 29110 is not intended to preclude the use of different life cycles such as: waterfall, iterative, incremental, evolutionary, or agile. The notation used in the diagrams of ISO/IEC 29110 does not imply the use of any specific process life cycle.

The following elements are used to describe the processes, activities, tasks, roles, and products of ISO/IEC 29110 (adapted from [ISO 11e]):

- Name of a process
  - A process identifier, followed by its abbreviation in brackets "()". For example, for the PM process, the notation is: PM process.
- Purpose of a process
  - General goals and results expected of the effective implementation of the process. The implementation of the process should provide tangible benefits to the stakeholders.
  - The purpose is identified by the abbreviation of the process name. For example, for the PM process:
    - PM purpose: The purpose of the PM process is to establish and carry out in a systematic way the tasks of the SI project, which complies with the project's objectives in regards to quality, time, and costs.
- Objectives
  - SG to ensure the accomplishment of the purpose of the process. The objectives are identified by the abbreviation of the name of the process, followed by the letter "O" and a consecutive number. For example, PM.O1.
  - Each objective is followed by the square box which includes a list of the chosen processes for the Basic profile from ISO/IEC/IEEE 12207 and its outcomes related to the objective. For example, for Project Objective 7 (PM.O7), the management and engineering guide describes the objective and a note is added followed by a square box:
    - PM.O7: SQA is performed to provide assurance that work products and processes comply with the Project Plan and Requirements Specification.
    - Note: The implementation of the SQA process is through the performance of the verifications, validations, and review tasks completed in the PM and SI processes.

7.2.3 Software Quality Assurance Process

– a strategy for conducting quality assurance is developed;

– evidence of software quality assurance is produced and maintained;

– problems and/or non-conformance with requirements are identified and recorded; and

– adherence of products, processes, and activities to the applicable standards, procedures, and requirements are verified.

ISO/IEC 12207:2008, 7.2.3

Figure 9.19 illustrates the graphical representation of the PM process of ISO/IEC 29110 as follows [ISO 11e]:

– the large round-edged rectangles indicate process or activities.

– the smaller square-edged rectangles indicate the products.

– the directional or bidirectional thick arrows indicate the major flow of information between processes or activities.

– the thin directional or bidirectional arrows indicate the input or output products.

Input work products

– Input work products are products required to perform the process and its corresponding source, which can be another process or an external entity to the project, such as the customer.

– Input work products are identified by the abbreviation of the process name and showed as a two column table of product names and sources. As an example, for the PM process, we have the following table:

Name	Source
Statement of work	Customer
Software configuration	Software Implementation
Change request	Customer Software Implementation

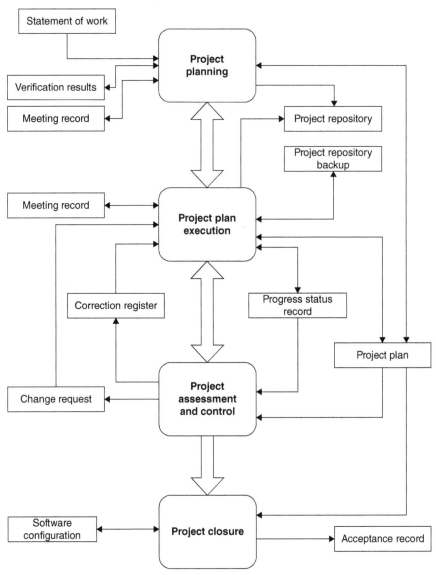

Figure 9.19 PM process diagram of ISO/IEC 29110 [ISO 11e].
Source: Standards Council of Canada.

The statement of work is defined as follows:

Name	Description	Source
Statement of work	Description of work to be done related to Software development. It may include: – Product description o Purpose o General customer requirements – Scope description of what is included and what is not – Objectives of the project – Deliverables list of products to be delivered to Customer The applicable status is: reviewed.	Customer

Output work products

– Work products generated by the process and its corresponding destination, which can be another process or an external entity to the project, such as the customer or organizational management.

– Output work products are identified by the abbreviation of the process name and showed as a two column table of product names and destinations. As an example, for the PM process we have the following table:

Name	Destination
Project plan	Software Implementation
Acceptance record	Organizational Management
Project repository	Software Implementation
Meeting record	Customer
Software configuration	Customer

Internal work products

– Work products generated and consumed by the process and not reviewed or approved by the customer.

– Input work products are identified by the abbreviation of the process name and showed as a one column table of the work product names.

– All work product names are printed in cursive and start with capital letters. Some products have one or more statuses attached to the product name surrounded by square brackets "[]" and are separated by a comma ",".

– The work product status may change during the process execution.

– The source of a work product can be another process or an external entity to the project, such as the Customer. As an example, for the PM process we have the following table:

Name
Change request
Correction register
Meeting record
Verification results
Progress status record
Project repository backup

Roles involved

– names and abbreviations of the functions to be performed by project team members;

– several roles may be played by a single person and one role may be assumed by several persons;

– roles are assigned to project participants based on the characteristics of the project;

– roles are defined in a table as follows.

Role	Abbreviation	Competency
Analyst	AN	Knowledge and experience eliciting, specifying, and analyzing the requirements. Knowledge in designing user interfaces and ergonomic criteria. Knowledge of the revision techniques. Knowledge of the editing techniques. Experience with software development and maintenance.

Activity

– An activity is a set of cohesive tasks.

– A task is a requirement, recommendation, or permissible action, intended to contribute to the achievement of one or more objectives of a process.

– A process activity is the first level of process workflow decomposition, the second of which is a task.

– Activities are identified by the process name abbreviation followed by consecutive numbers and the activity name. As an example, the activities of the PM process of the Basic profile are:
  o PM.1 Project planning
  o PM.2 Project plan execution
  o PM.3 Project assessment and control
  o PM.4 Project closure

Activity description

– Each activity description is identified by the activity name and the list of related objectives surrounded by brackets "()". For example, PM.1 project planning (PM.O1, PM.O5, PM.O6, PM.O7) means that the activity PM.1 project planning contributes to the achievement of the listed objectives: PM.O1, PM.O5, PM.O6, and PM.O7.

– The activity description begins with the task summary and is followed by the task descriptions table. As an example, the project planning activity of the PM process is illustrated as follows:

PM.1 Project Planning (PM.O1, PM.O5, PM.O6, PM.O7)

The project planning activity documents the planning details needed to manage the project. The activity provides:

– reviewed statement of work and the tasks needed to provide the contract deliverables to satisfy customer requirements;

– project life cycle, including task dependencies and duration;

– project quality assurance strategy through verification and validation of work; products/deliverables, customer, and team reviews;

– team and customer roles and responsibilities;

– project resources and training needs;

– estimates of effort, cost and schedule;

– identified project risks;

– project version control and baseline strategy;

– project repository to store, handle, and deliver controlled product and document versions and baselines.

Task description

– Tasks are described in tables of four columns:
  o Role—the abbreviation of roles involved in the task execution, for example, Project Manager (PM), Team Leader (TL), and Customer (CUS),

Table 9.9   Two Tasks of the Planning Activity [ISO 11e]

Role	Task list	Input work products	Output work products
PM TL	PM.1.1 Review the Statement of Work	Statement of Work	Statement of Work *[reviewed]*
PM CUS	PM.1.2 Define the Delivery Instructions of each one of the deliverables specified in the Statement of Work with the customer	Statement of Work *[reviewed]*	Delivery Instructions

- o Task—description of the task to be performed. Each task is identified by activity ID and consecutive number, for example, PM1.1,
- o Input work products—work products needed to execute the task, for example, statement of work,
- o Output work products—work products created or modified by the execution of the task.
- – The task description does not impose any technique or method to complete it.
- – The selection of the techniques or methods is left to the VSE or project team.

Table 9.9 describes the two tasks of the planning activity.

To further help VSEs that develop systems or software, a four-stage roadmap has been published by ISO as illustrated in Figure 9.20. The entry profile is for VSEs working on small projects (e.g., a project of up to six person-months) and for start-ups; the basic profile is for VSEs developing only one project at a time with a single team; the intermediate profile is for VSEs involved in the simultaneous development of more than one project with more than one team; the advanced profile is for VSEs wishing to significantly improve the management of their business and their competitiveness.

Readers are invited to download ISO/IEC 29110 technical reports, such as the management and engineering guides. ISO/IEC 29110 technical reports are

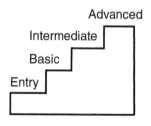

Figure 9.20   Four-stage roadmap of ISO/IEC 29110.

available at no cost from ISO. [1] Many of them have also been translated into Spanish, Portuguese, French, and Japanese. ISO/IEC 29110 has also been adopted by some countries, such as Brazil, Japan, Mexico, Uruguay and Peru, as a national standard.

9.8 CASE STUDY

Process description and process improvement are more than just technical activities. The management of change (i.e., organizational culture change) is, most of the time, the real challenge of a major process improvement initiative. An improvement process conducted over several years in a company in the defense sector has identified the following lessons learned [LAP 98]:

Lesson 1: Set realistic expectations for senior management

Appropriate expectations must be defined before starting a process improvement program. A risk, especially for a low process maturity level organization, is to communicate to senior management the idea that the process improvement initiative will be simple, quick, and inexpensive. This should be avoided, as it creates an unrealistic expectation. Here is a typical scenario: senior management learns the advantages that a higher process maturity level might offer for the competitiveness of the organization. A project manager, or an external consultant, promotes that these goals are easily achievable. Upper management mandates a team to achieve certification in a very short time frame. Subsequently, senior management discovers with surprise that process improvement goals will require much more time, more resources and significant changes to existing practices than they expected.

Lesson 2: Secure management commitment

For a low process maturity level organization, most of the findings associated with a process assessment will target PM shortcomings (look back at the CMMI process areas for level 2 displayed in Figure 4.9). It is therefore necessary to promote an attitude where management is willing to invest in the PM process rather than blame its personnel for current shortcomings. This is one other reason why it is necessary to frequently inform management so that they can be understanding and offer their commitment when such findings are made public in the organization.

"To reach a process maturity level 2, of CMMI, basically means getting rid of productivity inhibitors caused by a weak management processes, thus enabling skilled software engineers to be fully exploited."

Watts S. Humphrey

---

[1] http://standards.iso.org/ittf/PubliclyAvailableStandards/index.html

Lesson 3: Establish an improvement working group before a formal evaluation

It would be better if a small process improvement group be active a few weeks before a formal process assessment is begun by an external consultant. A process improvement group can take their time to familiarize themselves with the methods and tools associated with process improvement. Ideally, there should be a full-time person in this group, while other members could be involved on a part time basis. In addition to good technical skills, the members of this group must be selected according to their recognized expertise and their known enthusiasm for the improvement of project processes.

Lesson 4: Start improvement activities shortly after the first assessment

Regarding the development of an action plan, the organization should capitalize on the momentum created by the process assessment. The organization does not have to expect that a fully completed plan of action be finalized before starting process improvement activities. Some improvement activities can begin immediately after the first assessment. Executing small improvements is an important motivational factor for all technical personnel and management.

Lesson 5: Collect data to document improvements

Before and during the process assessment, it is recommended that quantitative and qualitative data be collected. These data will be used later to measure progress. One can obtain data such as budgets, schedules, quality, and the level of customer satisfaction. Given that management is investing in improvement, it will be important to be able to demonstrate that some gains are made.

Lesson 6: Train all concerned personnel in regards to the process, methods and tools

Once the processes are defined, it is essential to train personnel. Otherwise, it is likely that the process will end up unused (i.e., on the shelves). It is misleading to think that developers will learn new processes on their own in addition to their current project workload. Training sessions also send a strong message that the organization is moving forward and that its developers should use these processes. During training sessions, it is necessary to communicate to personnel that with their first use of a new process, mistakes and questions are likely to emerge. This message may help reduce the stress level of first time users. A contact person should be assigned to help (e.g., a hotline) and coach personnel when faced with obstacles in the execution of new processes.

Lesson 7: Managing the human dimension

We often underestimate the importance of the human dimension during a process improvement initiative. Those responsible for technological change are often extremely talented technically, but they are rarely well equipped in terms of change management approaches and techniques. The reason is simple: most of their training was focused on technology and not on the soft skills. However, the major difficulty of an improvement program is often managing the human dimension.

"BOEING noted early that technology transfer is almost as difficult as technology development."

ADA Strategy, June 1994

"The first lesson to use, from socio-technological research, is that social and technical changes must be managed jointly.

If we want that process innovation succeeds, then the human side of change cannot be left to itself. The organization and its human resources are more important than technology issues in order to obtain a behavioral change that is reflected in the process."

[DAV 93]

During the preparation of the technical part of an improvement action plan, change management elements must be planned as well. This implies, among other things, knowledge of: (1) the history of the organization with respect to similar efforts, successful or not; (2) the culture of the organization; (3) motivational factors, positive and negative, available to facilitate change; (4) the degree of perceived urgency communicated by management to conduct changes (look back at the case study presented in section 6.10).

Lesson 8: Process improvement requires additional people skills

As mentioned above, an organization that really wants to make substantial productivity and quality gains must manage cultural change. A culture change requires a special skill set. The profile of the process improvement coordinator and facilitator is a person with skills in sociology and psychology. This often requires both management and personnel to change/adapt their behavior.

With the formalization of its processes, management must change its authoritative management style to a more participatory style. For example, if the organization really wants to improve its processes, a major source of improvement ideas must come from those who work, on a daily basis, with these processes. This implies that management should encourage and listen to new ideas. This also implies that the decision process may have to change from an autocratic style of "Do as you are told" to a participatory style of "Let's talk about this idea." Similarly, the behavior of some employees who currently behave like "heroes" that can solve any problem should change to act as team members who can generate ideas, listen to the ideas of others and follow the process.

Professor Laporte was consulting at an organization in the public sector. During an informal discussion with a project manager, he stated that "in this organization you cannot make a mistake." A few weeks later, a member of the executive team, surrounded by six of his directors, approved a very important process improvement project at a meeting and assigned the same project manager. The meeting was to continue after lunch.

When the meeting reconvened, the project manager was no longer present. His secretary informs the meeting participants that he will be on sick leave for several months due to burn-out.

Process improvement is not an exact science, but rather an experimental approach that has its share of difficulties and mistakes. The project manager for process improvement, aware of the vice-president's low tolerance level for error, decided to leave the project to preserve his health.

In addition, in the first months after the introduction of a new process, a new practice or a new tool, management and employees must recognize that mistakes and questions will be inevitable. Unless a clear signal is sent by management that this is acceptable and a safety net is set-up to recognize this situation, employees will "cover up" their mistakes. The result is that not only does the organization not learn from these mistakes, but other employees will make the same mistakes again. For example, the main purpose of an inspection process is to detect and correct errors and defects as soon as possible in the life cycle of the project. Management must accept that to increase the error detection rate, the results of individual inspections have to remain known only to the author and the inspection process coordinator; only the *average* obtained from numerous inspections will be made public. When this rule is accepted by management, employees feel safe to identify errors and report them. Another advantage is that those who participate in an inspection will learn to avoid these mistakes in their own work.

Facilitating behavior change requires skills that are not taught in technical courses. It is highly recommended that those responsible for facilitating change receive appropriate training.

Professor Laporte recommends two books that can facilitate change management: the first book entitled "*Flawless Consulting: A Guide to Getting Your Expertise Used*" [BLO 11], advises anyone acting as an internal consultant; the second book, entitled "*Managing Transitions – Making the Most of Change*" [BRI 16], provides the steps required to develop and implement a change management plan.

Lesson 9: Choose pilot projects carefully

It is very important to carefully select pilot projects and pilot participants, as these projects, if completed successfully, will promote the adoption of new practices across the organization. Users of a new process will make mistakes, so it is imperative to train participants and provide them with a safety net. If participants see that their errors are used to learn and make improvements to the process instead of to blame them, the level of anxiety will be reduced and the participants will bring more suggestions.

Managing the human dimension is something that will not only facilitate the adoption of changes, but also create an environment where changes can be introduced at a faster rate.

When using a new process, a new practice, or a new tool, management and employees must recognize that mistakes are inevitable. When deploying the new process for documentation, the first engineer to use this process was reprimanded by his manager because he had made some mistakes.

Other engineers in the department, whose offices were nearby, heard the manager reprimand the engineer. The engineer was almost in tears. During the meal that followed this event, the other engineers in the department looked for excuses not to be the next one to use this new process!

Lesson 10: Conduct regular process audits

Process audits should be conducted on a regular basis for two main reasons: first, to ensure that practitioners use the process, and second, to discover errors, omissions or misunderstandings during their execution.

Lesson 11: Link process improvement activities with the organization's business objectives

It was observed that the improvement of software engineering processes really gains momentum when management realizes that the real benefit of process improvement is the improvement of the quality of products (e.g., it reduces time-to-market and costs). Consequently, it improves the ability of the organization to better compete.

A multi-year process improvement plan is a very important tool to illustrate the links between the objectives of the organization, the requirements of the organization's projects and process improvements. Essentially, this plan shows that process engineering is not a static exercise, but a central infrastructure component for the success of the organization's projects. Finally, a multi-year plan also shows practitioners that a long-term management commitment to process improvement activities is present.

Lesson 12: Adopt a common vocabulary

To succeed in any project, the use of common terminology is a fundamental requirement. In developing the software processes, it was observed that different participants had different meanings for the same word and the meaning of some words was unknown to some people. For example, prototyping had a meaning in systems engineering that was very different from that of software engineering. A terminology glossary was developed by a member of the process improvement project. The activity required collecting the vocabulary used by the participants and proposing definitions in order to gradually build a common glossary for all processes.

You can use the ISO/IEC/IEEE 24765 as a dictionary for software development processes. This will avoid lengthy and often unnecessary discussions between specialists. See the glossary at the following site http://pascal.computer.org/sev_display/index.action

9.9 PERSONAL IMPROVEMENT PROCESS

Although the SEI maturity models, such as the CMMI models, provide organizations with a proven framework for system and software process improvement, they describe "what" organizations should do and not "how" they should do it. But, software engineers also want to know how to do a process.

Since software development is a very complex process, it cannot be reduced to a cookbook of procedures. Watts S. Humphrey, the advocate of the maturity models, completed research in the early 1990s to show how process improvement principles can be applied to the daily work of software engineers. From this research, he concluded that the process management principles of Deming and Juran are also applicable to the individual processes used by software engineers, given that they apply to other areas of technology.

He used the fundamental principles of processes to show how engineers can define, measure and improve their own personal processes. Since each engineer is different, he must adopt his own practices to produce efficient and effective software. The Personal Software Process (PSP), developed by Humphrey, is a disciplined and structured approach to software development that enables all engineers to significantly increase the quality of their software products while increasing their productivity and respecting deadlines.

Personal Process

Set of steps or activities that guide individuals in doing their personal work. It is usually based on personal experience and may be developed entirely from scratch or may be based on another established process and modified according to personal experience. A personal process provides individuals with a framework for improving their work and for consistently doing high-quality work.

[SEI 09]

The PSP method is based on planning principles and the following quality principles [HUM 00]:

– Every software engineer is different. To be more effective, they must estimate and plan their work. They also have to develop this information using their personal data.
– To constantly improve their performance, engineers must use defined and measured processes.
– To produce quality products, engineers must feel personally responsible for the quality of their products. Premium quality products are not just produced with luck, engineers must strive to do quality work.
– It is less expensive to find and fix defects earlier than later in the process.
– It is more effective to prevent defects than to find and correct them.
– The right way to work is always the fastest and cheapest way to do work.

"According to the data of thousands of experienced engineers who have learned the PSP method, it was generally found that developers involuntarily inject about 100 defects in every 1000 lines of code they write."

Humphrey (2008) [HUM 08]

To properly complete a software engineering job, Humphrey said that engineers must plan their work before committing to or starting a job and they must use a defined process to do so. To understand their personal performance, they have to measure the time spent on each step of the work, count the defects they inject and correct and measure the size of the products they develop. To consistently produce quality products, engineers must plan, measure, and monitor the quality of the products and they must focus on quality early, at the beginning of a task. Finally, they must analyze the results of each task and use these results to improve their own processes [HUM 00].

"Since the PSP is a set of practices and methods that enable software developers to control their professional lives, when competent professionals learn and follow technical and scientific principles and when they are empowered to manage their own work, they do incredibly good work."

Watts S. Humphrey

The PSP method consists of the following five elements (scripts, forms, measurements, standards, and checklists) (adapted from [SEI 09]):

Scripts
Scripts are descriptions that guide the fulfillment of a personal process. They contain references to relevant forms, standards, checklists, sub-scripts, and actions. Scripts can be developed at a high level for a process or at a more detailed level for a particular phase of a process (e.g., a procedure). A script documents:

- the purpose or objective of a process;
- one or more entry criteria;
- general guidelines, considerations of use or constraints;
- phases or steps to be performed;
- the measures and process quality criteria;
- one or more exit criteria.

Forms
Forms provide an appropriate and coherent framework for the collection and retention of data. Forms indicate the necessary data and where to store it. When appropriate, forms define the necessary calculations and data definitions. Paper forms may be used if automated tools for data collection and storing are not readily available.

Measures
Measures are used to quantify the process and product. They allow for a better understanding of how the process works by allowing users to:

- elaborate data project profiles that can be used for planning and process improvement;

- analyze a process to determine how to improve it;
- determine the effectiveness of a change to a process;
- monitor the performance of their processes and take decisions for the next step;
- monitor the capacity to meet their commitments and take the necessary corrective measures.

Standards

The PSP method recommends the use of a number of standards such as coding standards, the guidelines for counting lines of codes and a defect classification standard.

Checklists

In the PSP method, checklists are specialized forms (or standards) used to guide the personal reviews of a software product. Each item in a checklist verifies that the product is correct or verifies its compliance with standards or specifications. A checklist includes a list of the most common defects that can be found by reviewing a specific software product. The product is fully reviewed using a single item, taken from top to bottom, from the checklist. When a list item is reviewed and completed, this item is marked as done. When all the items in the checklist have been checked and the list has been signed by the person who performed the review, it can be used as a quality record and proof that a review was done. Table 9.10 shows an example of a checklist. A line at the top of the checklist is available to indicate the title of the document/product verified. A column to the right of the checklist allows the auditor to note when an item is checked. Finally, a line at the bottom of the checklist identifies the person who carried out the review and indicates the date.

Figures 9.21, 9.22, and 9.23 present improvements to the process for effort estimation, quality, and productivity. These improvements are achieved when practitioners complete the 10 training sessions of the PSP method. Participants must write 10 programs and collect data during that time. In these three figures, the performance

Table 9.10   Example of a Partial Review Used by the PSP Method (adapted from [HUM 00])

No.	Name	Description	Verified
\multicolumn{4}{l}{Title of the document/product reviewed :}			
10	Documentation	Comment, messages	5
20	Syntax	Syntax problem	8
21	Typographic error	Spelling, punctuation	6
23	Start-end	Limits are not properly identified	12
52	I/O	File, display, printer, communication	70
70	Data	Structure, content	70
80	Function	Logic	
Date:		**Verified by:**	

during the development of program 1 is displayed to the left and the performance obtained from the tenth program is presented to the right. The resulting data are the average result of over 298 students that participated in this training over time [HUM 00].

Figure 9.21 shows improvements in the estimate of effort.

Figure 9.22 shows the quality improvements.

Figure 9.23 shows the productivity improvements.

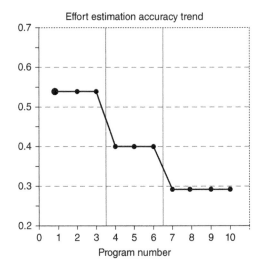

Figure 9.21   Effort estimation improvement (adapted from [HUM 00]).

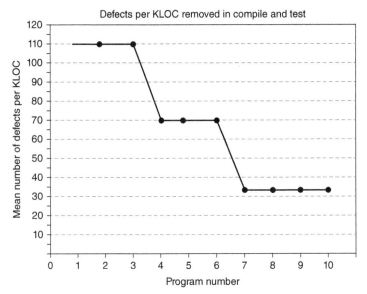

Figure 9.22   Quality improvement (adapted from [HUM 00]).

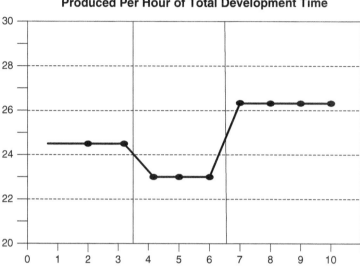

Figure 9.23   Productivity improvement (adapted from [HUM 00]).

After each exercise, the student takes the time to reflect on his performance to identify areas for improvement. As shown in these figures, the PSP method greatly improves the ability to estimate, improve quality, and improve productivity. After completing the PSP training, the student was better equipped to document his own estimates and more importantly, to defend them to his management team or to a future client.

9.10 POLICIES, PROCESSES, AND PROCEDURES IN THE SQA PLAN

The IEEE 730 standard begins with a statement that SQA procedures, practices, and policies should be developed and should conform to this standard.

This requires management to ensure that an organizational policy is established that defines and governs SQA roles and responsibilities in their organization. This is because the purpose of the IEEE 730 standard is to define the scope of SQA as:

– assessing the software development process;

– evaluating the conformance to software processes;

– evaluating the effectiveness of the software processes.

These processes include those that identify and establish the software requirements, develop the software product, and maintain the software product.

The newly created SQA function will then define, in collaboration with software developers and managers, the processes as well as the SQA function's role, concepts, methods, procedures, and practices to be applied in the organization. One of the first actions will be to define the organizational quality policy statement to be included in the future organizational quality management system. This organizational quality policy statement that defines the SQA process as an organization level process independent of SQA processes established for specific projects, will need to be supported by documented processes and procedures. In doing this, SQA must lead the identification of the standards, models, and procedures established by the project or organization. It is at that time that tasks are assigned to those responsible for SQA activities and for implementing the organizational quality policy statement.

Once these are defined and set up, adherence of products, processes, and activities to the applicable standards, procedures, and requirements are verified by project teams as well as independently by SQA.

SQA must also ensure that life cycle processes, models, and procedures for use by the organization have been defined, maintained, and improved. They must have helped in developing appropriate policies and procedures for the management and deployment of life cycle models and processes. These must have been made available, explained and provided to the personnel. Of particular interest to them is a need to ensure that processes and procedures exist related to reporting policy, process, and procedure non-conformances. Once this is available, SQA is responsible for monitoring the implementation of processes and procedures related to corrective and preventive action that can be used for the organizational repository and documentation.

The SQA function should regularly review the organizational quality policy and identify gaps and inconsistencies between the policy and proposed SQA roles and responsibilities.

Questions that project managers should ask themselves concerning policy, processes, and procedures are:

- what organizational reference documents (such as standard operating procedures, coding standards, and document templates) are applicable to this project?
- has adherence of products, processes, and activities to the applicable standards, procedures, and requirements been verified?

9.11 SUCCESS FACTORS

The following text box lists some factors that affect the development, implementation, and improvement of an organization's processes.

Factors that Foster Software Quality

1) A visible and sustained management commitment.

2) Defined business objectives.

3) Processes and process improvement objectives that support business goals.

Factors that may Adversely Affect Software Quality

4) No reasonable goals and plans.

5) Not tying processes and their improvement objectives to business objectives.

6) Inadequate resources and unrealistic expectations.

7) Thinking that institutionalization is the same as standardization [HEF 01]:
   - The CMM does not say that everyone has to do everything the same way, but we should understand when and where we need to be different. Institutionalization means that the essential practices are aligned with and reinforce the organization's infrastructure.

8) Ignoring middle managers.
   - In a low maturity organization, middle managers have the most to lose in a major cultural change. They are also the most effective group to resist change when they are not convinced that change is good. Consequently, they must recognize how to be effective in the new culture and we must give them the tools to help them make and sustain change.

9) Level 1 organizations often perform their improvement efforts as a maturity level 1 project. They do not have the discipline to manage the effort to improve as a mature project by defining the requirements, developing a good plan, and by tracking with a plan, etc. [HEF 01].

10) Organizations can also try buying a level of maturity by buying processes from a consulting firm. This approach runs the risk of alienating staff.

9.12 FURTHER READING

POTTER N. and SAKRY M. *Making Process Improvement Work*. Addison-Wesley Professional, 2002.

GARCIA S. and TURNER R. *CMMI Survival Guide*, Addison-Wesley Professional, 2007.

LAPORTE C. Y., BERRHOUMA N., DOUCET M., and PALZA-VARGAS E. Measuring the cost of software quality of a large software project at Bombardier Transportation, *Software Quality Professional Journal*, ASQ, vol. 14, issue 3, June 2012, pp. 14–31.

9.13 EXERCISES

9.1 Develop a checklist to determine whether a draft policy really is a policy. Provide five criteria.

9.2 Develop a list of criteria to guide a project manager to adapt the organization's process to the requirements of his project.

9.3 Develop a "scoring" grid to determine whether a process is consistent with a model (e.g., the CMMI-DEV).

9.4 Develop a list of criteria to help you determine whether a standard or model is compatible with the organization's culture and its way of working.

9.5 You have chosen the ETVX notation to document processes and business procedures. Your boss asks you to explain why. Give five reasons.

9.6 You have chosen the CMMI-DEV as a repository for documenting business processes. Your boss asks you to explain why. Provide five reasons.

9.7 You have decided not to use the CMMI-DEV as a repository for documenting business processes. Your boss asks you to explain why. Provide five reasons.

9.8 You have chosen the ISO 29110 standard for the development of software for your organization. Your boss asks you to explain why. Provide five reasons.

9.9 You decided not to use the ISO 29110 standard for the development of software for your organization. Your boss asks you to explain why. Provide five reasons.

Chapter 10

Measurement

After completing this chapter, you will be able to:

- understand the importance of measurement;
- understand the measurement process according to the ISO 12207 standard;
- understand the Practical Software and Systems Measurement method;
- understand the ISO/IEC/IEEE 15939 measurement standard;
- understand the measurement perspective according to the CMMI® for Development;
- understand the benefit of using surveys as a measurement tool;
- understand how to implement a measurement program;
- learn practical considerations for measurement;
- understand the ISO/IEC 29110 measurement perspective;
- learn the measurement requirements described in the IEEE 730 standard.

10.1 INTRODUCTION – THE IMPORTANCE OF MEASUREMENT

Software measurement has been a research topic in software engineering for over 30 years [FEN 07]. Sadly, many measurement programs can hardly report the most basic software measurement such as: schedule, costs, size, and efforts. This means that little recent and factual information is available to project teams and their management [LAN 08].

We recall here the definition of software engineering. This definition puts emphasis on the importance of measuring software activities and products.

Software Quality Assurance, First Edition. Claude Y. Laporte and Alain April.
© 2018 the IEEE Computer Society, Inc. Published 2018 by John Wiley & Sons, Inc.

Software Engineering
The systematic application of scientific and technological knowledge, methods, and experience to the design, implementation, testing, and documentation of software.
ISO 24765 [ISO 17a]

Today's organizations that develop software, either as standalone products or as components of systems, must continue to improve their performance and their software. Consequently, they must establish a performance target for their software development and maintenance processes. This would allow for better decision making and assessments of the rate of improvement compared with the needs of clients.

Victor Basili summarizes the many problems relative to measurement [BAS 10]. Organizations developing software experience many problems when trying to implement measurement programs. As an example, they often try to collect too much data where a good amount of it is not useful. They often do not implement a process to analyze data in a way that can help with strategic and tactical decision making. This situation leads to many problems such as a reduction of the benefits that can be achieved from a proper measurement program and disillusionment on the part of customers, management, and software developers. The inevitable result is the failure of the measurement initiative.

Watts S. Humphrey described the key roles related to software measurement, that is, understand and characterize, evaluate, control, predict, and improve [HUM 89]:

– understand and characterize: measures allow us to learn about software processes, products, and services. Measures also:
  o establish baselines, standards, and business and technical objectives;
  o document the software process models used;
  o set improvement objectives for software processes, products, and services;
  o better estimate effort and schedule costs for a specific project;
– evaluate: measures can be used to conduct cost/benefit analysis and determine if the objectives have been met;
– control: measures can help with project control of resources, processes, products, and services by sounding alarms when control limits are surpassed, performance criteria are not met and standards are not followed;
– predict: when software processes are stable and under control, measures can be used to predict budgets, schedules, resources needed, risks, and even quality issues;
– improve: measures allow us to identify the root causes of defects and other inefficiencies where improvements can be proposed.

What is the Average Cost of a Line of Source Code in a Military System?
"The average cost per line of source code, for a military system used for command and control varies between one and three hours per line, where an hour is labour time that is directly attributable and a line of code is defined as a logical line as described in the Software Engineering Institute measurement guide. In addition, the effort associated with this estimate includes requirements analysis, architectural design, development and integration of software and test tasks. The estimate does not include testing of the system or beta testing, but includes support for the requirements analysis."

Reifer (2002) [REI 02]

As an example, we illustrate the use of measurement to control the quality of a software project during development. Figure 10.1 presents the defect density of software components that have been inspected. The dotted line shows the level of quality required for a software component. For example, component number 10 should be inspected again after defects of the first inspection have been corrected.

Measurement, when available, allows the software project manager [SEI 10a]: to better plan and objectively assess the project state as well as the tasks that have been assigned to a supplier; to track the actual project performance against approved plans and objectives; to quickly identify process and product problems in order to act on them; and to collect baseline data useful for benchmarking future projects. We will also see that measures can be used to better estimate project schedules, requests

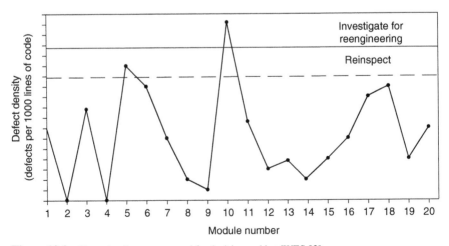

Figure 10.1 Example of a measure used for decision making [WES 03].

for proposals, answers from suppliers during the selection process, supplier and competitors' offers and proposed project schedules.

A measurement program is also helpful with the improvement of the quality of software acquisition, development, maintenance, and infrastructure processes. The program must use a measurement repository where data are collected, analyzed, and measurements are reported and available to all stakeholders within the organization. This measurement repository should be designed to answer all types of questions, for decision making and performance indicators. It should also allow for the coherent measurement of the software processes improving quality and allowing for efficient defect removal.

Measurement has been at the origin of all science and is partly responsible for all scientific advancements. Measurement contributes to the maturing of a concept by quantifying it. Using measurement allows software processes to move from an artisanal state to a controlled and repeatable state. Software engineers should design and use sound measures to improve the maturity of the software processes.

Measure (Noun)

Variable to which a value is assigned as the result of measurement.

> Note 1 to entry: The plural form "measures" is used to refer collectively to base measures, derived measures and indicators.

Measurement Process

Process for establishing, planning, performing, and evaluating measurement within an overall project or organizational measurement structure.

Measurement Process Owner

Individual or organization responsible for the measurement process.

ISO 15939 [ISO 17c]

Measuring allows us to understand the past, better inform current activities and try to predict the quality of developed products. This capacity will benefit projects where performance has not always been very precise historically with unclear schedules, budget overruns, and final products that include defects. In the past, measurement was considered as overhead to the project. Software development managers use measurement in their processes and set objectives. Measurement results are then used to take short term and active decisions during delivery and operation of systems. This helps identify and solve business-IT alignment issues or even with assigning work dynamically. For example, Google's rule of thumb is that a system reliability engineer must spend 50% of his time on software development/maintenance. To enforce

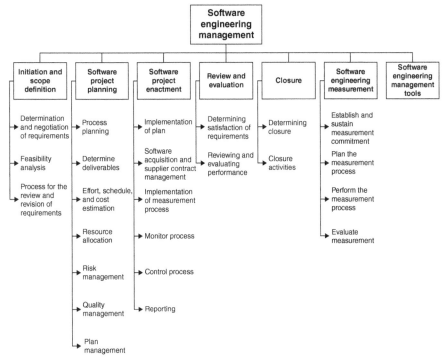

Figure 10.2   Software engineering measurement as proposed by the SWEBOK [SWE 14].

this threshold, they measure where time is spent and use this measure to ensure that teams consistently follow the proposed ratio.

Figure 10.2 describes the software engineering management knowledge area of the SWEBOK. On the right hand side of the figure, four software engineering measurement sub-topics are presented: (1) how to establish and sustain measurement commitment; (2) how to plan the measurement process; (3) how to perform the measurement process; and (4) how to evaluate measurement.

10.1.1 Standards, the Cost of Quality, and Software Business Models

The concepts of cost of quality and software business models were presented in a previous chapter. In terms of cost of quality, measurement is considered a preventive cost in the sense that a large part of measurement investment focuses on error prevention in all the stages of the software life cycle processes; for example, the cost of collecting, analyzing, and sharing these data. Table 10.1 presents different cost items with respect to preventive costs.

Table 10.1   Preventive Costs

Major category	Sub-category	Definition	Typical cost item
Preventive cost	Establish quality fundamentals	Efforts to define quality measures, establish objectives, standards and thresholds, and analysis required on data.	Definition of success criteria for acceptance testing and quality standards/guidance.
	Interventions toward projects and processes	Efforts to prevent bad quality or improve process quality.	Training, process improvement, measurement collection, and analysis.

Source: Adapted from Krasner (1998) [KRA 98].

Measures are often used in the following software business models: custom systems written on contract and mass-market software. In these business models, policies, processes, and procedures are often used and followed closely to control the development progress and minimize risks and the impact of defects.

In this chapter, the first topic described in detail is the measurement processes as described in ISO 12207 and ISO 9001. To illustrate how to implement these recommendations, we then present the practical software and systems measurement (PSM) which was initially developed to guide American Defense software projects and later became an influential component that led to the emergence of the ISO/IEC/IEEE 15939 standard on software measurement [ISO 17c]. The ISO 15939 standard is also summarized to provide an overview of the software measurement process. After this introduction to the topic, the CMMI point of view is then presented. Next, we discuss how the survey can be an efficient measurement tool. It is another illustration of a simple measurement process. Then, the use of measurement in very small entities is presented. Finally, as with all the other chapters of the book, the last section describes the measurement requirements of IEEE 730 that should be included in the software quality assurance (SQA) plan (SQAP) of a project. We conclude with a review of how to successfully implement a software measurement program in your organization as well as suggestions on how to avoid pitfalls.

10.2 SOFTWARE MEASUREMENT ACCORDING TO ISO/IEC/IEEE 12207

Measurement is one of the many processes described in the ISO 12207 standard. Its purpose is to collect, analyze, and report objective data and information to support effective software management and demonstrate the quality of the products, services, and processes [ISO 17].

As a result of the successful implementation of the measurement process [ISO 17]:

a) information needs are identified;

b) an appropriate set of measures, based on the information needs, is identified or developed;

c) required data is collected, verified, and stored;

d) the data is analyzed and the results interpreted;

e) information items provide objective information that supports decisions.

The project shall implement the following activities and tasks in accordance with applicable organizational policies and procedures with respect to the measurement process [ISO 17]:

– Prepare for measurement:
   1) define the measurement strategy;
   2) describe the characteristics of the organization that are relevant to measurement, such as business and technical objectives;
   3) identify and prioritize the information needs.
   4) select and specify measures that satisfy the information needs.
   5) define data collection, analysis, access, and reporting procedures.
   6) define criteria for evaluating the information items and the measurement process.
   7) identify and plan for the necessary enabling systems or services to be used.

– Perform measurement:
   1) integrate manual or automated procedures for data generation, collection, analysis, and reporting into the relevant processes.
   2) collect, store, and verify data.
   3) analyze data and develop information items.
   4) record results and inform the measurement users.

To obtain an overview of the measurement process, ISO 12207 refers the reader to the ISO 15939 standard that will be presented in a later section.

10.3 MEASUREMENT ACCORDING TO ISO 9001

ISO 9001 highlights the fact that a quality system needs a measurement component to be efficient. In addition to the typical process components, Figure 10.3 describes where performance measurement applies.

Clause 7.1.5 of ISO 9001 entitled "Monitoring and measuring resources" describes some measurement obligations [ISO 15]: "The organization shall determine and provide the resources needed to ensure valid and reliable results when

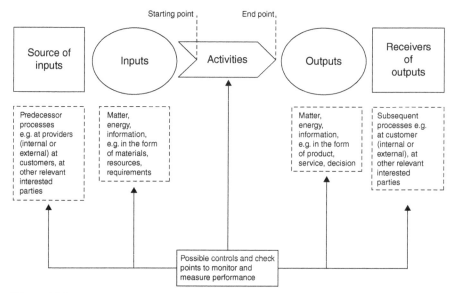

Figure 10.3   Measuring process performance according to ISO 9001 [ISO 15].

monitoring or measuring is used to verify the conformity of products and services to requirements."

Clause 9.1, entitled "Monitoring, measurement, analysis and evaluation," states that [ISO 15]: "The organization shall determine:

- what needs to be monitored and measured;
- the methods for monitoring, measurement, analysis, and evaluation needed to ensure valid results;
- when the monitoring and measuring shall be performed;
- when the results from monitoring and measurement shall be analyzed and evaluated."

Lastly, clause 10.3 of the ISO 9001 standard entitled "Continual improvement" describes another necessity of the use of measurement [ISO 15]: "The organization shall continually improve the suitability, adequacy and effectiveness of the quality management system."

10.4 THE PRACTICAL SOFTWARE AND SYSTEMS MEASUREMENT METHOD

The PSM methods were developed for the American defense industry [JON 03]. It served as a major input to the ISO 15939 standard on systems and software engineering measurements. Given that standards do not usually explain *how* things should be done, the PSM is useful for its practical examples.

The objective of the PSM is to provide measurement guidelines to software project managers. In it, directives, examples, lessons learned and case studies are presented. It provides a measurement framework that is ready to be used by software project managers. It also explains how to define and design a software measurement program to support the information needs of customers when they acquire software and systems from external providers.

The PSM covers three perspectives: (1) the project manager so as to provide a good understanding of the measures and how to use them to manage their project; (2) the technical staff that conducts measurement during planning and execution phases; and (3) the management team so that they can understand the measurement requirements associated with software.

The PSM is available for free at: http://www.psmsc.com/

The nine principles of the PSM are (adapted from [PSM 00]):

1) Use issues and objectives to drive the measurement requirements.
2) Define and collect measures based on the technical and management processes.
3) Collect and analyze data at a level of detail sufficient to identify and isolate problems.
4) Implement an independent analysis capability.
5) Use a systematic analysis process to trace the measures to the decisions.
6) Interpret the measurement results in the context of other project information.
7) Integrate measurement into the project management process throughout the life cycle.
8) Use the measurement process as a basis for objective communications.
9) Focus initially on project-level analysis.

As shown in Figure 10.4, quantitative project management includes the following specialities: risk management, measurement, and financial performance management. The PSM concentrates primarily on the measurement process but also includes the interface to other specialties like risk management and financial performance management.

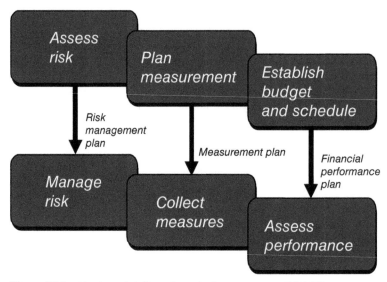

Figure 10.4   The three disciplines of quantitative management [PSM 00].

It has been observed that measurement is not as effective if it used in an independent and isolated process. Measurement can be effective in describing overall project challenges and also point to issues in interrelated systems. Measurement will be more effective when included in all aspects of project management. For example, it will be more effective when integrated with risk management and financial performance management [PSM 00]. Another chapter of this book discusses risk management in detail.

The PSM is composed of the following parts [PSM 00]:

– Part 1, The Measurement Process, describes the measurement process at a summary level and provides an overview of measurement tailoring, application, implementation, and evaluation. Part 1 explains what is required to implement the measurement process for a project.

– Part 2, Tailor Measures, describes how to identify project issues, select appropriate measures, and define a project measurement plan.

– Part 3, Measurement Selection and Specification Tables, provides a series of tables that help the user select the measures that best address the project's issues. These tables support the detailed tailoring guidance of Part 2.

– Part 4, Apply Measures, describes how to collect and process data, analyze the measurement results, and use the information to make informed project decisions.

- Part 5, Measurement Analysis and Indicator Examples, provides examples of measurement indicators and associated interpretations.
- Part 6, Implement Process, describes the tasks necessary to establish the measurement process within an organization.
- Part 7, Evaluate Measurement, identifies assessment and improvement tasks for the measurement program as a whole.
- Part 8, Measurement Case Studies, provides three different case studies that illustrate many of the key points made throughout the Guide. The case studies address the implementation of a measurement process on a DoD weapons system, an information system, and a government system in the operations and maintenance life cycle phase.
- Part 9, Supplemental Information, contains a glossary, list of acronyms, bibliography, project description, comment form, and an index.
- Part 10, Department of Defense Implementation Guide, this addendum provides information specific to implementing the PSM guidance on Department of Defense programs. It addresses implementation issues of particular concern to DoD acquisition organizations.

The PSM Insight tool, which can be found at www.psmsc.com/PSMI.asp, has been designed to run on PC's. This tool automates the PSM measurement process and can be adapted. It contains three modules:

- customization;
- data entry; and
- analysis.

The PSM approach to software measurement, as illustrated in Figure 10.5, addresses the following four key measurement activities (adapted from [PSM 00]):

1) Tailor Measures

The objective of this activity is to define the set of software and system measures that will provide the best understanding of challenges in the project at the lowest cost. A measurement plan documents the result of this first activity.

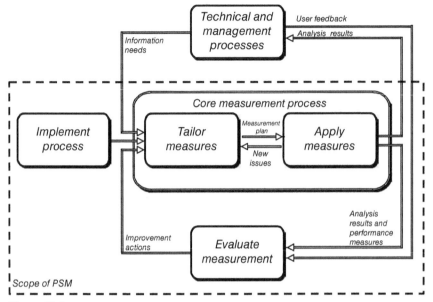

Figure 10.5 Measurement process activities proposed by the PSM [PSM 00].

2) Apply Measures

During this activity, measures are analyzed in order to provide the feedback necessary for effective decision making. Information on risks and financial performance can also be taken into account for decision making.

3) Implement Process

This activity consists of three tasks:

– obtain organizational support: including the right to measure at all organizational levels;

– define responsibilities concerning measurement;

– provide resources, purchase required tools, and recruit personnel for this process.

4) Evaluate Measurement

This activity includes four tasks:

– evaluate measures and indicators as well as their results;

– evaluate the measurement process according to three perspectives: (1) the quantitative performance evaluation of the measurement process; (2) a conformity assessment of the process executed versus the one that was planned; and (3) the measurement capability as compared to a standard recommendation;

– update the experience base with lessons learned;

– identify and implement improvements.

The key roles and responsibilities associated with measuring are (adapted from [PSM 00]):

- executive manager: is typically a manager responsible for more than one project. This manager defines the expected high level performance and the business objectives. He ensures that individual projects align with the general measurement policy. He uses the measurement outputs to make decisions;
- project or technical manager: this individual or group identifies the project challenges, reviews the measurement analysis, and acts on the information. In the case of the acquisition of complex software, the customer and the external provider will have a dedicated project manager that will use this information to make joint decisions.
- measurement analyst: this role can be assigned to a person or a group. The responsibilities include the design of the measurement plan, data collection and analysis, as well as the presentation of results to all the stakeholders. Typically, in large and complex software acquisition projects, both the external provider and the customer have a measurement analyst assigned to the project;
- project team: is the team responsible for the acquisition, development, and maintenance/operation of the software and systems. This team can include government or industry organizations as part of an Integrated Product Team (IPT). The project team collects measurement data periodically and uses it to orient engineering decisions.

The PSM defines seven categories of information that should be produced for software projects (adapted from [MCG 02]):

1) schedule: this measurement category aims at tracking the progress of the project at each step and milestone. A project that experiences delays will have a hard time meeting its delivery objectives. The project manager may have to make decisions such as reducing the functionality to be delivered or sacrificing its quality;
2) resources and costs: this measurement category evaluates the balance between the work to be done and the availability of human resources to do this work. A project that overruns its personnel budget will have difficulty completing the work unless some functionality is dropped or the quality reduced;
3) software size and stability: this measurement category addresses the stability of the progress made with respect to the delivery of functional and non-functional requirements. It uses the delivered and tested functional size to assess the delivery trend. The stability measurement considers functional change rates. Scope creep is characterized by a growing number of change requests being submitted. This situation will likely extend the schedule and increase the human resource costs;

4) product quality: another dimension of a software project that needs to be controlled is the product quality. This measurement category considers the current state of the defect removal trend for both functional and non-functional requirements. When a defective product is delivered to the customer acceptance testing step, it generates a large number of defect reports. Forcing delivery in this condition will drastically impact maintenance efforts;

5) process performance: this measurement category assesses the ability of the external providers to meet both the contract clauses as well as the requirements identified in the attachments of the contract. An external provider with weak control of his processes or experiencing weak productivity is an early sign of possible delivery problems;

6) effectiveness of technology used: this measurement category measures the effectiveness of the technology chosen to be used by the project to address the requirements. Relatively technical measures assess the software engineering techniques, like reuse, development methods and frameworks and software architectural concerns. It aims at discovering the use of risky technologies or those that have not been mastered;

7) user satisfaction: this last category assesses how the customers feel about the progress of the project and how it meets their requirements.

The SMERFS Tool

The SMERFS (*Statistical Modeling and Estimation of Reliability Functions for Systems*) tool is used to analyze software, hardware and system data with the help of reliability modeling. This tool, available at www.slingcode.com/smerfs/, is free and included with the PSM Insight tool. SMERFS tries to help answer the following questions:

- Is the software ready for release to the customer?
- How many more tests will be required before delivery?
- Does this software require significantly more rework?

This tool was developed by Dr. William Farr. To use it, five steps have to be followed:

- Step 1. Record failure data.
- Step 2. Draw a failure graph.
- Step 3. Identify a curve that best matches the observations.
- Step 4. Assess the precision of the proposed curve.
- Step 5. Use the prediction model.

The next section presents the ISO/IEC/IEEE 15939 standard, which is the current international standard for software process measurement.

10.5 ISO/IEC/IEEE 15939 STANDARD

This section presents four key software measurement activities of ISO 15939 as well as examples of measures. ISO 15939 defines a software measurement process that applies to both the systems and software engineering disciplines for software suppliers and acquirers.

The ISO 15939 standard is aligned with the measurement requirements of ISO 9001. It elaborates the measurement process for software projects as described in ISO 15288 and ISO 12207.

Measure (Verb)
Make a measurement.
Measurement Experience Base
Data store that contains the evaluation of the information products and the measurement process as well as any lessons learned during the measurement process.

ISO 15939 [ISO 17c]

In this standard, the measurement process is represented by a model that describes the activities and tasks to specify, implement, and interpret results. It does not describe how to perform these tasks nor give examples of measures. Its purpose is to describe the activities and tasks that are necessary to successfully identify, define, select, apply, and improve measurement within an overall project or organizational measurement structure. It also provides definitions for measurement terms commonly used within the system and software disciplines [ISO 17c].

As a result of the successful implementation of the measurement process, you can expect the following outcomes [ISO 17c]:

a) information needs are identified;

b) an appropriate set of measures, based on the information needs, is identified or developed;

c) required data is collected, verified, and stored;

d) the data is analyzed and the results interpreted;

e) information items provide objective information that supports decisions;

f) organizational commitment for measurement is sustained;

g) identified measurement activities are planned;

h) the measurement process and measures are evaluated;

i) improvements are communicated to the measurement process owner.

Note that the first five outcomes presented above are the same as those described in ISO 15288 and ISO 12207.

10.5.1 Measurement Process According to ISO 15939

A software measurement process should detail its activities and tasks to achieve its goal. Figure 10.6 presents the reference process model. It contains four activities, where each has a certain number of tasks [ISO 17c]:

1) Establish and sustain measurement commitment;

2) Prepare for measurement;

3) Perform measurement;

4) Evaluate measurement.

The model includes a feedback loop to the information technology life cycle processes and assumes that the organization has formalized them (i.e., technical and management processes). The activities are represented by an iterative cycle allowing for continuous feedback and improvement. This is an adaptation of the Plan-Do-Check-Act model widely used in process improvement.

The measurement repository described in Figure 10.6 collects data during a project iteration and stores historical data for all projects and software engineering processes.

The typical functional roles that are mentioned in the ISO 15939 standard are: stakeholder, sponsor, measurement user, measurement analyst, data provider, and measurement process owner.

10.5.2 Activities and Tasks of the Measurement Process

The measurement process is launched by the measurement requirements, also known as the technical and management information needs of the organization. The activities and tasks are described in Figure 10.7.

10.5.3 An Information Measurement Model of ISO 15939

Annex A of the ISO 15939 standard is only informative. It presents the model that links the information needs to the measures. This model shows what the

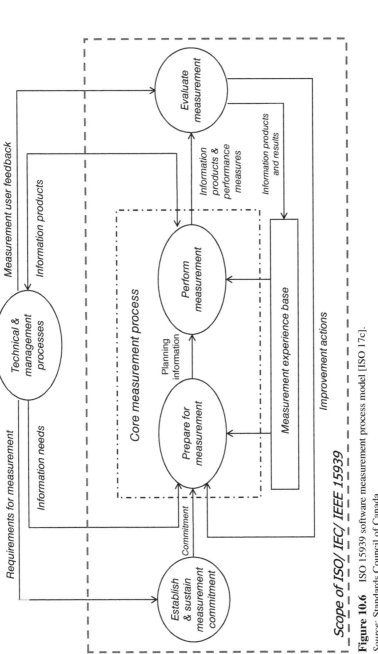

Figure 10.6 ISO 15939 software measurement process model [ISO 17c].

Source: Standards Council of Canada.

- Activity 1: Establish and sustain measurement commitment
- Accept the requirements for measurement
- Assign resources
- Activity 2: Prepare for measurement
- Define the measurement strategy
- Describe the characteristics of the organization that are relevant to measurement.
- Identify and prioritize the information needs
- Select and specify measures that satisfy the information needs
- Define data collection, analysis, access, and reporting procedures
- Define criteria for evaluating the information items and the measurement process
- Identify and plan for the enabling systems or services to be used
- Review, approve, and provide resources for measurement tasks
- Acquire and deploy supporting technologies
- Activity 3: Perform measurement
- Integrate procedures for data generation, collection, analysis, and reporting into the relevant processes
- Collect, store, and verify data
- Analyze data and develop information items
- Record results and inform the measurement users
- Activity 4: Evaluate measurement
- Evaluate information products and the measurement process
- Identify potential improvements.

Figure 10.7   The software measurement process activities and tasks [ISO 17c].
Source: Standards Council of Canada.

measurement planner has to design during the planning, execution and evaluation stages. Three types of measures are presented: base measures, derived measures, and indicators. In this section, the measurement model is explained and followed by an example of its use.

Attribute
Property or characteristic of an entity that can be distinguished quantitatively or qualitatively by human or automated means.

Indicator
Measure that provides an estimate or evaluation of specified attributes derived from a model with respect to defined information needs.

Scale

Ordered set of values, continuous or discrete, or a set of categories to which the attribute is mapped.

Note 1 to entry: The type of scale depends on the nature of the relationship between values on the scale. Four types of scale are commonly defined:

- nominal: the measurement values are categorical;

- ordinal: the measurement values are rankings;

- interval: the measurement values have equal distances corresponding to equal quantities of the attribute;

- ratio: the measurement values have equal distances corresponding to equal quantities of the attribute, where the value of zero corresponds to none of the attribute.

Base Measure

Measure defined in terms of an attribute and the method for quantifying it.

Note: A base measure is functionally independent of other measures.

Derived Measure

Measure that is defined as a function of two or more values of base measures.

Unit of Measurement

Particular quantity defined and adopted by convention, with which other quantities of the same kind are compared in order to express their magnitude relative to that quantity.

ISO 15939

Figure 10.8 presents the model and its components. Components are explained from the top to the bottom of the figure [ISO 17c]:

- An entity is an object (e.g., a process, product, project, or resource) that is to be characterized by measuring its attributes. Typical engineering objects can be classified as products (e.g., design document, network, source code, and test case), processes (e.g., design process, testing process, and requirements analysis process), projects, and resources (e.g., the systems engineers, the software engineers, the programmers, and the testers). An entity may have one or more properties that are of interest to meet the information needs. In practice, an entity can be classified into more than one of the above categories.

- An attribute is a property or characteristic of an entity that can be distinguished quantitatively or qualitatively by human or automated means. An entity may have many attributes, only some of which may be of interest for measurement. The first step in defining a specific instantiation of the measurement information model is to select the attributes that are most relevant to the measurement user's information needs. A given attribute may be incorporated in multiple measurement constructs supporting different information needs.

- A measure is defined in terms of an attribute and the method for quantifying it. A measure is a variable to which a value is assigned. A base measure is

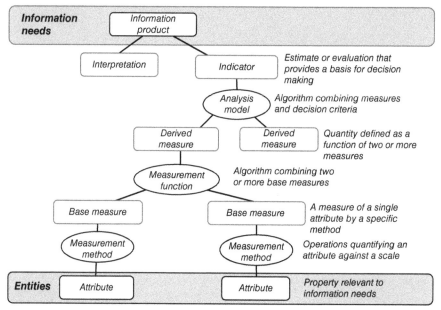

Figure 10.8   Information measurement model [ISO 17c].
Source: Standards Council of Canada.

functionally independent of other measures. A base measure captures information about a single attribute. Data collection involves assigning values to base measures. Specifying the expected range or type of values of a base measure helps to verify the quality of the data collected.

– A measurement method is a logical sequence of operations, described generically, used in quantifying an attribute with respect to a specified scale. The operations may involve activities such as counting occurrences or observing the passage of time. The same measurement method may be applied to multiple attributes. However, each unique combination of an attribute and a method produces a different base measure. Some measurement methods may be implemented in multiple ways. A measurement procedure describes the specific implementation of a measurement method within a given organizational context:

o The type of measurement method depends on the nature of the operations used to quantify an attribute. Two types of method may be distinguished:

□ Subjective: quantification involving human judgment.

□ Objective: quantification based on numerical rules such as counting. These rules may be implemented by human or automated means.

– A derived measure is a measure that is defined as a function of two or more values of base measures. Derived measures capture information about more than

- Characterization of the organizational unit;
- Business and project objectives;
- Prioritized information needs, and how they link to the business, organizational, regulatory, product or project objectives;
- Definition of the measures and how they relate to the information needs;
- Responsibility for data collection and sources of data;
- Schedule for data collection (e.g., at the end of each inspection, monthly);
- Tools and procedures for data collection (e.g., instructions for executing a static analyzer);
- Data storage;
- Requirements for data verification;
- Data entry and verification procedures;
- Data analysis plan including frequency of analysis and reporting;
- Necessary organizational or process changes to implement the measurement plan;
- Criteria for the evaluation of the information products;
- Criteria for the evaluation of the measurement process;
- Confidentiality constraints on the data and information products, and actions/precautions necessary to help ensure confidentiality;
- Schedule and responsibilities for the implementation of measurement plan including pilots and organizational unit wide implementation;
- Procedures for configuration management of data, measurement experience base, and data definitions.

Figure 10.10 Examples of information included in a measurement plan [ISO 17c].
Source: Standards Council of Canada.

areas of the model: in the generic goals (GG), in the generic practices (GP), in the specific goals (SG), and in the specific practices (SP) of specific process areas like "project planning," "project monitoring and control," "organizational process definition," and "quantitative project management."

Measures collected by maturity level 1 organizations are often of poor reliability because at that maturity level, their processes are often chaotic and not documented. At maturity level 2, also referred to as "managed," organizations have processes that are planned and executed. Therefore, at that level, it is possible to measure processes and software products. We recall here that one of the generic practices of maturity level 2 is "GP 2.8 Monitor and Control the Process," and it refers to some process attributes such as the percentage of projects that use progress and performance measures and the number of outstanding open and closed corrective actions [SEI 10a].

Concerning software products developed by suppliers, the CMMI-DEV recommends that the acquirer needs to closely follow the project quality, schedule, and costs. Measurement and data analysis are key activities of project monitoring.

The ISO 15939 standard was used by the CMMI-DEV "measurement and analysis" process area. This allows both the systems engineering and software engineering communities to share the same measurement recommendations. The next text box describes the level 2 objectives and specific practices of the measurement and analysis process area.

Measurement and Analysis

The purpose of measurement and analysis (MA) is to develop and sustain a measurement capability used to support management information needs.

SG 1 Align Measurement and Analysis Activities

Measurement objectives and activities are aligned with the identified information needs and objectives.

- SP 1.1 Establish measurement objectives: establish and maintain measurement objectives originating from identified information needs and objectives;
- SP 1.2 Specify measures: specify measures to address measurement objectives;
- SP 1.3 Specify data collection and storage procedures: specify how measurement data are obtained and stored;
- SP 1.4 Specify analysis procedures: specify how measurement data are analyzed and communicated.

SG 2 Provide Measurement Results

Measurement results, which address identified information needs and objectives, are provided.

- SP 2.1 Obtain measurement data: obtain specified measurement data;
- SP 2.2 Analyze measurement data: analyze and interpret measurement data;
- SP 2.3 Store data and results: manage and store measurement data, measurement specifications, and analysis results;
- SP 2.4 Communicate results: communicate results of measurement and analysis activities to all relevant stakeholders.

CMMI-DEV

This process area is used by many other process areas of the model. For example, for measuring project performance, the project monitoring and control process area should be consulted; for controlling software products, refer to the configuration management process area; for requirements traceability, the requirements management process area contains measurement guidelines; for organizational measurement, refer to the organizational process definition process area. To learn more about the appropriate use of statistical methods, the quantitative project management process area of CMMI-DEV provides more guidance.

SEI Measurement Program

The *"Software Engineering Measurement and Analysis"* (SEMA) program aims at studying trends, improving current solutions and promoting underutilized technologies. It also develops and improves software measurement and analysis methods and tools. To accelerate the adoption of proven methods and tools, this program also provides case studies, training and consulting.

www.sei.cmu.edu/measurement/index.cfm

The Software Engineering Information Repository (SEIR)

The objective of the SEIR is to offer a free and open forum for exchange and contributions of process improvement information.

https://seir.sei.cmu.edu/seir/

10.7 MEASUREMENT IN VERY SMALL ENTITIES

Worldwide, there are a large number of Very Small Entities (VSEs) that develop and maintain software. These are organizations, companies, departments, and projects involving up to 25 people. In an earlier chapter, the ISO 29110 was introduced. ISO 29110 proposes a four-stage roadmap referred to as a profiles. The profiles apply to VSEs in start-up mode; small projects that have a limited duration of six person-months; those that create only one product with only one team; VSEs that have more than a project with more than one team; and a VSE that wants to improve the management of its business and its competitiveness.

The activities of the ISO 29110 project management process that are related to measurement are: project planning activity, where size, effort, calendar, and resources are estimated and used in the preparation of the project plan, as well as in the project assessment and control activity, where progress is evaluated against the project plan.

The tasks of the software implementation process of ISO 29110 related to measurement are mainly those related to defects identified and corrected during reviews and testing.

10.8 THE SURVEY AS A MEASUREMENT TOOL

Surveys are used by organizations to obtain an overview of complex questions, aid problem resolution, and support decision making. Surveys are tools that allow information to be collected quickly and anonymously. It can be done during meetings and, most often, using an internet survey tool that sends questionnaires or invites to participants to answer a survey by clicking on a web link.

Concerning SQA, surveys can be used to obtain service satisfaction information from individuals and organizations like developers, project managers, testers, configuration management, and sometimes suppliers. For example, a few weeks after the deployment of a new measurement program, surveys can be prepared by SQA to assess the level of satisfaction of customers of the organization concerning its products and services.

In this section, two case studies are presented: one survey conducted by the SEI concerning measurement and one survey conducted by the ISO working group for VSEs.

What is a survey? According to Kasunic, of the SEI [KAS 05], a survey is a collection of data and an analysis method where solicited individuals answer questions or comment on declarations previously elaborated.

The SEI developed a survey process with seven steps [KAS 05]:

1) identify the research objectives;

2) identify and characterize the target audience;

3) design the sampling plan;

4) design and write the questionnaire;

5) pilot test the questionnaire;

6) distribute the questionnaire;

7) analyze results and write a report.

According to Kasunic, a good survey has to be systematic, impartial, representative, quantitative, and repeatable. Although surveys show good results compared with other data collection techniques, they have limitations [KAS 05]:

– To generalize for a population, a survey must follow strict procedures in defining which participants are studied and how they are selected.

– Following the rules and implementing the survey with the rigor that is necessary can be expensive with respect to cost and time.

– Survey data are usually superficial. It is not typically possible to go into any detail—that is, we are not capable of digging deeply into people's psyches looking for fundamental explanations of their unique understandings or behaviors.

– Surveys can be obtrusive. People are fully aware that they are the subjects of a study. They often respond differently than they might if they were unaware of the researcher's interest in them.

An Example of a Survey Questionnaire

The Acme Corporation is committed to satisfying their clients. We would like to obtain your opinion regarding our ABC software. Please indicate your level of satisfaction as well as the level of importance of each of the product characteristics.

On a scale of 1–5, circle the appropriate value that indicates your level of satisfaction for each item. A score of 1 indicates that you are not satisfied and a score of 5 indicates that you are very satisfied.

On a scale of 1–5, circle the appropriate value that indicates the level of importance for each item. A score of 1 indicates that the item is not very important and a score of 5 indicates that the item is very important.

For each question, do not hesitate to add comments concerning your satisfaction or dissatisfaction. If you have specific examples, please include them and describe any suggestions you may have.

Ease of installation Comments and suggestions	
Ease of use Comments and suggestions:	

	Satisfaction					Importance				
	NS				VS	NI				VI
Ease of installation	1	2	3	4	5	1	2	3	4	5
Comments and suggestions:										
Ease of use	1	2	3	4	5	1	2	3	4	5
Comments and suggestions:										

Legend: Very important (VI), Very satisfied (VS), Not satisfied (NS), Not important (NI)

Adapted from Westfall (2002) [WES 02]

The SEI conducted this survey to understand the state of the practice of software measurement. The following text box describes the results.

An SEI Survey of the State of the Practice of Software Measurement

The Software Engineering Institute (SEI) conducted an initial survey to understand how widely software measurement was used in the industry. This survey contained 17 questions and was distributed randomly to 15,180 software practitioners.

The survey results can be used to determine: (1) what measurement definitions and implementation methods are used, (2) the types of measures that are most widely used, and (3) what attitudes prevent the use of measurement.

The survey results indicate that, in an organization, managers and staff do not have the same perception of measurement. Managers had a stronger response than staff regarding the following items:

– understanding of the need for measuring;

– measurement allows teams to obtain better results;

– a documented process for collecting and reporting measures needs to be followed;

– definitions of measurement are generally well understood by their organization;

– measurable criteria exist for their products and services;

– corrective measures are taken when a limit has been breached.

Results show that the size of the organization has an impact on several survey items. The table below presents the percentage of respondents that have answered "often" to questions. This percentage increases slightly with the size of the organization.

Question	Number of people in the organization		
	≤100	101–499	≥500
My team follows a documented process for transmitting measurement data to management.	37.0%	46.4%	54.7%
I use measurement to understand the quality of the products and/or services on which I am working.	38.4%	42.0%	52.8%
My team follows a documented process for collecting measurement data.	42.3%	46.2%	53.1%
Corrective measures are taken when a limit has been breached.	35.1%	41.1%	46.2%
I understand the reasons why I collect data.	65.7%	71.6%	72.1%

Answers to some questions of the SEI survey on the use of measurement [KAS 05]
    The complete survey questionnaire can be found in Appendix A of the SEI report.
[KAS 05]

A Survey of SEI on the State of the Practice of Measurement

Methods Used

The CMMI MA process area was identified by respondents as the measuring method most often used to identify, collect, and analyze measurement data. About 56% of respondents said they only use this process area, while 27.4% of respondents indicated that this was the only method they used. As mentioned in the introduction of the survey, the population for the survey consisted of those who had contacted the SEI. Therefore, these people may have had prior knowledge of SEI's products and services. A correct interpretation of the results must take into account the possibility that the results were biased in favor of SEI products and services.

About 41% of respondents said they only used one method to identify, collect, and analyze measurement data, while 59% used two or more methods. About 21% reported not using any measurement method.

Measures Used

The schedule and task effort were the measures which the respondents indicated were used most often. Most respondents (97%) indicated that progress in regards to the schedule was the most commonly used measure, while 93% indicated that the effort (time) applied to tasks was the most used. The rate of increase of code, capacity, and stability were measures which the respondents indicated that they used least often.

The frequency of measurement reports varied depending on the measure with most respondents indicating that measures were reported on a weekly, monthly, or daily basis.

Adapted from Kasunic (2005) [KAS 05]

10.9 IMPLEMENTING A MEASUREMENT PROGRAM

First of all, we would think that measurement is a quick and easy thing to do. But there are many obstacles to successfully implementing a measurement program:

- we do not know why we are collecting measures;
- we intend to collect too many measures;
- some measures are not collected in the same manner in other projects;
- there are no adequate tools to easily collect and analyze measurements;
- measures are deployed without having been tested in pilot projects;
- measurement adds to the current workload;
- we think measures will not be used;
- we believe measures will be used to assess our individual performance;
- there is no commitment to measurement from the organization;
- there is little support for the measurement program.

Impacts of Low Data Quality

Some problems related to the use of low quality data are:

- low quality estimates of project costs and schedules for future projects;
- difficulty tracking project costs and progress;

– inappropriate salary levels;

– inefficient testing processes;

– inferior quality systems moved to production;

– inefficient adjustments/changes to processes.

To overcome these potential obstacles, a seven-step implementation approach has been created, tested, and is recommended [DES 95]:

1) demonstrate the value and potential of the measurement program to upper management to gain their support;

2) implicate the delivery personnel early in the design of the program;

3) identify the key processes to be improved where most benefits would arise;

4) identify the measurement goals and objectives for these key processes;

5) design and publish the measurement program for comments;

6) identify the tools/processes to be used for measurement and test them;

7) launch a first pilot and then extend gradually.

10.9.1 Step 1: Management Commitment Build-Up

Senior managers do not readily see the relevance of initiating measurement programs in software engineering since they perceive them to be expensive and bureaucratic. They also mention significant time delays prior to obtaining the expected results and the limited impact of these measurements which is limited to only a few sub-groups within the whole software engineering department. Furthermore, they often get contradictory advice from experts on the strategies for initiating a measurement program.

To address the issue of the relevance of measurement with respect to management concerns, the benefits and alignment to organizational strategy must be identified for the measurement program. This first step consists of finding the necessary information that will help the managers make a decision on the relevance of implementing a measurement program within the organization. Demonstration of the benefits of a software engineering measurement program is challenging because many results are not tangible and are realized over a long period of time.

"A large number of measurement programs die after their creation, usually because they did not provide relevant information to users."

Jones (2003) [JON 03]

10.9.2 Step 2: Staff Commitment Build-Up

It appears that there is almost always major reluctance from staff to accept a measurement program. Project managers do not usually like control and productivity measures. On the one hand, nobody likes to be "measured." On the other hand, when measurement programs are implemented, they are often labor-intensive data collection processes. To address these issues, we must offer useful tools to automate the data collection process. We must at the same time, find ways to help project managers control the data collection process and develop analytical skills to extract information from the data and measures available.

This step consists in finding the necessary arguments that will lead the staff involved in the data collection process to accept and support the measurement program.

10.9.3 Step 3: Selection of Key Processes to be Improved

This step consists in evaluating the maturity level of the software development organization. The CMMI model and assessment results provide much more information than the well-publicized single-digit maturity level. Based on this assessment of the organizational maturity level, multiple key candidates for process improvements are provided. Furthermore, the CMMI models helps with the selection of the priorities to be given to the key processes targeted for improvement programs.

10.9.4 Step 4: Identification of the Goals and Objectives Related to the Key Process

The purpose of this step is to determine the goals and objectives of the measurement program. For each CMMI process area, there is one or more goals. A goal describes the purpose of what should be achieved (e.g., improve the estimation of a development project). An objective is the wording of a goal to be reached (with or without specifying the achievement conditions) through measurable behavior over time.

Organizational goals must also correspond to the ability to achieve them. The selection of key processes considering the organization's maturity level is not enough. Processes already in place are also important. From this perspective, an organization should not have too many goals, and they must be prioritized. An organization cannot achieve all of its goals within the first year if this organization is just embarking on a measurement program.

10.9.5 Step 5: Design of the Measurement Program

This step consists in designing a measurement program that will allow management not only to see if the objectives have been reached, but also to understand why if

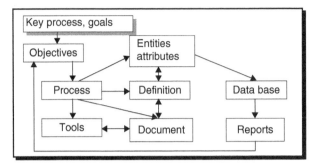

Figure 10.11 Components of a software measurement program [DES 95].

they have not been reached. Figure 10.11 suggests the components of a measurement program: tools, standards, definitions, and a choice of measures. The implementation of this design will vary from one organization to another.

10.9.6 Step 6: Description of the Information System to Support Measurement

This step consists in modeling all the measures to be collected to meet the objectives. These measures must specify measurement units and, whenever feasible, be based on standards.

This step must also define the validation process and the control reports.

10.9.7 Step 7: Deployment of the Measurement Program

This step consists in deploying a measurement program through:

– selection of a pilot site;
– personnel training;
– assigning responsibilities and tasks;
– setting-up the measurement group.

The responsibilities for the measurement objectives are distributed at different levels for different types of staff: senior management, measurement program manager, experts, and developers. This is illustrated in Table 10.2.

The success of creating and sustaining a software measurement program relies on constant support from higher management. The known presence of a leader is essential for other staff to contribute to measurement. The leader will often be part of senior management and motivate personnel.

Table 10.2 Responsibility Matrix [DES 95]

Level/personnel	Upper management	Measurement program manager	Experts	Developers
Strategic	Define strategic objectives	Supply information about the measurement program Ensure objectives are consistent	Ensure the consistency of resources	
Tactical	Endorse and promote the measurement program Approve tactical objectives	Track/validate the coherence of objectives	Assist the leader Using the objectives, define the entities, attributes and measures Assist delivery personnel Create reports and have them approved Define and document resources	Participate in the definition of tactical objectives
Operational	Supply needed resources Approve operational objectives	Manage the measurement program Implement the program and tools Provide status/recommend adjustments to the measurement program	Follow-up and implement tools Conduct statistical analysis Design the measurement repository	Participate in the definition of operational objectives Participate in data collection

There is probably a link between process maturity and the success of this program. Only more mature organizations will support process improvement in a structured way by clarifying goals and objectives. Chances for success are better than if this commitment is not communicated.

Published studies show that when measurement programs are not supported by leadership, they do not last very long.

The next text box presents common errors of measurement.

Common Errors of Measurement:
– missing or unclear measurement objectives;
– absence of adequate resources and training;
– more than one operational definition;
– difficulty with the measurement method itself;
– undisciplined measurement process;
– conflicting motivations;
– low priority/interest for measuring and analyzing;
– absence of variation analysis;
– data capture errors.

Kasunic et al. (2008) [KAS 08]

10.10 PRACTICAL CONSIDERATIONS

This section presents base measures recommended by the SEI. In every software project, management usually wants the same type of information (adapted from Carleton et al. (1992) [CAR 92]):

– what is the size of the product to be developed?
– do we have enough qualified/available personnel for the task ahead?
– can we meet the schedule?
– what level of quality is expected of this product?
– where are we according to plans?
– how are we doing on costs?

The base measures proposed by the SEI are: a size measure, an effort measure, a calendar measure, and a quality measure. For size, the SEI recommends the number of lines of source code for the following reasons (adapted from [PAR 92]):

- it is easy, simply count the end of line markers;
- counting methods do not greatly depend on the programming language;
- it is easy to automate the counting of physical lines of code;
- the majority of the data used to create cost estimation models like COCOMO [BOE 00] used lines of code.

With regards to the effort measure, the SEI recommends using the number of staff-hours. The organization should track both normal and overtime hours, whether they are paid or not. Ideally, hours dedicated on important activities like requirements, design, and testing should be calculated separately.

Effort
The number of labor units required to complete a schedule activity or work breakdown structure component. Usually expressed as staff hours, staff days, or staff weeks.

Staff-Hour
An hour of effort expended by a member of the staff.

ISO 24765 [ISO 17c]

The SEI recommends also adopting structured methods to define two important aspects with respect to a measure for schedules: the dates and the exit criteria. It is recommended to compare dates (planned and actual) associated with project milestones, reviews, and audits.

Examples of calendar measures or schedule completion criteria are: milestones, end of phase reviews, audits, and deliverables approved by the client.

With regards to quality, the SEI recommends measuring problems and defects. It is recommended that the problems and software defects be used to help determine when products will be ready for customer delivery and to provide data for the improvement of both processes and products.

Measures Available for all the Employees

In the audit chapter, we presented a case study for the Bombardier Transportation Company. It presented data measured during on-site evaluations. The evaluation team found that this data was regularly produced for each of the organization's projects. Measures were used by project managers and displayed on a wall located near the cafeteria so that all employees could view project data. Measures that were displayed included project status and progress indicators, risk and project schedule, among others.

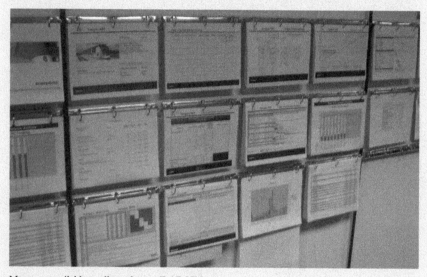

Measures available to all employees [LAP 07b]

Laporte et al. (2007) [LAP 07b]

10.10.1 Some Pitfalls with Regards to Measurement

In the 1990s, Professor Austin of the Harvard Business School [AUS 96] warned organizations about the unexpected side effects of measurement. Austin showed how a measurement program may cause dysfunctional behavior and could even affect organizational performance. Most of the literature on software measurement focuses on its technical aspects and ignores the cultural or human side of it [MCQ 04]. For example, the following text box describes how human behavior can be modified when observed and measured.

The Hawthorne Effect

The Hawthorne effect (also referred to as the observer effect) is a type of reactivity in which individuals modify or improve an aspect of their behavior in response to their awareness of being observed.

The original research at the Hawthorne Works in Cicero, Illinois, on lighting changes and work structure changes such as working hours and break times were originally interpreted by Elton Mayo and others to mean that paying attention to overall worker needs would improve productivity. Later interpretations such as that done by Landsberger suggested that the novelty of being research subjects and the increased attention from such could lead to temporary increases in workers' productivity. This interpretation was dubbed "the Hawthorne effect."

Wikipedia

We have seen that developing a measurement program is not easy. There are many drawbacks that can lead to its failure. The first pitfall is to use the measure to evaluate the performance of a developer instead of measuring the performance of the process and the tools he is using. For example, data from an inspection process should not be used to measure the productivity of the author of the document. A second pitfall is to develop an ambitious action plan which attempts to measure everything (as described in the following text box).

A Grand Measurement Plan

In a company that aimed to reach the SEI maturity level 2, the manager of the software engineering department was proud to write a 45-page measurement plan. According to this plan, dozens of measures would need to be collected on a regular basis. After a discussion with the manager, it was agreed to start measuring in a more modest way. He was encouraged to start with some basic measures such as size, effort, the number of defects.

Another pitfall is to develop a measurement plan without involving developers. Often when the measurement plan is shared, a wave of resistance may occur and could result in its abandonment. You will also want to avoid developing measures that are not used for decision making. In such cases, without understanding the use of the proposed measures, developers will be less motivated to collect them in a precise manner. For example, an organization may invest a large amount of time to measure the size of its software without measuring other key aspects such as effort or quality. This size measurement, in and of itself, is not sufficient to make decisions. The following text box shows other common pitfalls to measurement.

Measurement Pitfalls to Avoid

– Launching a measurement program that tries to measure too many elements (there are dozens and dozens of items that can be measured); when launching a program, it should be limited to a few basic measures, such as the size of a product, the effort to produce it, product quality, project status, and the satisfaction of clients and developers;

– Launching an organization wide measurement program without having tested the program with a pilot project;

– Launching a measurement program without having involved those who will be subject to measurement: most people do not like being observed—a way to get staff commitment is to ask them to participate with the development of the measurement objectives, the identification of measures that will be collected, and how the measures will be collected, stored, analyzed, used and published;

– Allowing some projects not to collect measures that are required of other projects;

– Not using the collected measurements: staff will gradually lose the motivation to collect measures in their future project if they see that their efforts result in measurements that are not used;

– Asking to collect measurements, such as the size of a software, without having precisely defined what is included and excluded in the measure: the result can neither be compared to similar results from other projects nor used in decision making;

– Using the measure to assess, reward or punish the performance of staff: although standards, such as IEEE 1028, clearly specify that you should not make this mistake, managers still insist on using measures to assess individual performance, grant salary increases or promotions (staff could react to this situation by no longer providing measures, trying to sabotage the measurement program or providing measures that cannot assess their performance);

– Producing beautiful color graphs that are not used in decision making;

– Not measuring the right things: not identifying the objectives of the organization and then launching a measurement program that does not measure what might help it make better technical and management decisions;

– Using measurement definitions that are inconsistent from one project to another;

– Predicting the results of a project (e.g., effort, schedule, quality) without ensuring that the organization's processes are stable.

10.11 THE HUMAN SIDE OF MEASUREMENT

The attitude of the people involved in measurement, that is to say the definition of measures, data collection, and analysis, is an important criterion to ensure that the program is useful to the organization. Measurement can affect the behavior of the individual observed. When something is measured, it is implied that it is important.

A software manager of a group of about 25 software engineers was briefed about the cost and benefits of software reviews. When he was presented with the description of the metrics that will be collected for each software review, he was very happy to discover that he would have plenty of measures (e.g., number of defects injected, effort to correct defects) to evaluate the performance of each software engineer. He told Professor Laporte that, equipped with such measures, he would be able to compute the pay raise, identify candidates for promotions, etc.

Professor Laporte explained to the manager that, once software engineers learn that review measures would be used for performance evaluations, they would find many creative ways to influence the metrics collected (e.g., perform a first "informal" review i.e., not known by the manager, to detect defects and correct them, then perform the "official" review where just a few defects would be detected).

(DILBERT: © Scott Adams/Distributed by United Feature Syndicate, Inc.)

After a short discussion, the manager agreed not to access software metrics collected from individual reviews. He agreed that a monthly report listing averages (e.g., mean number of defects detected, mean number of hours to correct defects) would meet his needs as process owner of the software development process. He also agreed, during a meeting with all software engineers, that he would not access the metrics since he wanted only to get the big picture about the effectiveness of the review process. After this meeting, the review process was deployed successfully. Hundreds of software reviews have been conducted since then and the confidentiality of the review metrics have never been breached.

Every staff member wants to look good and therefore would like the measures to help him look good. When developing a measurement program, think of the behaviors we want to encourage and the behaviors that we do not want to encourage.

For example, if you measure software productivity in terms of lines of code per hour, developers will target that goal to meet the objective of productivity. They will perhaps focus on their own work to the detriment of the team and the project. Worse, they can find ways to program the same functionality with additional lines of code to impact the measure.

The Defect Injection and Detection Rate Measurement at the Rolls-Royce Company

The Rolls-Royce Company produces aircraft engines. The software development department has developed a methodology to estimate the rate of injection of defects and the defect detection rate for each of its developers. While the best organizations reach a level of quality of one defect per 1000 lines of code, Rolls-Royce achieved a level of 0.03 defects per 1000 lines of code.

The many measures taken in their development process allow them to estimate the rate of injection and detection of each developer. The results of measuring the effectiveness of developers have shown that the difference in injection fault defect rates between the best and worst varies by one order of magnitude, that is, the best developer produces an average of 0.5 defects per 1000 lines of code, while the worst produces 18 defects per 1000 lines of code.

Another study showed a factor of 10 in defect detection efficiency, even if all developers use exactly the same process and exactly the same checklists.

Adapted from Nolan et al. (2015) [NOL 15]

There is only one way to avoid this behavior and it is to focus measurement on processes and products instead. Here are some suggestions to promote the implementation of effective measures (adapted from [WES 05]):

- once measures are collected, they should be used to make decisions. One sure way to undermine the measurement program is to accumulate them in a database and ignore them for decision making;
- given that software development is an intellectual task, it is recommended to develop a set of measures to fully capture the complexity of the task. As a minimum, quality, productivity and project schedule should be measured;
- to gain developer commitment to this program, they must have a sense of belonging (ownership). Participation in the definition, collection, and analysis of the measures will improve this sense of belonging. People working with

a process on a daily basis have an intimate knowledge of this process. They can help suggest ways to better measure the process, ensure the accuracy of valid measurements and propose ways to interpret the results to maximize their utility;

– provide regular feedback to the team about the data collected;

– focus on the need to collect data (when team members see the data actually used, they are more likely to view the collection activity as important);

– if team members are kept informed of how the measures are used, they are less likely to become suspicious of the measurement program;

– benefit from the knowledge and experiences of team members by involving them in the analysis of data for process improvement efforts.

At Siemens, over 50% of sales were based on products or systems that included software and more than 27,000 software engineers were employed all over the world (about 10% of employees). Some very large projects involved 2,000 developers in 13 countries.

For the engineering business unit of Siemens, objectives were established and issues that needed to be addressed were defined according to Basili's Goal-Question-Measurement (Goal-Question-Metric) approach [BAS 10]. For example:

– Objective: Reduced turnaround time
  o Question: How was productivity affected?
  o Question: Is quality the same as before?

– Objective: Increase quality
  o Question: Has customer satisfaction been affected?

– Goal: Improving process maturity
  o Question: Have measurements had a significant effect on productivity or quality?

– Objective: Introduction of new technologies
  o Question: Has a significant reduction in development time been observed?

For this business unit, Siemens has implemented a set of six measures: customer satisfaction, quality, development cycle, productivity, the maturity of its processes, and the maturity of its technology as tools.

Adapted from Geck et al. (1998) [GEC 98]

An anecdote about the invention of the inspection method by Fagan while he was working at IBM [BRO 02] was presented earlier in the chapter on reviews. In the following text box, we continue this story with a brief description of the difficulties he encountered.

The History of the Inspection Process at IBM

Upon his arrival as head of a software development department, Fagan notes that development was chaotic. There were no appropriate measures that allowed one to understand what was going on or how to do better. There was pressure to reduce the number of defects that were delivered to customers. Estimates of the percentage of rework ranged from 30% to 80%. The two most obvious ways to solve these problems were to reduce the number of defects injected during development, as well as finding and fixing those that were injected as close to their originating point in the process as possible.

The atmosphere at that time was not conducive to improvement. We had to "deliver this functionality now!" They developed software with a brute force approach, intellectual of course, but with heroic efforts. Today, we would call this an SEI level 1 organization.

One of the first actions taken was the establishment of measurable output criteria for all key activities of the development process. Fagan had decided to develop and implement a new review method, called inspection, to detect early design errors before coding. These reviews helped reduce rework significantly. In addition, inspections did not delay delivery and customers noticed an improvement in the quality of products they received.

Since he was a manager and that he felt his duty was to provide high quality software on time and within budget, Fagan had taken the liberty and the risk of implementing an inspection process for source code as well. In the early years of inspections, even with results that seemed convincing, Fagan faced derision and exasperation, but very little acceptance from his peers. He did not receive much support—in fact, he was ridiculed. He was often asked to stop this "nonsense" and refocus on project management as everyone else did.

Even when those who had no interest in the methodology confirmed how inspections had helped them, others were still reluctant to try it. Resistance to change was an obstacle to the spread of the use of inspections.

After the inspection process had proved, beyond a shadow of a doubt, to be an effective way to reduce defects and improve product quality and productivity, IBM asked him to deploy this process in its other divisions. Again, Fagan had to convince non-believers and change their work habits in software development.

Having saved the company millions of dollars, Fagan was awarded the largest single prize ever awarded by IBM.

Adapted from Broy and Denert (2002) [BRO 02]

10.11.1 Cost of Measurement

Obtaining precise, complete, and analyzed measures add a non-negligible cost to the IT budget. For example, a measurement program for software products can cost between 2% and 3% of a project's budget. According to Grady (1992) [GRA 92], organizations that use measurement obtain a competitive edge compared to their competitors who do not. Alternatively, the cost associated with not having any measures can be seen by all of the software projects that fail to meet their budget, schedule, and quality objectives. Those with measures have the advantage of making sound decisions that will allow them to obtain greater customer satisfaction.

Lessons learned from an Improvement Activity led over a Number of Years in an Organization in the Defense Industry

Collect Data to Document Improvements

Given the investment in process improvement, it becomes important to demonstrate the benefits. It is recommended that quantitative and qualitative data be collected before and during the process improvement activity so as to be used later to measure progress. Project data such as initial and final budgets, initial and final schedule, planned versus actual quality and the customer level should be collected.

Evaluate Work Groups

From time to time, the members of a work group evaluated their effectiveness. A questionnaire [ALE 91] was distributed at the end of a meeting. Members filled out the questionnaire individually and sent their answers to the work group animator. Consequently, even the shyest member could reveal his thoughts about the group dynamics. The questionnaire addressed the following topics: goals and objectives, use of resources, level of confidence between participants, conflict resolution, leadership, control and procedures, interpersonal communications, problem resolution, experimentation and creativity. For example, the following question addressed leadership:

Leadership

One individual dominates the scene. Leadership functions are neither undertaken nor shared.				There is total leadership participation. Members share leadership functions.		
1	2	3	4	5	6	7

For each question, survey participants answer on a scale from 1 to 7. Difficulties highlighted by the team members were considered to propose improvements.

Laporte and Trudel (1998) [LAP 98]

10.12 MEASUREMENT AND THE IEEE 730 SQAP

Measurement, described in the IEEE 730 standard, is helpful in demonstrating that the software processes can and do produce software products that conform to requirements. This confirmation includes evaluating intermediate and final software products along with methods, practices, and workmanship. Evaluation further includes measurement and analysis of a software process as well as product problems and related causes and provides recommendations about ways to correct current problems. IEEE 730 also explains that the measurement activities and tasks can supply objective data to improve an organization's life cycle management process. Similarly, evaluating software products for compliance can identify improvement opportunities.

This standard provides a number of questions that the project team should use during planning and execution in order to ensure its conformity to the measurement requirements. For example, [IEE 14]:

- Are requirements specific, measureable, attainable, realistic, and testable?
- Have the information needs required to measure the effectiveness of technical and management processes been identified?
- Has an appropriate set of measures, driven by the information needs, been identified and developed?
- Have appropriate measurement activities been identified and planned?
- Is the review process of the project measured and effective?
- Have all corrective actions that were implemented proven to be effective as determined by effectiveness measures?
- Have the measurement process and specific measures been evaluated?
- Have improvements been communicated to the measurement process owner?

Of the 16 SQA activities recommended by this standard, activity 5.4.6 describes the measurement of software products while activity 5.5.5 describes the recommended process measurement of a software project.

10.12.1 Software Process Measurement

Effective SQA processes identify what activities to do, how to confirm the activities are performed, how to measure and track the processes, how to learn from measures to manage and improve the processes, and how to encourage using the processes to produce software products that conform to established requirements. SQA processes are continually improved based on objective measures and actual project results. During SQA planning, the project team will define specific measurements for assessing project software quality and the project performance against project and organization quality management objectives. The following activities are recommended [IEE 14]:

1) Identify applicable process requirements that may affect the selection of a software life cycle process.
2) Determine whether the defined software life cycle processes selected by the project team are appropriate, given the product risk.

3) Review project plans and determine whether plans are appropriate to meet the contract based on the chosen software life cycle processes and relevant contractual obligations.

4) Audit software development activities periodically to determine consistency with defined software life cycle processes.

5) Audit the project team periodically to determine conformance to defined project plans.

6) Perform Task 1 through Task 5 above for subcontractor's software development life cycle.

Performing these activities should provide the following outcomes [IEE 14]:

– Documented software life cycle processes and plans are evaluated for conformance to the established process requirements.

– Project life cycle processes and plans conform to the established process requirements.

– Non-conformances are raised when software life cycle processes and plans do not conform to the established process requirements.

– Non-conformances are raised when software life cycle processes and plans are not adequate, efficient, or effective.

– Non-conformances are raised when execution of project activities does not conform to software life cycle processes and plans.

– Subcontractor software life cycle processes and plans conform to the process requirements passed down from the acquirer.

10.12.2 Software Product Measurement

Measurement, from a product perspective, determines whether product measurements demonstrate the quality of the products and conform to standards and procedures established by the project. This is even more important when a supplier is involved. Prior to delivery, determine the degree of confidence the supplier has that the established requirements are satisfied and that the software products and related documentation will be acceptable to the acquirer. The project will then collect measurement data sufficient to support these satisfaction and acceptability decisions. A contract may demand that the acquirer, prior to delivery, determine whether software products are acceptable. The following activities are recommended [IEE 14]:

1) Identify the standards and procedures established by the project or organization.

2) Determine whether proposed product measurements are consistent with standards and procedures established by the project.

3) Determine whether the proposed product measurements are representative of product quality attributes.

4) Analyze product measurement results to identify gaps and recommend improvements to close gaps between measurements and expectations.

5) Evaluate product measurement results to determine whether improvements implemented as a result of product quality measurements are effective.

6) Analyze product measurement procedures to confirm they are sufficient to satisfy the measurement requirements defined in the project's processes and plans.

7) Perform Task 1 through Task 6 above for software products developed by all subcontractors.

Performing these activities should provide the following outcomes [IEE 14]:

– Software product measurements conform to the project's processes and plans, and conform to standards and procedures established by the project or organization.

– Software product measurements accurately represent software product quality.

– Software product measurements are shared with project stakeholders.

– Software product measurements are performed on software products developed by the supplier as well as all of the supplier's subcontractors.

– Software product measurements are presented to management for review and potential corrective and preventive action.

– Non-conformances are raised when required measurement activities are not performed as defined in project plans.

Finally, IEEE 730 refers to the ISO 15939 measurement recommendations, presented in section 10.5, as measurement recommendations to be implemented in a SQAP.

Tips Concerning Measurement

– start small; identify major issues and select only a few measures, for example, five to seven measures;

– be aware that measurement can be threatening for many people. Ensure that the staff understands why we measure, what will be measured, how it will be done and what the measures will be used for;

– ensure that the measurement process is effective and does not hinder personnel's current work;

– make measurement results accessible and visible;

– use measurement to make decisions;

– communicate clearly to what purpose measurements will and will not be used.

10.13 SUCCESS FACTORS

Following are the factors that help or adversely affect software quality in an organization.

Factors that Foster Software Quality

1) A measurement program that supports business objectives.

2) Measures used to improve processes and the products of the organization.

3) Developers that are involved in the planning and implementation of the measurement program.

4) Results of measurement analysis are communicated to developers.

5) Project managers use measures to make decisions.

Factors that may Adversely Affect Software Quality

1) Measures are used to assess the performance of developers.

2) Measures are used to negatively motivate personnel (as a stick).

3) Confidential data are revealed.

4) Data are badly interpreted.

5) Process measures are not documented.

6) Data collected are not used.

7) Data are not communicated to those who are affected by the measurement.

8) Ignoring the cultural and human aspects of measurement.

9) Measurement is perceived as an overhead that can be cut when the schedule is tight or budgets are overspent.

10.14   FURTHER READING

FLORAC W. A. and CARLETON A. D. *Measuring the Software Process*, Addison-Wesley, Boston, MA, 1999.

HUMPHREY W. S. *Managing the Software Process*, Addison-Wesley, Boston, MA, 1989.

IISAKKA J. and TERVONEN I. The darker side of inspection. In: First Workshop on Inspection in Software Engineering (WISE'01), Paris, July 2001.

WEINBERG G. M. *Quality Software Management*, Volume 2: First Order Measurement. Dorset House, New York, 1993.

10.15 EXERCISES

10.1 You need to measure the size of a software. Before programming a measurement tool, you will need to specify what will and what will not be measured for a specific programming language. Choose a programming language and write the specifications for this measurement tool.

10.2 In the same organization, many software have been developed using different programming languages. You have access to size, effort, and quality measures (e.g., the number of defects). How will you proceed to compare the productivity and the quality of these software?

10.3 List criteria that would allow you to choose measures for a specific project.

10.4 What are the principal questions that a project manager should ask and for which a good measurement program will provide answers?

10.5 Write the task description to hire someone that will be responsible for the measurement analysis for your organization.

Chapter 11

Risk Management

After completing this chapter, you will be able to:

- understand risk management;
- know the main standards and models that include requirements for risk management;
- understand the risks that can affect the quality of a software;
- understand the techniques used to identify, prioritize, document, and mitigate risks;
- understand the roles of participants in risk management;
- understand the human factors involved in risk management;
- understand how to conduct risk management for very small entities;
- recognize the requirements for risk management in a software quality assurance plan.

11.1 INTRODUCTION

Software engineers and project managers are eternal optimists. When planning a project, they often assume that everything will go as planned. Reality is very different as every software project includes risks. Risk management is recognized as a proven practice in the software industry. According to Charrette (1992) [CHA 99], many software professionals have the wrong perception of risk management. They see it as a necessary but uninteresting task to be done before the really interesting coding work begins. It is perceived as over management or as another bureaucratic activity that prevents the organization from achieving its objectives.

Software Quality Assurance, First Edition. Claude Y. Laporte and Alain April.
© 2018 the IEEE Computer Society, Inc. Published 2018 by John Wiley & Sons, Inc.

"Risk, in and of itself, is not bad. Risk is essential to progress. Failure is often a key element of learning. We need to learn to balance the negative consequences of risk with the potential advantages associated with an opportunity."

Van Scoy (1992) [VAN 92]

In some organizational cultures, those that raise the flag to indicate a new risk are often perceived as negative or as trouble makers. Management will often react by attacking these individuals instead of attacking the risks highlighted. These organizations are often in reactive mode. When a risk becomes a real problem, they try to mitigate it and then manage it by adding personnel to a project that is already late, for example. When these strategies fail, the organization goes into crisis management. Now it needs to put out fires.

"With risk management, the focus is moved from crisis management to anticipatory management."

Down et al. (1994) [DOW 94]

There is a large number of sources of risk that are both external and internal to a software project. Figure 11.1 illustrates some of these sources.

With the growing complexity of software and the demand for even better, bigger, and more performing software, the industry is now becoming a high risk endeavor. When software development project teams do not manage their risks, they become vulnerable to major rework, additional costs, late delivery, or simply project failure.

Figure 11.2 illustrates the context within which software is developed. Risk management, as presented in this chapter, covers project management, that is, the risks related to the developed processes and products, although the organizational context can also present risks.

At the beginning of a software project, there are things that you know, things that you know you do not know (but that you need to understand) and unknown–unknowns, which are things that you do not know at all. The ones that you should worry about are the unknown–unknowns. These are the items that come as surprises and are unpredictable. Risk management as presented in this chapter, aims at managing the first two types of unknowns listed above. With respect to the

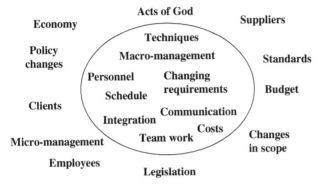

Figure 11.1 Some software project internal and external risk sources.
Source: Adapted from Shepehrd (1997) [SHE 97].

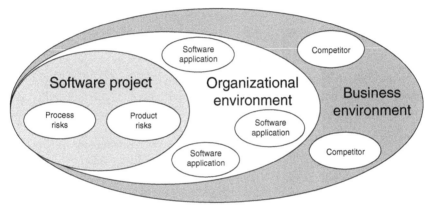

Figure 11.2 Software projects—many surrounding contexts [CHA 99].

unknown–unknowns, you will become quicker at identifying them with experience and through the use of risk management over a number of years.

The following text box describes some of the definitions of risk.

Risk
Effect of uncertainty.

Note 1: An effect is a deviation from the expected - positive or negative.

Note 2: Uncertainty is the state, even partial, of deficiency of information related to understanding or knowledge of an event, its consequence, or likelihood.

Note 3: Risk is often characterized by reference to potential events and consequences, or a combination of these.

Note 4: Risk is often expressed in terms of a combination of the consequences of an event (including changes in circumstances) and the associated likelihood of occurrence.

Note 5: The word "risk" is sometimes used when there is the possibility of only negative consequences.

Note 6: This constitutes one of the common terms and core definitions for ISO management system.

ISO 9000

An uncertain event or condition that, if it occurs, has a positive or negative effect on one or more project objectives.

PMBOK® Guide [PMI 13]

The combination of the probability of an event and its consequence.

ISO/IEC/IEEE 16085 [ISO 06a]

For the PMI, the risk management objectives of a project are: to augment the probability and the impact of positive events as well as reduce the probability and impact of negative events in the project [PMI 13]. In this chapter, we use a more widely accepted definition of risk, which is more pessimistic.

Risk management allows for raising the awareness of a doubt before it becomes a crisis. This technique improves the chances of successfully completing the project and reduces the impact of risks that cannot be avoided. Effective risk analysis and its management, for a given project, helps in identifying the hypothesis, constraints, and objectives that may change for the worse.

Boehm states four good reasons to use software risk management techniques (adapted from Boehm (1989) [BOE 89]):

– avoid catastrophic events in software projects, including budget and schedule overruns, defective software products and production failure;

– avoid rework caused by requirements, by design or by source code that is wrong, missing or ambiguous which generally accounts for 40–50% of the total cost of software development;

– avoid overkill by using detection and prevention techniques when the risk is minimal or non-existent;

– encourage a win–win software solution where the customer gets the product he needs and the supplier receives the expected benefits.

Figure 11.3 illustrates the typical progress of a project. We see that, in the early analysis phase, a very small part of the budget is spent. On the other hand, a large proportion of the budget is already committed before this phase is even completed. This situation raises the risk profile of the project as there may be budget overruns.

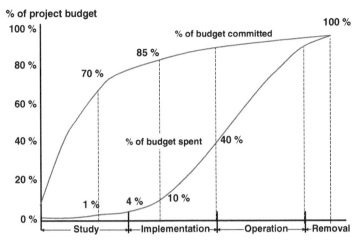

% of project budget

Figure 11.3 Typical expense curve of a project [FOR 05].

We cannot only be involved with low risk projects. A risk represents a competitive advantage or a deciding factor. Since every software project is unique, there is no miracle recipe for a development project where some, often unpleasant, surprises occur. By definition, a software development project always includes some risks. The saying "forewarned is forearmed" applies perfectly to risk management. It is better to be proactive than reactive with respect to software development.

Risk management is only one of the elements of the project decision process in an organization. Risk can appear at software, system, or organizational levels. These three risk management levels are intimately linked. To fully understand these links, it is necessary to identify what is valued in the organization or the project. For example, is innovation and risk taking well regarded or frowned upon? The organizational culture will impact the tolerance for risk in software projects.

The second issue to clarify and take into account is the risk management process itself. It needs to be documented, known, reproducible, measurable, deployed, and used.

Behaviors should also be considered. For example, is the organization's communication about risks open and honest? When there is a risky situation in the project, we need to think about what we want the project team members to do or not to do. What would you like the individuals to do when they are faced with a risky situation or when faced with a situation they feel they may fail?

Professor Laporte was consulting at an organization in the public sector. During an informal discussion with a project manager, he stated that "in this company, you cannot make a mistake." A few weeks later, a member of the executive team approved a very important project at a meeting and assigned the same project manager. The meeting was to continue after lunch.

When the meeting reconvened, the project manager was no longer present. At the end of the day, we heard that the project manager would be on sick leave due to burn-out.

It is clear that senior management of that organization had a low tolerance for error. Unfortunately, managers and employees of the organization quickly learned not to take risks, to cover up mistakes or to blame someone else. Such an organizational culture does not invite anyone to innovate or to look for opportunities to increase quality and productivity!

Interestingly, risk management may seem easy to do at first. In practice it can become a complex process, because risks are not tangible, only the resulting problems are. Risks are potential problems. When you try to lower their probability or consequences, the question is: will the investment in managing this risk improve the chance of success for the project?

Most project risks are a combination of political, social, economic, environmental, and technical factors. It can be difficult to isolate the factors and quantify the risk when they depend on individual perceptions. Given that risk is generally perceived, its estimation and probability is also generally perceived and can be biased. In organizations that have been managing risk for a long time, information accumulated from many software projects allow managers to anticipate and manage risk better with the use of quantitative measures.

Other difficulties may arise due to the varying levels of tolerance to risk from different stakeholders (e.g., marketing team, development team, or the client). For example, a lending agent in a bank and a professional sports player have different tolerance levels depending on the requirements of their profession. If they exchange jobs they will have to adapt their levels of tolerance to risk.

The three main risk factors in software development are (adapted from Charette (2006) [CHA 06]):

- an unrealistic attitude: in the software industry, it is common to make promises and under estimate effort. Unrealistic objectives are especially observed in complex projects;

- lack of discipline: we note, with regards to the CMMI, that the majority of software organizations are not disciplined or use deficient development practices, such as a bad project management approach that will destroy the project faster than any other risk factor, apart from unrealistic attitudes;

- political games: some projects are not planned or executed in a completely objective or rational way. They are part of other, more organizational, political issues. Most software project managers have difficulty managing these situations, as well as understanding their influence on the success of the project.

Professor Laporte had a mandate in an organization where he witnessed political games.

At that time, the organization did not have a tool to perform configuration management activities. The organization's configuration management expert had proposed to senior management to assess the tools available in the market so that the organization could rapidly acquire and deploy the tool and use it in their daily activities. To his surprise, senior management opted for the development of such a tool by the internal IT division. The CM expert estimated that it would take at least 2 years of work, by the internal IT team, to develop and deploy a tool with a minimal set of features. In addition, the organization would have to invest IT resources to maintain the tool, resources that could not be used for important activities of the organization.

Hallway conversations helped to understand the politics that led to this decision. Without this development project, the director of the IT division would have had to reduce the size of his division. The Director of Engineering had to return a favor to the IT manager. At the meeting, during which the decision was taken to develop internally, the engineering manager voted to develop the CM tool internally.

After over a year of development, the organization decided to buy a commercial tool!

Risk management is a good tool to continuously review the feasibility of project plans, identify and resolve problems that could impact the life cycle processes of the project, product quality, and performance with the objective of improving project management processes.

Software development risk management does not guarantee project success or the total absence of risks. It does not relieve stakeholders of their social, moral, financial, or legal obligations.

Figure 11.4 outlines the SWEBOK body of knowledge for software management. Among the major categories, risk management is highlighted in the left hand side of the diagram.

11.1.1 Risk, the Cost of Quality and Business Models

The importance of software business models and the cost of quality have already been discussed. In regards to the cost of quality, risk is considered as an element of prevention costs, that is, the costs incurred by an organization to prevent defects in different phases of its software life cycle. With respect to risk management, prevention costs are identification costs, along with the analysis and execution of risk mitigation measures. Table 11.1 describes the different prevention elements.

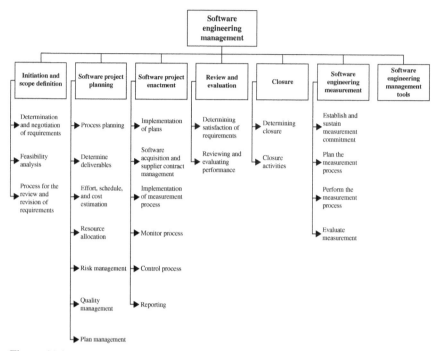

Figure 11.4   Risk management in the SWEBOK [SWE 14].

Table 11.1   Risk Management Prevention Costs

Major category	Sub categories	Definition	Typical elementary costs
Prevention costs	Establish quality fundamentals	Efforts to define quality, set quality objectives, standards and checklists. Analysis of compromises linked to quality.	Definition of success criteria, acceptance tests and quality standards.
	Interventions oriented toward projects and processes	Efforts dedicated to prevent bad quality and improve the quality of the process/project.	Training for process improvement, measurement and analysis. Risk identification, analysis, and mitigation.

Source: Adapted from Krasner (1998) [KRA 98].

Risks are at the foundation of all the business models. Software development practices should be chosen in response to the inherent risks of each model. To minimize losses and errors, developers must make a careful selection of software quality assurance (SQA), verification and validation, and risk management practices.

11.1.2 Costs and Benefits of Risk Management

The costs and benefits of risk management can vary greatly: from not managing risks at all to controlling all the project risks. The objective is to strike a balance where risk is minimized at an optimum cost.

All new approaches, like implementing risk management for the first time, will require an initial investment to document the activities of this process, training personnel, and ramping up its use in all projects. The most important investments are made when projects apply the process for mitigating the probability and consequences of risk, for example, with the development of a prototype to better understand the customer requirements or with the additional hiring of an external consultant to conduct a feasibility study.

Risk management offers a structured mechanism to give better visibility to threats perceived by the project team. It also allows for the quantifying of the schedule slippage if some risks materialize. Project performance becomes more predictable and project reviews are more effective. It helps the project manager to plan a contingency budget (e.g., in money and time) to avoid errors made in the past (e.g., overconfidence). Risk management activities should start with the request for information, before an acquisition project is initiated. This technique can also be used to assess the capability of a supplier to deliver a critical component, on time and at the quality level specified.

The advantages associated with the effective use of risk management in projects are [ISO 17 and SEI 10a]:

- a defined and executed risk management strategy;
- potential problems, that is, risks that could influence the success of the project are identified;
- the probability of risk and their consequences are understood;
- risks are ranked by priority highlighting those that will be tracked closely;
- appropriate mitigation solutions are developed proactively, taking into account the project context, diminishing crisis situations where risks become problems;
- mitigation solutions are chosen for risks that have surpassed threshold limits;
- project risk information is captured, analyzed, and exploited with the objective of improving the risk management procedures and policies.

11.2 RISK MANAGEMENT ACCORDING TO STANDARDS AND MODELS

This section briefly presents the standards and models that describe risk management requirements. First, we present the requirements of ISO 9001. Then we discuss the ISO/IEC/IEEE 12207 that describes all the life cycle processes including the recommended risk management process. Next, a section is dedicated to the ISO/IEC/IEEE 16085 standard [ISO 06a]. Risk management is also covered by the CMMI. Given that software development is almost always associated with a project, the point of view of the PMBOK® Guide of the Project Management Institute (PMI) is presented. The following section presents a discussion on how to apply risk management to very small entities using the ISO 29110. Finally, the requirements for risk management included in a SQA plan are described.

"If you do not have time to mitigate the risk now, it is certain that you will make the time to resolve the problem later, when it arises."

11.2.1 Risk Management According to ISO 9001

It is important to point out that ISO 9001 uses a risk-based thinking approach, among others [ISO 15]. A risk-based approach allows the organization to identify factors that may create a gap between its processes and its quality management system (QMS) and the expected results. It also allows for the implementation of a preventive process to limit any negative effects and to capitalize on improvement opportunities.

Clause 6.1 describes the actions to address risks and opportunities, the quality objectives of the QMS and their achievement, and modifications to the QMS. It has a number of requirements that are listed in the next text box.

6.1.1 When planning for the quality management system, the organization shall consider the issues referred to in 4.1 and the requirements referred to in 4.2 and determine the risks and opportunities that need to be addressed to:

– give assurance that the quality management system can achieve its intended result(s);

– enhance desirable effects;

– prevent, or reduce, undesired effects;

– achieve improvement.

6.1.2 The organization shall plan:

– actions to address these risks and opportunities;

– how to:

o integrate and implement the actions into its quality management system processes;
o evaluate the effectiveness of these actions.

Actions taken to address risks and opportunities shall be proportionate to the potential impact on the conformity of products and services.

Note 1: Options to address risks can include avoiding risk, taking risk in order to pursue an opportunity, eliminating the risk source, changing the likelihood or consequences, sharing the risk, or retaining risk by informed decision.

Note 2: Opportunities can lead to the adoption of new practices, launching new products, opening new markets, addressing new customers, building partnerships, using new technology and other desirable and viable possibilities to address the organization's or its customers' needs.

11.2.2 Risk Management According to ISO/IEC/IEEE 12207

The purpose of the risk management process, according to the ISO 12207 [ISO 17] standard, is to identify, analyze, treat, and monitor the risks continually. The risk management process should be a continuous process for systematically addressing risk throughout the life cycle of a system, software product, or service. It can be applied to risks related to the acquisition, development, maintenance, or operation of a system.

As a result of the successful implementation of the risk management process [ISO 17]:

– risks are identified;

– risks are analyzed;

– risk treatment options are identified, prioritized, and selected;

– appropriate treatment is implemented;

– risks are evaluated to assess changes in status and progress in treatment.

11.2.2.1 Activities and Tasks of the Risk Management Process

In accordance with the policies and procedures of the organization concerning risk management, the project shall implement the following activities [ISO 17]:

– plan risk management;

– manage the risk profile;

– analyze risk;

– treat risk;

– monitor risk.

11.2.3 Risk Management According to ISO/IEC/IEEE 16085

Risk management according to the ISO 16085 [ISO 06a] standard supports the acquisition, supply, development, operation, and maintenance of products and services by providing a series of process requirements that can address a wide variety of risks. The purpose of this standard is to provide suppliers, acquirers, developers, and managers with a single set of process requirements suitable for the management of a broad variety of risks [ISO 06a].

This standard does not describe risk management techniques. It defines a risk management process that can be used with many different techniques. The use of this standard does not require a specific life cycle process. The measurement process, described within ISO 15939 [ISO 17c] and described in an earlier chapter, works closely with the risk management activities described in the ISO 16085 standard to both identify and quantify risks.

Risk Management Process

A continuous process for systematically identifying, analyzing, treating, and monitoring risk throughout the life cycle of a product or service.

Risk Management Plan

A description of how the elements and resources of the risk management process will be implemented within an organization or project.

ISO/IEC/IEEE 16085 [ISO 06a]

The ISO 16085 risk management process, as illustrated in Figure 11.5, is continuously executed during all the activities of the life cycle of the product. It is recommended that this process include the following activities [ISO 06a]:

– plan and implement risk management;

– manage the project risk profile;

– perform risk analysis;

– perform risk monitoring;

– perform risk treatment;

– evaluate the risk management process.

The risk management process is initiated using the information requested by the stakeholders of the technical and management process of an organization (see

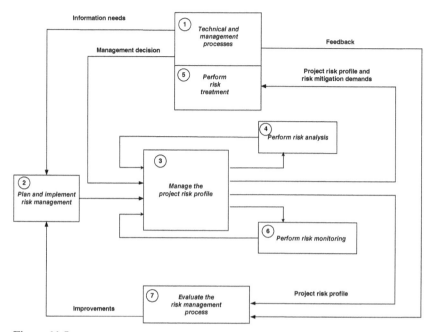

Figure 11.5   ISO 16085 recommended risk management process [ISO 06a].
Source: Standards Council of Canada.

rectangle number 1 of Figure 11.5) to make decisions that include risks. The risk management process activities are [ISO 06a]:

- during the execution of the activity entitled "Plan and implement risk management" (rectangle number 2 of Figure 11.5), the policies regarding the general guidelines under which risk management will be conducted, the procedure to be used, the specific techniques to be applied, and other matters relevant to risk planning are defined. The risk management plan (RMP) can include the following topics:
  o overview;
  o scope;
  o reference documents;
  o glossary;
  o risk management overview: describe the specifics of risk management for this project or organization's situation;
  o risk management policies: describe the guidelines by which risk management will be conducted;
  o risk management process overview;
  o risk management responsibilities: define the parties responsible for performing risk management;

     o risk management organization: describe the function or organization assigned responsibility for risk management within the organizational unit;

     o risk management orientation and training;

     o risk management costs and schedules;

     o risk management process description;

     o risk management process evaluation;

     o risk communication: describe how risk management information will be coordinated and communicated among stakeholders and interested parties (i.e., those who are interested in the performance or success of the project or product but not necessarily of the organization) such as what risks need reporting to which management level;

     o RMP change procedures and history.

Risk State

The current project risk information relating to an individual risk.

     Note: The information concerning an individual risk may include the current description, causes, probability, consequences, estimation scales, confidence of the estimates, treatment, threshold, and an estimate of when the risk will reach its threshold.

Risk Profile

A chronological record of a risk's current and historical risk state information.

Risk Threshold

A condition that triggers some stakeholder action.

     Note: Different risk thresholds may be defined for each risk, risk category or combination of risks based upon differing risk criteria.

Risk Treatment

The process of selection and implementation of measures to modify risk.

     Note:

   – The term "risk treatment" is sometimes used for the measures themselves.

   – Risk treatment measures can include avoiding, optimizing, transferring or retaining risk.

Risk Action Request

The recommended treatment alternatives and supporting information for one or more risks determined to be above a risk threshold.

                                        ISO 16085 [ISO 06a]

     – during the execution of the activity entitled "Manage the project risk profile" (rectangle number 3 of Figure 11.5), the context of current and historical risk management as well as the risks states are documented. The risk profile of the overall project is the sum of all the individual risk profiles;

– information on the project risk profile is constantly updated by the "Perform risk analysis" activity (rectangle number 4 of Figure 11.5) that identifies risks, determines their probability, lists their consequences, assesses the risk of exposure and prepares the risk action requests for risks that cross their established thresholds;

– the risk mitigation recommendations, the state of other risks, and their mitigation proposals are sent to management to be reviewed (rectangle number 5 of Figure 11.5). Management decides whether to approve that a mitigation of critical risks should be performed. The mitigation plans are then developed. These plans will be included in the project management activities to be coordinated with the other project plans and current activities;

– all the risks are then monitored until they stop posing a threat. For example, they are removed during the activity entitled "Perform risk monitoring" (rectangle number 6 of Figure 11.5). New potential risks are then investigated;

– a periodic assessment of the risk management process is necessary to ensure its effectiveness. During the activity entitled "Evaluate the risk management process" (rectangle number 7 of Figure 11.5), information, such as feedback, is documented to improve the process or the organizational or project's capacity to better manage risks. The improvements identified following a risk assessment are then used by the process entitled "Plan and implement risk management" (rectangle number 2 of Figure 11.5).

11.2.4 Risk Management According to the CMMI Model

The CMMI®-DEV contains many process areas that discuss risks. As illustrated in Figure 4.9, in the staged representation of the CMMI® model, risks are discussed in two separate process areas of maturity level 2 [SEI 10a]:

– project planning: one of the project planning practices, SP 2.2, is listed as identifying and analyzing project risks. Four sub-practices are:
  o identify risks;
  o document risks;
  o review and obtain agreement with relevant stakeholders on the completeness and correctness of documented risks;
  o revise risks as appropriate.

The typical outputs of these practices are:

  o identified risks;
  o risk impacts and probability of occurrence;
  o risk priorities.

The CMMI model also proposes examples of risk identification and analysis tools such as: risk taxonomy for determining the source and categories of risks, checklists, and brainstorming sessions;

- project monitoring and control: an appreciation of the project progress is obtained allowing corrective actions to be taken when the project performance diverges significantly from its plan. One of the specific practices, SP 1.3, discusses the need to monitor the identified risks. Three sub-practices state that:
  o periodically review the documentation of risks in the context of the project's current status and circumstances;
  o revise the documentation of risks as additional information becomes available;
  o communicate the risk status to relevant stakeholders.

An example work product of this SP is the records of project risk monitoring.

At maturity level 3, the "risk management" process area focuses on the prevention of potential problems before they appear. The purpose of the risk management process area is to identify potential problems before they occur so that risk handling activities can be planned and invoked as needed across the life of the product or project to mitigate adverse impacts on achieving objectives. This process area includes the following specific objectives and practices [SEI 10a]:

- SG 1 Prepare for risk management
  o SP 1.1 Determine risk sources and categories
  o SP 1.2 Define risk parameters
  o SP 1.3 Establish a risk management strategy
- SG 2 Identify and analyze risks
  o SP 2.1 Identify risks
  o SP 2.1 Evaluate, categorize, and prioritize risks
- SG 3 Mitigate risks
  o SP 3.1 Develop risk mitigation plans
  o SP 3.2 Implement risk mitigation plans

We can see that at maturity level 2, two of the process areas, that is, project planning and project monitoring and control, aim at risk identification and mitigation when they appear, whereas at maturity level 3, the risk management process proposes practices to ensure a systematic and continuous predictive practice for the planning, anticipation, and analysis of risk.

CMMI also addresses agile topics [SEI 10a]: some risk activities are already part of agile methodologies, for example, some technical risks can be addressed by early experimentation and experimenting early failures or by executing a spike outside the scope of the current iteration. However, the risk management process suggests a more systematic approach of technical management of risks. Such an approach can be included in agile iterations and meetings, as well as in iteration planning and task assignments.

11.2.5 Risk Management According to PMBOK® Guide

The Project Management Body of Knowledge (PMBOK® Guide) of the Project Management Institute [PMI 13] includes nine knowledge areas and the management of project risk is one of them.

Project risk, according to the PMBOK® Guide, is an uncertain event or condition that, if it occurs, has a positive or negative effect on one or more project objectives, such as schedule, costs, content, or quality (where the schedule objective is to deliver the product within the agreed upon delay and the cost objective is to deliver the product within the agreed upon budget, etc.) [PMI 13].

Figure 11.6, of the PMBOK® Guide, describes the difference between the influence of the stakeholders' risks and uncertainties and the costs of modifications as the project progresses.

The PMBOK® Guide, proposes that risk management includes six processes [PMI 13]:

1) Plan risk management
   – The process of defining how to conduct risk management activities for a project.
2) Identify risks
   – The process of determining which risks may affect the project and documenting their characteristics.

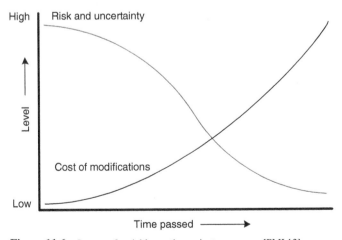

Figure 11.6   Impact of variables as the project progresses [PMI 13].

3) Perform qualitative risk analysis
   – The process of prioritizing risks for further analysis or action by assessing and combining their probability of occurrence and impact.

4) Perform quantitative risk analysis
   – The process of numerically analyzing the effect of identified risks on overall project objectives.

5) Plan risk responses
   – The process of developing options and actions to enhance opportunities and to reduce threats to project objectives.

6) Control risks
   – The process of implementing risk response plans, tracking identified risks, monitoring residual risks, identifying new risks, and evaluating risk process effectiveness throughout the project.

The Project Management Institute and the IEEE Computer Society published the "Software Extension to the PMBOK® Guide Fifth Edition" to complement the PMBOK® by describing the recognized project management practices that apply to software projects. In this guide, one chapter of more than 20 pages addresses risk management for software projects.

11.2.6 Risk Management According to ISO 29110

Very Small Entities (VSEs) manage software project risks "in the small." For example, the projects are executed on short schedules and we may not have had the time to think about risks and the ways to mitigate them. It is therefore necessary to be alert because we need to react very quickly when a risk emerges and rapidly becomes a problem because the schedule of the project is short. In these VSEs, the development teams are very small and when a problem occurs there is often no one available to address it.

Additionally, the VSE may already be in crisis mode and team members may not necessarily have the expertise or authority to address a risk. In VSEs, the authority often resides with the same individual that solves all the problems.

The authors of the VSE standard have included some risk management tasks to help. The following text describes risk management as presented in the Basic profile. The Basic profile refers to VSE who develop one project at a time with only one

Table 11.2   Risk Management Task During Project Planning [ISO 11e]

Role	Task list	Input work product	Output work product
PM TL	PM.1.9 Identify and document the risks which may affect the project	All elements previously defined	Project Plan – Identification of project risks

development team. The Intermediate and Advanced Profiles for VSEs require more involved risk management processes as they develop more than one project at a time with several teams.

The Basic profile of ISO 29110 [ISO 11e] has selected some of the expected outcomes from the ISO 12207 [ISO 17]. One objective of the project management process is "Risks are identified as they develop and during the conduct of the project." With these objectives in mind, tasks, roles, inputs, and outputs were described in an ISO 29110 management and engineering guide [ISO 11e].

Table 11.2 describes the risk management tasks suggested during the project planning activity of the project. Roles in these activities are: the project manager (PM), the technical lead (TL), and the work team (WT).

Table 11.3 describes the risk management tasks to be done during the project implementation.

Table 11.4 presents the three risk management tasks during the monitoring and control activity of the project (only the elements related to risk management are listed).

Software work products related to these tasks have also been developed. Table 11.5 describes the proposed content of the project plan and of progress reports (only the elements related to risk management are listed).

Table 11.3   Risk Management Task List During Project Implementation [ISO 11e]

Role	Task list	Input work products	Output work product
PM TL WT	PM.2.3 Conduct revision meetings with the work team, identify problems, review risk status, record agreements, and track them to closure	Project Plan Progress Status Record Correction Register Meeting Record	Meeting Record [updated]

Table 11.4 Risk Management Tasks List During the Assessment and Control Activity [ISO 11e]

Role	Task list	Input work products	Output work products
PM TL WT	PM.3.1 Evaluate project progress with respect to the Project Plan, comparing: – actual risk against previously identified	Project Plan Progress Status Record	Progress Status Record [evaluated]
PM TL WT	PM.3.2 Establish actions to correct deviations or problems and identified risks concerning the accomplishment of the plan, as needed, document them in Correction Register and track them to closure.	Progress Status Record [evaluated]	Correction Register
PM TL WT	PM.3.3 Identify changes to requirements and/or Project Plan to address major deviations, potential risks or problems concerning the accomplishment of the plan, document them in Change Request and track them to closure.	Progress Status Record [evaluated]	Change Requests [initiated]

Table 11.5 Proposed Content of the Project Plan and the Progress Status Records [ISO 11e]

Name	Description	Source
Project Plan	Presents how the project processes and activities will be executed to assure the project's successful completion, and the quality of the deliverable products. It includes the following elements which may have characteristics as follows: – Identification of project risks The applicable statuses are: verified, accepted, updated, and reviewed.	Project Management
Progress Status Record	Records the status of the project against the project plan. It may have the following characteristics: – Status of actual risk against previously identified The applicable status is: evaluated	Project Management

11.2.7 Risk Management and the SQA According to IEEE 730

We have already discussed that SQA ensures that processes are established, managed, maintained, and applied by skilled and qualified staff and that the activities and tasks performed are commensurate with product risk. Software systems are increasingly developed to perform tasks that can cause harm to living things, physical structures, and the environment. A fundamental principle of this standard is to first understand software product risk and then to ensure that the planned SQA activities are appropriate for the product risk. This means that the breadth and depth of SQA activities defined in the SQA plan are determined by and derived from software product risk.

The risk management descriptions for a project can be documented separately, as part of the project plan or as a section of the SQA plan. What is important is that it be present, complete, and available for reviews and audit. Following is a list of issues that the project manager and the SQA should consider [IEE 14]:

– prepare the SQA plan and identify the SQA activities and tasks for the project consistent with the software product risks established for the project;

– the SQA function will analyze product risks, standards, and assumptions that could impact quality and identify specific SQA activities, tasks, and outcomes that could help determine whether those risks are effectively mitigated;

– analyze the project and adapt the SQA activities to correspond to the risk;

– identify and track project changes that require further SQA planning, including changes to requirements, resources, schedules, project scope, priorities, and product risk.

– determine whether the defined software life cycle processes selected by the project team are appropriate, given the product risks.

For IEEE 730, software product risk refers to the inherent risks associated with use of the software product (e.g., safety risk, financial risk, security risk). Software product risk is differentiated from project management risk. Techniques for addressing software product risk are discussed in section 4.6.2 of this standard and in Annex J of this standard [IEE 14]:

– are potential product risks known and well documented?

– are potential product risks understood so that SQA activities can be planned in a manner appropriate to the product risk?

– has the scope of product risk management to be performed been determined?

– have appropriate product risk management strategies been defined and implemented?

– will a software integrity/criticality level be established, if appropriate?

- does the project team have adequate training in product risk management techniques?
- is the project team planning to adjust their activities and tasks in a manner consistent with product risk?
- are the breadth and depth of planned SQA activities proportional to product risk?
- are risks identified and analyzed as they develop?
- has the priority in which to apply resources to the treatment of these risks been determined?
- are risk measures appropriately defined, applied, and assessed to determine changes in the status of risk and the progress of the treatment activities?
- has appropriate treatment been taken to correct or avoid the impact of risk based on its priority, probability, and consequence or on the defined risk threshold?
- based on product risk, do the project tools, especially those used for the product SQA, require validation before they can be used on this project?
- have additional project risks appeared that could prevent SQA from accomplishing their project responsibilities?

Finally, we would like to note that high risk industries, such as medical devices, transportation, and nuclear energy have additional risk management recommendations that originate from national monitoring bodies. For example, for medical devices, risk management is not the same as the risk management defined in IEEE 730. Please refer to these additional guidelines when working within these high risk industries.

"If you do not actively attack risks, they will actively attack you."

Gilb (1988) [GIL 88]

11.3 PRACTICAL CONSIDERATIONS FOR RISK MANAGEMENT

In this section, we discuss a practical risk management approach step by step. This risk approach has been adapted from Boehm (1991) [BOE 91]. To facilitate its implementation, we have added a few tools that are easy to use. As illustrated in Figure 11.7, risk management has two major steps: risk evaluation followed by risk control. We have added another activity entitled "Lessons learned" for analyzing risks

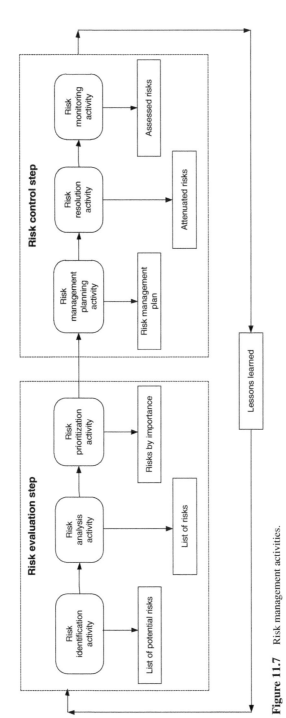

Figure 11.7   Risk management activities.

once the project is completed in order to update the risk process (e.g., checklist) of the organization.

The risk evaluation step consists of three activities: risk identification, risk analysis, and risk prioritization. The risk control step consists of three activities: risk plan development, risk mitigation, and risk monitoring.

The following text describes each activity along with helpful tools for their implementation.

Risk Management—Questions for a Software Project Manager:
– how do you identify project risks?
– are the risks of the project assessed and prioritized using their probability of occurrence and their potential impact?
– how do you mitigate a risk?
– how much budget and work days were placed in the risk reserve for your project?
– what risks are considered show-stoppers for a project?
– can you describe the most recent risks of your project?
– what is your mitigation plan for high risks?
– what risks impact the delivery of the software product?

11.3.1 Risk Evaluation Step

The risk evaluation step consists of three activities: risk identification, risk analysis, and assigning priorities.

11.3.1.1 Risk Identification Activity

Risk identification first produces a list of potential risks that are specific to the project and are susceptible of compromising its success. To do this, the following tools and techniques can be used: documentation review, using the organizational risk checklist, brainstorming with the project team, conducting interviews, strengths and weaknesses, opportunities and threats analysis (SWOT), using past experience, project lessons learned reviews and cause and effect diagrams.

Experienced personnel are often available in an organization. These individuals are well placed to propose ideas to resolve project problems and identify other potential problems that the project team did not consider. Table 11.6 describes the most common software project risks that are reported according to Boehm.

Table 11.6 List of Most Common Risks

Risk element	Risk management techniques
Personnel shortfall	Attract talented personnel (increase salaries), team training and cross functional training.
Schedules and budgets that are too optimistic	Estimates made using different techniques, incremental development, reuse, project hypothesis analysis (e.g., technology selected is adequate and available).
Incorrect functionalities and properties developed	User involvement, prototype, task analysis, user survey, user guide available early in the project.
Incorrect user interfaces developed	Prototyping, scenarios, task analysis, user participation.
Gold plating	Requirements scrubbing, prototype, cost-benefit analysis, fixed budget.
High requirements churn and requirements creep	Incremental development (to push forward changes to later iterations), stricter control from the configuration control board.
Defective components originating from outside the project	Benchmarking, inspection, reference verification, compatibility analysis.
Shortfalls occurring in the project tasks outside the project	Reference verification, audit, CMMI evaluation, award-fee contracts, increase the contract terms.
Real-time performance problems	Simulation, benchmarking, modeling, prototyping.
Capacities, both human and technical, pushed to their limits	Prototype, technical analysis, cost-benefit analysis, technology readiness level evaluation.

Source: Adapted from Boehm (1991) [BOE 91].

A software project can face different types of risks: technical, management, financial, personnel, and other resources (adapted from Westfall (2010) [WES 10]):

- technical risks include problems associated with project size, project functionality, platforms, methods, standards, or processes. These risks can originate from excessive constraints, inexperience, wrongly defined parameters or dependencies with organizations that are out of the control of the project team;
- management risks include not enough planning, inexperience in management and training, communications problems, a lack of authority, and financial control problems;
- contractual risks and judicial risks include changes to requirements, market influences, health and safety issues, government regulation, and product warranty;

– personnel risks include late acquisition of personnel, inexperienced and untrained personnel, ethical and moral issues, personnel conflicts, and productivity issues;

– other resources of risks originate from the unavailability or late delivery of equipment, tools, and environment configurations as well as slow response time.

NASA has developed a tool, called Technology Readiness Levels (TRLs), to help assess the risk involved when a project wants to include hardware or software technology that could pose major technological risks. TRLs have been developed to assess software risks. The following text box describes the TRLs.

Technology Readiness Assessment

A technology readiness assessment is a formal, systematic, metrics-based process and accompanying report that assesses the maturity of critical hardware and software technologies to be used in systems.

A technology readiness assessment is conducted by an independent review team of subject matter experts that uses Technology Readiness Levels (TRLs) to assess the maturity of a technology.

The TRL scale ranges from one through nine. The definitions are as follows:

– TRL 1: Basic principles observed and reported;

– TRL 2: Technology concept and/or application formulated;

– TRL 3: Analytical and experimental critical function and/or characteristic proof of concept;

– TRL 4: Component and/or breadboard validation in a laboratory environment;

– TRL 5: Component and/or breadboard validation in a relevant environment;

– TRL 6: System/subsystem model or prototype demonstration in a relevant environment;

– TRL 7: System prototype demonstration in an operational environment;

– TRL 8: Actual system completed and qualified through test and demonstration;

– TRL 9: Actual system proven through successful mission operations.

Department of Defense (2009) [DOD 09]

Many techniques can help with risk identification such as interviews, brainstorming, decomposition, project assumption analysis, documentation about the unknowns of a project, critical path analysis, reviewing the risk list generated by end of project reviews, and using risk taxonomies and checklists. In addition, a proper work environment facilitates the communication of risks.

Brainstorming
A general data gathering and creativity technique that can be used to identify risks, ideas, or solutions to issues by using a group of team members or subject matter experts.
PMBOK® Guide [PMI 13]

A risk statement typically includes two parts: the risk condition and its potential consequences. The condition is a statement of the potential problem that "describes the main circumstances, the situation generating doubt, anxiety or uncertainty" [DOR 96]. A consequence is a brief description explaining the potential loss or negative outcome if this condition appears during the project execution. For example, if the team does not deliver their components in compliance with the quality level expected by the customer, there will be a need to raise the effort required, with overtime, for the next three weeks.

We cannot expect that all the risks will be identified by the project manager. Project participants can identify other potential risks. Risk identification should be a team effort.

A simple and easy tool to use for identifying risks is a checklist. The following list includes typical risks (see the following text box).

Risks Related to Requirements
– no clear vision of the software product to develop;
– insufficient participation of customers during requirements gathering;
– disagreement concerning the software product requirements;
– requirements are not prioritized;
– a new market where requirements are unclear;
– rapid change in requirements;
– no requirements change management process;
– insufficient impact analysis of a requirements change.

Management Related Risks

- weak estimate of the software product size;
- poor task planning;
- lack of visibility of project progress;
- weak engagement toward project objectives;
- unrealistic expectations from the customer or management;
- personality conflicts between team members.

Once risks are identified, the next step is to document them. Table 11.7 provides an example of a risk documentation grid. The column on the left is entitled "Risk identification number" and is a number assigned to each risk by the project manager (e.g., using a simple sequence). The column entitled "Risk description" describes the risk by using the following formulation: "if event X happens then its consequence will be Y." For example, "if the estimation of the effort is incorrect by 10%, then the product delivery could be two weeks late."

11.3.1.2 Risk Analysis Activity

Once the risks are defined and documented, we proceed with the analysis of each risk. The probability of each risk and the impact is identified as well as the possible interactions between risks. The tools and techniques for this activity are: cost models, quality factor analysis (e.g., reliability, availability, security), sensitivity analysis, and decision trees [BOE 91].

Here is a list of questions that can facilitate the analysis:

- when could the event happen?
  o under what circumstances?
  o when should you act to avoid or lessen the consequences?
  o what could happen afterward?
- what is the probability of occurrence?
- what is the consequence?

Table 11.7   Risk Documentation Grid

Risk identification number	Risk description	P	C	E	Risk mitigation
1					
2					
3					

Source: Adapted from Wiegers (1998) [WIE 98].

– in what way can we quantify the consequence?
– what can we control or influence?
  ○ the probability of this event occurring?
  ○ the probability of possible results?
  ○ the consequence of the result?

A project is high risk if three or more of the following criteria exist:

– a new application domain;

– documentation is not updated;

– lack of experienced personnel;

– inflexible schedule;

– changing requirements;

– a new customer;

– a software defect that would lead to injury, financial loss, or environmental impacts.

Table 11.7 above indicates how to document a risk analysis. Column P is the probability of the risk occurrence, on a scale of 1 (not very probable) to 5 (almost certain to occur). Alternatively, you can also express the probability by the rating of low, medium or high. Column C describes the consequence if the risk becomes a problem, expressed on a scale of 1 (of little consequence) to 5 (catastrophic consequence), or with a rating of low, medium or high. Column E indicates the exposure to the risk. If numerical values were used to estimate the probability of the risk and its consequence, then the exposure is equal to $P \times C$. If relative interval rating values have been used (e.g., low, medium, high), we can estimate the risk exposure using Table 11.8.

Figure 11.8 presents a risk description template originally presented by Wiegers in [WIE 98]. It contains more information than the risk classification grid above.

Table 11.8 Risk Classification Grid

	Consequence		
Probability	Low	Medium	High
Low	Low	Low	Medium
Medium	Low	Medium	High
High	Medium	High	High

Source: Adapted from Wiegers (1998) [WIE 98].

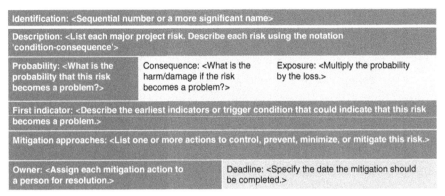

Figure 11.8   A risk documentation template [WIE 98].

A field entitled "first indicator" documents the trigger that would cause this risk to become a problem. Another field identifies the individual responsible for acting on that risk and another indicates the time when risk mitigation should be completed.

11.3.1.3 Risk Prioritization Activity

This activity produces a prioritized list of risks, such as the top-ten risks, that have been analyzed. Priority-setting techniques include: risk analysis, cost-benefit analysis, and the Delphi technique [BOE 91]. Regarding priorities, there are two simple questions to be asked:

     – what is going to hurt the project the most?
     – what is going to hurt the project the soonest?

If numerical values are used to determine the likelihood and consequence of a risk, one can then calculate risk exposure using the simple calculation "probability × consequence." For example, for a probability of 3 and a consequence of 4, we will get a priority of 12. You could add a column, to the right of the grid presented in Table 11.7 to document the resulting exposure. Having estimated the exposure of each risk, it is now easy to prioritize them.

Now that the risks have been assessed and prioritized, we can proceed with the risk control activities.

11.3.2 Risk Control Step

The risk control step involves three main activities: RMP, risk resolution, and risk monitoring.

11.3.2.1 RMP activity

Risk management planning is the selection of a risk management technique for each identified risk within the context of the project. The techniques and tools include risk control lists, cost–benefit analysis and the description of the contents of the RMP according to the standard used.

Each risk identified has its own mini action plan. The RMP consists in integrating each mini action plan. Some parts of the RMP may appear in other documents such as the project plan. For a large project, the table of contents of a plan may include the elements listed in the table of contents of ISO 16085 presented above. The RMP, once approved, is baselined and stored in the organizational repository. The RMP and it should follow, like any other document of a project, the organization's configuration management process.

Risk Mitigation

A risk response strategy whereby the project team acts to reduce the probability of occurrence or impact of a risk.

PMBOK® Guide [PMI 13]

Risk Treatment

Process to modify risk.

Note 1: Risk treatment can involve:
o avoiding the risk by deciding not to start or continue with the activity that gives rise to the risk;
o taking or increasing risk in order to pursue an opportunity;
o removing the risk source;
o changing the likelihood;
o changing the consequences;
o sharing the risk with another party or parties (including contracts and risk financing);
o retaining the risk by informed decision.

Note 2: Risk treatments that deal with negative consequences are sometimes referred to as "risk mitigation," "risk elimination," "risk prevention," and "risk reduction."

Note 3: Risk treatment can create new risks or modify existing risks.

[ISO Guide 73]

The column "risk mitigation" in Table 11.7 (also called "risk reduction") indicates, for each negative risk, an approach to avoid the risk, to transfer, check, accept, or monitor the risk. The risk mitigation actions must produce tangible results that will determine whether the risk of exposure changes [SEI 10a]:

– risk avoidance refers to the elimination of the risk from the project. For example, this can be done by not developing a risk component;

Table 11.9 Expanded Risk Document Grid [WIE 98]

Risk identification number	Risk description	P	C	E	Risk mitigation	Person responsible for risk	Risk mitigation completion date	Status (P/C)
1								
2								
3								

- risk transfer is to divert to a third party, such as a supplier, the risk and the responsibility of its resolution. Transferring the risk does not eliminate it;
- risk acceptance means that no action will be taken regarding the risk;
- risk control means that certain actions are taken, between now and the time when the risk can occur by reducing the probability and/or impact or its consequence;
- risk monitoring means observing and periodically re-evaluating the risk to detect changes in parameters.

Contingency planning means that preparations are made before the time when the risk can materialize, which define actions to be taken should the risk occur.

It is possible to add additional columns to the grid. For example, one might add the name of the person responsible for a risk to the right of the grid shown in Table 11.9. We could add a column to the right to indicate when the risk mitigation actions should have been established. Finally, we could add a column to show the status of the actions to reduce each risk as follows: P for in progress and C for completed.

For small projects, we could add, as an annex, the form illustrated in Figure 11.9 or the individual risk forms illustrated in Figure 11.8.

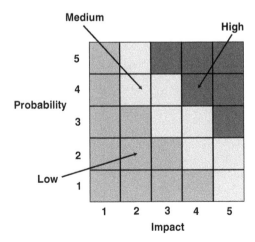

Figure 11.9 Grid illustrating risk exposure using three categories.

11.3.2.2 Risk Resolution Activity

The resolution of risks produces a situation in which identified risks are eliminated or otherwise resolved (e.g., by easing requirements) using the risk mitigation techniques documented in the project plan developed in the previous activity.

11.3.2.3 Risk Monitoring Activity

Monitoring risk involves monitoring project progress to address risk factors and take corrective action if necessary. The techniques and tools include: risk audits, gap analysis and trends, project reviews, monitoring milestones, and the list of the most significant risks [BOE 91]. We can use the form illustrated in Figure 11.9 or the set of individual forms of Figure 11.8 during the project progress meetings to track each risk and, if necessary, make changes to documents (e.g., probability, consequence, status).

Keep looking for new risks after the start of a project. The project conditions may change and risks that had not been identified at the beginning of a project, or for which the probability seemed very small, can now become part of the threats to the project.

It should not be assumed that risk is controlled simply because the selected mitigation measures were carried out. By conducting periodic risk control, we may need to change the control strategy of a particular risk if it is ineffective.

11.3.3 Lessons Learned Activity

The elements of the risk management process, as shown in Figure 11.7, can be improved by conducting lessons learned sessions, as discussed in Chapter 5, at the end of a development project to identify weaknesses and propose possible improvements.

When holding a lessons learned session, the project manager and his team could discuss the project risks, the risks identified and described in the project plan and unidentified risks that came up during the project.

Regarding the risks identified (i.e., the known risks) and documented in the project plan, the team could discuss the probabilities, consequences, and mitigation measures that were satisfactory. Otherwise, the team could suggest improvements to the process.

For risks that were not identified (i.e., the unknown risks), were these risks on the list of potential risks of the project management process but were not identified during the development of the project plan? In this case, the project manager should analyze the assumptions he used for the preparation of the project plan and decide whether to amend the risk management tasks of the planning process. If the risks that were not identified were not on the list of potential risks, the project manager should perhaps add them to the list of risks. The new list of risks will be used in a future project.

During a major hardware and software installation, system managers had indicated that the system was not in operation from midnight to 5 A.M. The manager for the supplier, who was responsible for the installation, prepared a project plan that allowed employees to install hardware and software from midnight to five in the morning.

Upon arrival on site, the maintenance manager indicated that, even if the system was not in operation from midnight to 5 A.M., employees used the system to perform daily maintenance operations. The actual time left for the supplier's installation task was now down to only 2 hours a day. The system manager of the supplier had to rework the installation plan by adding many days to the initial schedule and, since this facility was located more than 500 kilometers from their office, he had to significantly increase the budget for living expenses and add air travel costs for his employees so that they could return to their families on weekends and holidays.

11.4 RISK MANAGEMENT ROLES

A risk management process requires the participation of several project stakeholders such as the project manager, the development team, marketing and customers. In a small entity, one person may play many of these roles:

- the project manager is responsible for managing the risks associated with the development and maintenance of the system and ensures that risk management is conducted in accordance with the organizational process;
- the risk manager (a role of large organizations or projects): the project manager of a large project can choose to play this role. This role must perform the risk management process and serve as a "facilitator" for the risk analysis activity with other stakeholders;
- developers participate in the risk identification, analysis, documentation, and monitoring;
- SQA periodically reviews risk management activities to ensure that they are carried out as they were planned by the project. The SQA specialist can also participate in the risk identification and lessons learned by playing the role of a facilitator;

– configuration management can also play a role in risk monitoring and reporting. For example, the CM manager could be responsible for determining the risk status;
– risk monitor or risk owner is the person responsible for monitoring the evolution of a specific risk.

11.5 MEASUREMENT AND RISK MANAGEMENT

Assessing risk requires the following measures at the least:

– probability: a measure of the likelihood of a threat occurring. For example, this measure can be a value from 0 to 5 or 0 to 10 or a qualitative value such as low, medium, high. Table 11.10 shows an example;
– a measure of the extent of the potential impact or consequence of the risk: a measure of the loss than could occur if a threat materializes. For example, this measure can be a value from 0 to 5 or 0 to 10 or also a qualitative value, such as low, medium, or high. Table 11.11 shows an example;
– risk exposure: a measure of the magnitude of a risk based on the probability and potential impact. It is easier to provide this if the probability and impact measures were calculated numerically. Otherwise, a grid, as shown in Figure 11.9, may be used to illustrate the risk exposure using a scale of low, medium or high. The portion of the grid which is located at the bottom left shows the low exposure area, the upper right region indicates high exposure, and the intermediate zone indicates an average exposure to risk.

Figure 11.10 shows three examples of risks. Risk 1 is a medium risk since it is likely that the schedule is acceptable, risk 2 is low and risk 3 is a high risk.

Table 11.10   Example of Risk Probability Categories

Value	
1	Not likely
2	Low likelihood
3	Likely
4	Highly likely
5	Near certainty

Source: Adapted from Shepehrd (1997) [SHE 97].

Table 11.11 Example of Risk Consequence Categories

Level	Technical	Schedule	Cost
1	Minimal or no impact	Minimal or no impact	Minimal or no impact
2	Minor performance shortfall, same approach retained	Additional activities required; able to meet key dates	Budget increase or unit production cost increase <1%
3	Moderate performance shortfall, but workarounds available	Minor schedule slippage; will miss need date	Budget increase or unit production cost increase <5%
4	Unacceptable, but workarounds available	Critical path affected	Budget increase or unit production cost increase <10%
5	Unacceptable, no alternatives exist	Cannot achieve key milestone	Budget increase or production cost increase >10%

Source: Adapted from Shepehrd (1997) [SHE 97].

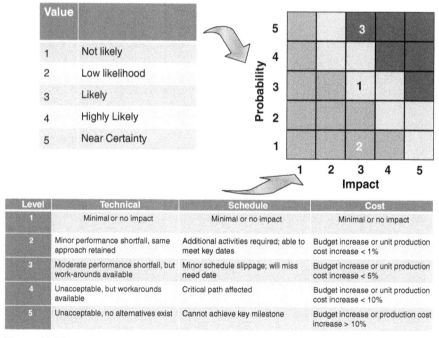

Figure 11.10 Example of three project risks.
Source: Adapted from Shepehrd (1997) [SHE 97].

For low risks, usually we will not take any specific action. For medium risks, close monitoring will be sufficient, while for high risk, action must be taken as soon as possible.

We can also measure different risk management elements during a project, for example:

- the number of risks identified;
- the number of active risks;
- the number of risks by exposure category (e.g., low, average, high);
- the effort for risk management (e.g., in person-hours);
- the number of risks identified, managed, and monitored;
- the number of risks that were not identified;
- the number of risk management process audits;
- the number of closed risks since the project was started compared to the number initially identified;
- the percentage of the budget dedicated to risk management activities.

Note that you should be prudent with these measurements since risk management is already a delicate subject to manage.

The following text box describes an industry application of a risk management approach.

Risk Management Applied to the Re-Engineering of a Weapon System

Oerlikon Contraves (now Rheinmetall Canada) was the integrator of an air defense missile system. The system consists of a missile launcher mounted on a tracked vehicle or a fixed platform, together with radar and optical sensors, electronic control systems, and communication equipment.

The corporate systems engineering process had been applied to the re-engineering of two sub-systems: the launcher control electronics and the radar and electro-optical operator consoles. The table below lists only the four activities and 12 tasks of the systems engineering process related to the management of risks.

Perform risk analysis	Identify potential risks
	Identify potential loss and consequences
	Analyze risk dependencies
	Identify risk probability of occurrence
	Prioritize risks
	Identify risk aversion strategies for each risk
Review risk analysis	Review risk analysis
	Identify risks to be part of the risk management plan (RMP)
Plan risk aversion	Define a risk monitoring approach
	Estimate risk aversion strategy cost and schedule
	Recommend risk aversion strategies
Commit to strategy	Obtain stakeholders commitment

A two-section RMP was developed. The first section described the overview of the project and defined terms such as:

– type of risk (cost, program, schedule, supportability, technical)

– assessment of risk impact (catastrophic, critical, marginal, negligible)

– overall categorization of risk (high, moderate, low)

The RMP specified who was responsible for the risk management and how the risks were to be managed.

The second section of the RMP was mainly composed of a matrix that lists all of the identified risks. The risk identification process was performed through brainstorming sessions with both the development team members and stakeholders. Along with the list of risks, in the same matrix, were the following elements of information:

– type of risk (i.e., cost, schedule, program, and technical)

– probability of occurrence (i.e., very low, low, medium, high, and very high)

– impact (i.e., negligible, marginal, critical, catastrophic, and cost)

– overall risk (i.e., low, medium, high, and cost)

– identification of impact on other projects

– brief resolution plan

– drop-dead date

– person(s) responsible (e.g., member of the project team, functional manager, project manager, and director of engineering)

– hours or resources required to perform the project

– resolution Status (i.e., open and close)

Once the RMP was approved, actions and status of the risks were then reviewed on a weekly basis during the project reviews. When costly mitigation actions were required, for example, special resources and a considerable amount of hours, then specific risk activities were directly integrated in the detailed work breakdown structure (WBS) of the project plan and scheduled like any other major development items.

Adapted from Boucher (2003) [LAP 03]

11.6 HUMAN FACTORS AND RISK MANAGEMENT

Risk management is not a purely rational process. It includes a strong human and cultural component regarding motivations, perceptions, interactions between roles, communication, decision making, and risk tolerance.

For example, a corporate culture that values and rewards heroes and fire fighters and that does not value people who can solve issues proactively before they become problems, will have a hard time implementing an effective risk management process. To change this culture, the organization will need to reward people who know how to identify, address, and avoid risks before they become problems. The organization will also have to accept that occasionally there will be fires to put out and that firefighters will still be required.

The next text box presents some guidelines that allow a person to anonymously report a risk in an organization that is wary of risks.

Anonymous risk reporting process in an organization that does not have a high risk tolerance:

– Anyone, involved or not involved with a project should be allowed to report what is perceived as an unreported risk or potential problem.
– Reporting can be submitted to an anonymous email address where messages are subsequently sent to the project manager.
– Risks can be examined in a meeting, led by the project manager, where this information is shared with every team member and stakeholders.

Table 11.12 lists some attitudes that, if present in an organization, will make it difficult to implement an effective risk management process.

Table 11.12   Attitudes that Make it Difficult to Implement Risk Management

---

We blame someone who has committed an error.
Information is not shared because information is power.
Lone rangers and fire fighters are promoted.
Failure is not an option and should not be possible.
We never talk about risks.
We look for scapegoats.
We never reflect on past projects.
We shoot the messenger when he comes with bad news.
We never address real problems in meetings.
We hide risks, because no contingency budget was approved.
We (i.e. management) do not want to hear about problems if the solution does not
   come with it. "Don't bring me problems, bring me solutions!"
We make decisions only when the problem has erupted into a crisis.
A quiet meeting is a sign that everything is under control.
We think that success comes with hard work.
We believe in miracles.

The next text box presents a list of excuses used to avoid implementing risk management.

Excuses for not Using Risk Management
The "Software Project Manager Network" provides this list of excuses that managers and developers have used to justify not using risk management:
1) We have no risks in this project.

2) Publicly discussing the risks will kill the project.

3) The customer gets anxious when he hears about a potential problem.

4) My client does not want to hear that he is a source of risk.

5) Risk identification is bad for my career.

6) This is a development project; why should we be concerned about the risks of maintenance?

7) How can you predict what will happen a year from now?

8) We expect to begin implementing risk management next year, after we have defined the process and trained the staff.

9) Our job is to develop software and not to fill out bureaucratic forms.

10) The commercial software industry cannot waste time implementing risk management.

11) We do not need a risk management program because we have frequent technical discussions and meetings.

12) If I gave a realistic effort/schedule estimation, no one would listen.

13) The use of this tool is not a risk, the vendor told me!

14) This method is proven and is therefore not a risk. The speaker at the conference said so!

15) The project is too small for risk management.

16) There is no risk to costs or schedule because the new technology will dramatically increase our productivity by five or ten times.

17) New technologies, which we have never used before, will be used to mitigate the risk.

18) We need to make the most economically advantageous proposal to win. We will concern ourselves with the job (e.g., risks) when we get the contract.

19) We have to cut corners to win this contract.

20) We do not need risk management since this software is only one component of a subsystem.

21) Our development approach is rapid application development so we do not need risk management.

Adapted from SPMN (2010) [SPM 10]

11.7 SUCCESS FACTORS

The following text box lists some of the factors that help or prevent effective risk management.

Factors that Foster Risk Management

1) A commitment from senior management.

2) Risk management is part of every project plan.

3) A risk management process that is documented and approved by management.

4) There is a risk reserve for contingencies.

5) Measures are collected and analyzed.

6) Risk management begins at the start of a project or when preparing a proposal.

7) Anticipated risks are openly discussed.

8) A RMP is used.

9) The list of risks is analyzed and updated regularly.

10) Project performance measures are used.

When developing a project plan, any assumption that we make, often unconsciously, is a risk that we accept. For example, an assumption used by most organizations when they underestimate effort is that the productivity of their programmers is above average and therefore will require less effort. If you consider your developers to be above average, while they are realistically average or even below average, your risk increases right at the beginning of a project.

Factors that may Adversely Affect Risk Management

1) An organizational culture where the project will not be approved if one acknowledges the risks.

2) An organizational culture that does not want to acknowledge risks.

3) A risk management culture that seeks to blame people who have made mistakes.

4) A RMP that stays on the shelves once approved.

5) Customers or managers engaged in tasks, milestones, technical commitments or unrealistic delivery deadlines.

6) Unplanned contingencies and unknown potential impacts.

7) No contingency built into schedules.

11.8 CONCLUSION

We can summarize risk management as taking into account these basic rules:

– always have an alternative;

– be prepared to manage crises;

– increase the probability of acceptable results;

– reduce the impact of less desirable results;

– reduce the probability of the risk event itself;

– identify parallel tracks, delivery deadlines, decision points, and action plans;

– have a clear mechanism to solve problems and communicate results.

Do not forget that despite it all, risks can also be opportunities.

Software Engineering Institute:
http://www.sei.cmu.edu/programs/sepm/risk/risk.mgmt.overview.html
Society for Risk Analysis: a forum for individuals interested in risk analysis.
http://www.sra.org/
A public forum on risks caused by computers. Incidents have been inventoried since 1986.
http://catless.ncl.ac.uk/risks

11.9 FURTHER READING

CHARETTE R. N. *Software Engineering Risk Analysis and Management*. McGraw-Hill, New York, 1989.

CHARETTE R. N. *Applications Strategies for Risk Management*. McGraw-Hill, New York, 1990.

DEMARCO T. and LISTER T. *Waltzing with Bears: Managing Risk on Software Projects*, Dorset House Publishing, New York, 2003.

MCCONNELL S. *Rapid Development: Taming Wild Software Schedules*. Microsoft Press, Redmond, WA, 1996.

HALL E. M. *Managing Risk – Methods for Software Systems Development*. Addison-Wesley, London, UK, 1998.

OULD M. *Strategies for Software Engineering: The Management of Risk and Quality*. John Wiley & Sons, Ltd, Chichester, UK, 1990.

POULIN L. *Reducing Risk in Software Process Improvement*. Auerbach Publications, Boca Raton, FL, 2005.

11.10 EXERCISES

11.1 Develop and draw, using the ETVX notation and the ISO 16085 standard, a risk management process for an organization with fewer than 10 employees.

11.2 List five risks for each phase of a typical development project.

11.3 List five potential risks when reusing components already developed in your organization.

11.4 List five potential risks when a project intends to acquire Commercial Off-The-Shelf (COTS) software components.

11.5   An emergency plan is a plan that is implemented when a risk becomes a problem. Give examples of emergency plans.

11.6   A supplier cannot deliver the software at the required level of reliability and consequently the reliability of the system may not meet the customer performance specifications. Describe the possible risk management measures to be taken in such a case.

11.7   The interface with a new control device is not yet defined. The software driver may take longer to develop than initially estimated. Describe the possible risk management measures to be taken in such a case.

Chapter 12

Supplier Management and Agreements

After completing this chapter, you will be able to:

- understand the importance, as well as the effect, of including SQA in projects that involve external suppliers;
- know the requirements of the ISO 9001, ISO/IEC/IEEE 12207 standard, and the CMMI® model for the management of agreements with suppliers;
- recognize the difference between suppliers and external participants;
- communicate and manage the risks associated with external participants;
- be aware of the two main software contract reviews;
- understand the requirements of the IEEE 730 standard regarding the monitoring of suppliers in the quality assurance plan of the project.

12.1 INTRODUCTION

When software work involves external suppliers, the software quality assurance (SQA) staff and project managers should be knowledgeable in the management of suppliers and agreements/contracts. The quality of a relationship between partners is a complex concept and it is key to the success of the project. We believe that adequate preparation, the choice of an adequate agreement or contract type, frequent reviews, and follow-up are fundamental for a good relationship. The development of contractual clauses that apply the knowledge described in this book is also key for delivering quality software in this complex situation.

Ensuring quality results in this type of project requires that the supplier's personnel are involved and knowledgeable regarding SQA processes. To ensure this, we expect that the supplier provides a SQA plan (SQAP) (in addition or included

Software Quality Assurance, First Edition. Claude Y. Laporte and Alain April.
© 2018 the IEEE Computer Society, Inc. Published 2018 by John Wiley & Sons, Inc.

in the project and technical plans) before we can finalize contractual negotiations. This allows SQA, as well as client's project manager, to assess the SQAP intentions of the external supplier for this project. Thus, the acquirer SQA function has the important and arduous task to ensure early discussion on this type of software project.

We have observed that customer satisfaction will be higher when the supplier adopts a collaborative strategy as well as simple and straightforward language. The following sections present an overview of the required knowledge in the management of suppliers and contracts.

12.2 SUPPLIER REQUIREMENTS OF ISO 9001

Clause 8.4 of the ISO 9001 standard describes the control of externally provided processes, products, and services, whether through:

– purchasing through a supplier;
– an arrangement with an associate company; or
– outsourcing to an external provider.

The controls required for external provision can vary widely depending on the nature of the services or products acquired. The organization can apply risk-based thinking (that was covered in a previous chapter) to determine the type and extent of controls appropriate to particular suppliers and externally provided services and products.

Customer
Person or organization that could or does receive a product or a service that is intended for or required by this person or organization.

Example: Consumer, client, end-user, retailer, receiver of product or service from an internal process beneficiary and purchaser.

Note 1 to entry: A customer can be internal or external to the organization.

Supplier
Organization that provides a product or a service.

Note 1 to entry: A provider can be internal or external to the organization.

Note 2 to entry: In a contractual situation, a provider is sometimes called "contractor".

ISO 9000

Many topics presented in ISO 9001 are helpful in ensuring that a supplier delivers quality software. For example, clause 4.4 insists that processes needed (i.e., inputs and outputs), their interactions, and responsibilities and authorities be clear to ensure that the quality system works effectively. Clause 5.1 goes further and recommends that the supplier demonstrate leadership and commitment. Clause 6.2 discusses the establishment of quality objectives and the planning to achieve them. We have already covered this topic in an earlier chapter.

Clause 8.4 of ISO 9001 describes the controls required to manage a supplier's processes, products, and services: "The organization shall ensure that externally provided processes, products and services conform to requirements.

The organization shall determine the controls to be applied to externally provided processes, products, and services when:

- products and services from suppliers are intended for incorporation into the organization's own products and services;
- products and services are provided directly to the customer(s) by suppliers on behalf of the organization;
- a process, or part of a process, is provided by a supplier as a result of a decision by the organization.

The organization shall determine and apply criteria for the evaluation, selection, monitoring of performance, and re-evaluation of suppliers, based on their ability to provide processes or products and services in accordance with requirements. The organization shall retain documented information of these activities and any necessary actions arising from the evaluations." [ISO 15]

Additionally, clause 8.5 of ISO 9001 specifies the responsibilities of the procuring organization regarding intellectual property [ISO 15]: "The organization shall respect the property of customers or external service providers when it is under their control or being used. The organization shall identify, verify, protect, and safeguard the property that customers or external service providers have provided them for use or incorporation into their products and services."

12.3 AGREEMENT PROCESSES OF ISO 12207

There are two agreement processes in ISO 12207: the acquisition process and the supply process: "These processes define the activities necessary to establish an agreement between two organizations. If the acquisition process is invoked, it provides the means for conducting business with a supplier. This may include products that are supplied for use as an operational software system, services in support of operational activities, software elements of a system, or elements of a software system being provided by a supplier. If the supply process is invoked, it provides the means for an

agreement in which the result is a product or service that is provided to the acquirer."
[ISO 17].

Acquisition
Process of obtaining a system, product or service.
Agreement
Mutual acknowledgement of terms and conditions under which a working relationship is conducted.
    Example: Contract, memorandum of agreement.

                                                                ISO 12207 [ISO 17]

According to ISO 12207, the purpose of the acquisition process is to obtain a product or service in accordance with acquirer's requirements. The process begins with the identification of the customer needs and ends with the acceptance of the product and/or service needed by the acquirer. The following activities are described in the standard [ISO 17]:

a) prepare for the acquisition: the acquirer defines a strategy, describes what is to be acquired in enough detail (i.e., system/software requirements) so that the supplier can understand it;

b) advertise the acquisition and select the supplier: the acquirer communicates the request for the supply of a product and/or services, then using an established procedure, the acquirer evaluates and selects one or more suppliers;

c) establish and maintain an agreement: the acquirer prepares and negotiates an agreement with the supplier. During this activity, he identifies the necessary changes and their impact on the agreement with the supplier including: acquisition requirements, costs and schedule, and many other topics like acceptance criteria, warranty, and licensing;

d) monitor the agreement: the acquirer assesses the execution of the supplier's activities in accordance with the software review and audit processes (as seen in earlier chapters). He also provides data needed by the supplier and resolves issues in a timely manner;

e) accept the product or service: the acquirer conducts the acceptance review and testing of the deliverable. During this process the acquirer confirms that the delivered product or service complies with the agreement, provides payment or other agreed consideration, takes on the ownership of the configuration management (as discussed in a previous chapter) and finally closes the agreement.

According to the ISO 12207, the purpose of the supply process is to provide an acquirer with a product or service that meets the agreed requirements. The following activities are described in the standard [ISO 17]:

a) prepare for the supply: the supplier determines the existence and identity of an acquirer who has a need for a product or service. Once done he defines a supply strategy;

b) respond to a request for supply of products or services: the supplier evaluates a request to determine its feasibility and how to respond. Then he prepares a response that satisfies this solicitation;

c) establish and maintain an agreement: the supplier negotiates an agreement with the acquirer that includes acceptance criteria. The supplier may identify necessary changes to the agreement and their impact as part of a change control mechanism. Then a negotiation will take place and the final agreement is made official;

d) execute the agreement: the supplier executes the agreement according to the established project plans. The most important agreement execution activities are summarized as: reviews, choosing an appropriate software life cycle model, detailing project management plans, including an SQAP, performing V&V, assessing the execution and quality and managing subcontractors;

e) deliver and support the product or service: the supplier delivers the software product or service as specified in the agreement. He also needs to provide assistance to the acquirer in support of the delivered software.

A software quality management process suited for the project is established. This is the mechanism to assure quality and conformance to established plans. As many software engineering processes work together, the SQA process should be coordinated with the software V&V, review and the audit processes. It also requires that a SQAP (software quality assurance plan) be developed and typically include the following:

– quality standards, methodologies, procedures, and tools for performing the SQA activities;

– procedures for contract review and coordination thereof;

– procedures for identification, collection, filing, maintenance, and disposition of quality records;

– resources, schedule, and responsibilities for conducting the SQA activities;

Once a plan is in place, it will be necessary to schedule the SQA activities and execute the plan. A problem resolution process will be used between the acquirer and the supplier during this period to solve all outstanding issues. For this process to work correctly, individuals performing SQA functions should have a position within the organization that provides an unimpeded communication mechanism with management. This will allow free circulation of the information for problem resolution. Before delivery, software products will be verified to ensure they have fully satisfied their contractual requirements and are acceptable to the acquirer.

12.4 SUPPLIER AGREEMENT MANAGEMENT ACCORDING TO THE CMMI

We know that the CMMI® for Development (CMMI-DEV) is a descriptive model in that it describes the essential attributes (or major attributes) that are expected from the processes to be executed in an organization that is working, in this case, in the Supplier Agreement Management process area of maturity level 2 of the staged representation. This model is also used as a normative model because the objectives and practices describe the practices that the acquirer expects from the in-house or external suppliers undertaking projects in a contractual context. CMMI proposes that project managers, in the acquirer's organization, must master the agreement management process with suppliers.

Acquisition Strategy
The specific approach to acquiring product and services that is based on considerations of supply sources, acquisition methods, requirements specifications types, agreement types, and related acquisition risks.

CMMI-DEV

Agreement management involving external suppliers is set at level 2 of the CMMI-DEV. This is mainly due to the fact that acquiring software products and services is much more common nowadays. CMMI describes the specific goals (SG) and specific practices (SP) required as [SEI 10a]:
SG 1 Establish supplier agreements:

- SP 1.1: Determine acquisition type
- SP 1.2: Select suppliers
- SP 1.3: Establish supplier agreements

SG 2 Satisfy supplier agreements:

– SP 2.1: Execute the supplier agreement

– SP 2.2: Accept the acquired product

– SP 2.3: Ensure transition of products

Figure 12.1 shows the interaction between the SP of this process area.

In addition, according to the CMMI, the organization should commit that its software projects follow a written policy for the management of software acquisition. Furthermore, a contract manager should be assigned for the establishment and management of contract activities. CMMI also requires that these contract management activities are reviewed with the project manager on a periodic basis rather than on an event-driven basis.

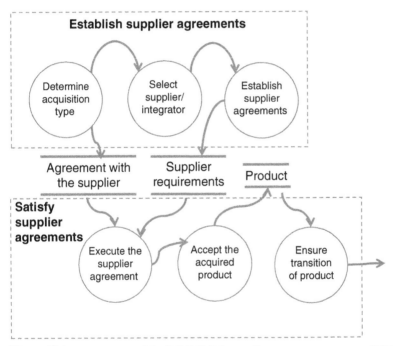

Figure 12.1   Interpretation of the CMMI-DEV for supplier agreement management [KON 00].

In addition to activities directly carried out regarding the project, CMMI suggests that the organization has implemented project verification procedures. It is therefore expected that the contract management activities be reviewed with management on a periodic basis. To support this activity, SQA and project managers should carefully review the activities and work products described in the agreement and/or perform third-party audits, as described in an earlier chapter.

The following anecdote describes how a public service manages its suppliers using the CMMI model.

The Assessment of an Underground Transportation Equipment Supplier

A Canadian public service division, responsible for underground transport systems, added a requirement that all its suppliers, located in many countries, demonstrate their software process maturity level against the CMMI requirements in the tendering process for the supply of a new subway car fleet and monitoring systems.

The contract of the customer stated that an independent assessment of the prime contractor and all of its suppliers would be required within 90 days of signing a contract. The contract would state that all of the suppliers who had not met a certain level of maturity would need to develop action plans to improve their situation and report their improvement progress on a monthly basis.

The contract also mandated the prime contractor to produce a specific action plan showing how each supplier would reach the CMMI level of maturity required within 24 months of the signature of the contract.

12.5 MANAGING SUPPLIERS

As we have seen, the need to manage suppliers during a software project is presented in many standards. The number of external providers that contribute to a software project can be important and are often working in the background. There can also be a partnership between suppliers. The bigger the project, the more complex this situation can become. The types of suppliers can be:

- subcontractors: who take responsibility for a specific part of a contract. We will call on contractors when seeking specific software expertise and also to ensure the timely availability of experts when needed;
- software package suppliers: offering off-the-shelf software ready to be adapted and implemented. These external providers are becoming more popular because they offer proven software packages that can reduce new software development costs and delays;
- consultants: who are hired on the project to help with specific tasks, such as explain the requirements and business rules, develop interfaces with other internal software, evolution and maintenance staff, and IT infrastructure staff. These additional resources, coming from outside the organization, bring considerable expertise to the project's success but must be coordinated properly.

As the number of people involved grows, the more coordination and quality issues can arise:

– deadlines: the orchestration of the various stakeholders may become more difficult and many meetings are needed to address the issues, verify each intermediary deliverable and coordinate work. A problem with a key supplier will have a direct effect on the project;

– quality of deliverables: we saw that quality issues can be varied, for example: defects, failure to comply with established rules, incomplete intermediate deliverables and misunderstanding of a requirement. These problems will create rework and additional testing effort to re-verify deliverables;

– transition difficulty: during the transition, the software operation becomes the responsibility of the customer. Customer software maintainers and IT operations can also identify non-conformities and ask the external providers to rework their deliverables before accepting them. It is therefore important that the SQA function help project managers that are involved with several external providers in understanding the complex process that a project will require. It can support this by describing the SQA tools available to help:

  o prevent delays and make sure to prevent the arrival of impending problems;
  o ensure early assessment of the quality of deliverables (preventive approach);
  o pay attention to potential downstream requirements of IT operations and software maintenance/support requirements;
  o keep an eye on the performance of each stakeholder involved.

A solid SQAP (discussed in the next chapter) as well as a clear process map of the software acquisition process are two key elements of success for these complex software projects. The quality plan will bring more precision to the obligations of each of the stakeholders and the process mapping will clarify the roles, activities, and deliverables of the project from beginning to end.

12.6 SOFTWARE ACQUISITION LIFE CYCLE

Software projects that involve suppliers are often more complex than traditional software development projects. To illustrate this, we will describe an acquisition process specially designed for the acquisition of off-the-shelf products like SAP/R3 solutions, Oracle Financials and similar software packages. There are several factors to consider before purchasing such software solutions. The software acquisition life cycle description will define and clarify the activities and the roles involved in the software procurement process. It is an essential part of SQA for this particular situation.

A supplier management (SM) process requires goals, applicability, objectives, and a detailed process mapping of the activities. The example that follows shows a high level vendor management process for a real company.

Supplier Management Process

Purpose of the Process

The goal of the software SM process is to select qualified software suppliers and manage them effectively.

Applicability

The software vendor supplier process applies to all software products that are obtained from suppliers, developed and maintained by a supplier including commercial off-the-shelf software (COTS).

Objectives

- The acquirer selects qualified suppliers.

- The acquirer and the supplier of software products agree upon their mutual commitments.

- The acquirer and the software product supplier maintain communication throughout the project.

- The acquirer tracks the results of the software product provider and his performance against commitments.

Overview of the SM Process

The process consists of two sets of activities: the SM-100 series, shown in the figure below, defines the steps that start when a project requires software products that cannot be developed internally and stops when a contract is awarded to a supplier.

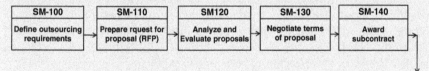

Select the supplier and award the contract.

The SM-200 series, shown in figure below, defines the steps that begin after a contract award and ends when the software products are delivered according to the requirements of the contract or when the contract with the supplier must be terminated. When the contract closing activities have been completed, a review is usually performed to discuss improvements in the SM process.

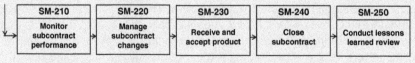

Perform supplier monitoring and close the contract.

Adapted from Laporte and Papiccio (1997) [LAP 97]

The software acquisition life cycle is complex and needs to be described in more detail so that external suppliers clearly understand what is expected of them. For a large Canadian equipment distributor, three detailed process maps were developed to better explain the detailed activities to the supplier. These maps describe the following three steps (see Figure 12.2):

– the request for information (RFI) stage (stage 1);

– the supplier selection stage (stage 2);

– the adaptation and implementation of the software package (stage 3).

The RFI stage represented here aims at identifying and selecting the best candidates to lead a complex software project. The first activity of this process is to document business requirements and existing processes/systems. The result of this activity is high-level requirements. This is used as an entry point to develop the RFI, which, once approved by the legal department, will be sent out to potential suppliers. Each supplier response will then be evaluated and potentially used to help develop a request for proposal (RFP).

As we can see from Figure 12.2, a number of roles and responsibilities as well as templates are described and greatly clarify all the activities. When the customer takes the time to provide this level of information, the quality of the project increases. The two other process maps of this process (i.e., stage 2 and stage 3) are available on this book's website. The next section presents software contract types that can be considered for software suppliers.

12.7 SOFTWARE CONTRACT TYPES

We have seen that the software acquisition life cycle processes require choosing a contract type. Contracts, in the software domain are complex and involve legal, project management, and technology-related clauses. There are many types of contracts. The simplest types involve consultancy and contract hire where expertise is needed: for example, to document requirements, to write specifications, or convert data. They are relatively simple because the sums of money involved are small and when the work is finished, the deliverables are verified by the customers. For larger endeavors where the customer would like a turnkey solution, this contract type is not fit for that purpose.

Another type of contract is the fixed-cost contract. When should this type of contract be used? Projects with clear tasks are good candidates for a fixed price contract. Some organizations are forced to use this type of contract by law. To clarify deliverables, during requirements generation, a column can be added to the WBS definition of the tasks identifying each status task as clear/not risky or unclear/risky/surge. For example, consider whether the daily operations and maintenance of a system, as well as regularly scheduled maintenance requirements, are clearly identified and their costs reasonably estimated. An unclear/surge task, like unscheduled maintenance or

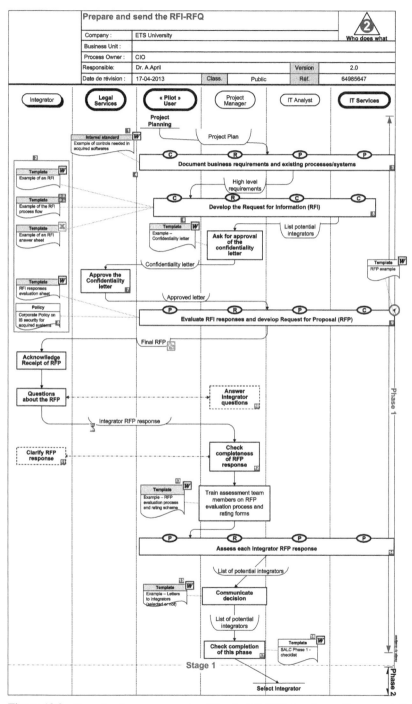

Figure 12.2   Example of a process map describing the RFI-RFQ stage for a software acquisition.

repair, could be contracted separately as time and materials, labor hours plus other fixed costs.

Four popular contract types that are commonly used for software services acquisition are listed below. Each type offers a different risk profile for the external provider and the customer:

- fixed priced contract;
- cost plus percentage of cost;
- cost plus fixed fee;
- risk sharing.

For all these types of contracts, it is necessary that SQA takes the time to propose appropriate contract clauses to the project manager. This will ensure the quality of results (i.e., ensure alignment of the contract clauses with the process, project plan, and SQAP). In addition, it is important to describe, among other things, how cost overruns will be handled as well as the supplier cost control process.

12.7.1 Fixed Price Contract

This type of contract is an agreement where the supplier undertakes the work at a specific price. The supplier assumes the greatest share of the risk. However, the profit margin can be very advantageous. Indeed, the supplier's interest is to reduce his costs and be efficient because the margin between the agreed price and the actual cost of the project will be his profit. The only way this can be done is if the supplier controls the process. The contractor cannot control costs effectively without controlling the processes, inputs, and outputs. This type of contract is rigid and uses a change control process when unforeseen activities need to be completed. This type of software contract is usually used when the project specifications are well defined, contractors are experienced and market conditions are stable. The main risk to the customers are that software suppliers will typically provide a low bid to get the business and then, once the work is started, will identify missing specifications and send change requests. Enhancements are one of the most common sources of software supplier claims for additional compensation with this type of contract.

When this type of contract has been chosen for a software project, SQA will want to help the project management team to reduce uncertainty. A clear WBS will be required from the supplier to verify that detailed activities required have been already identified as part of the initial bid. When this cannot be done with assurance then hybrid contracts should be considered. Currently, for turnkey delivery of complex software solutions, a very popular approach consists of contracting an integrator. An integrator is a special type of supplier that takes on the responsibility of coordinating the overall responsibility for all the subcontractors (i.e., hardware, software package, data conversion, and others). In this situation, the integrator assumes a lot of responsibilities. Consequently, the contract and SQAP will be more complex.

12.7.2 Cost plus Percentage of Cost

This type of contract will reimburse the supplier for allowable expenses specified in the software requirements. In addition, the supplier receives a percentage, defined in the contract, to reflect his profit. From the customer's point of view, this type of contract is riskier because there is little incentive for the supplier to cut costs. In fact, the supplier will tend to increase costs since this will increase his profits.

With this type of contract, the project manager will want to focus particular attention on the control of worked hours and the cost of materials to ensure that the supplier will not increase costs only for the purpose of increasing his profit.

12.7.3 Cost plus Fixed Fee

In this type of contract, the supplier charges back allowable expenses for performing the software contract. The fixed fee is how the supplier makes a profit. Unless there is a change to the contract, this fee remains constant throughout the contract.

The customer still assumes a high share of the risk. However, compared with the previous type of contract, the supplier is encouraged to complete the work as fast as possible to get his fee. Still, the supplier will try to lower his own costs and ensure a profit margin on every activity. The project manager will have to ensure tight control of worked hours and the quality of supplier personnel.

12.7.4 Risk Sharing

This type of contract is well adapted for complex software acquisitions. With this type of contract, the supplier is reimbursed for allowable expenses for the execution of the contract. In addition, the supplier has the opportunity to receive a bonus if the work is completed early. This bonus will be paid if the final cost is less than the agreed upon estimated price. The savings will be shared between the supplier and the customer. Both the supplier and the customer share the advantage of completing the project ahead of the deadline. On the other hand, missing the deadline is also a shared risk.

Here is an example of the use of a risk sharing approach used as part of a large financial software replacement project. In Figure 12.3, the costs borne by the customer are in black and the costs borne by the integrator are shown in white. The supplier agreed to implement this software package for $1,174,902 within 42 working days (i.e., 8 hours per day). This estimate includes some uncertainty and both partners are willing to share the risk associated with missing this deadline but the

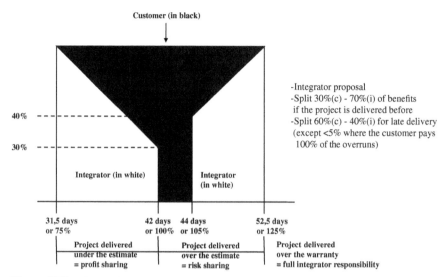

Customer (in black)

-Integrator proposal
-Split 30%(c) - 70%(i) of benefits
   if the project is delivered before
-Split 60%(c) - 40%(i) for late delivery
   (except <5% where the customer pays
      100% of the overruns)

40%

30%

Integrator (in white)   Integrator
                                (in white)

31,5 days   42 days 44 days   52,5 days
or 75%   or 100% or 105%   or 125%

Project delivered   Project delivered   Project delivered
under the estimate   over the estimate   over the warranty
= profit sharing   = risk sharing   = full integrator responsibility

Figure 12.3   Graphical representation of risk sharing contract.

customer wishes to obtain a guarantee of a fixed budget ceiling in the contract. Given that the supplier that was chosen has a very good track record of executing this type of project, this estimate is solid.

The stakeholders (i.e., the customer and supplier) agree that after a 5% overrun, the costs will be shared in this way—60% by the supplier and 40% by the customer. In addition, the customer wishes to establish a limit (a guarantee) because a fixed budget needs to be approved. This limit is set at 25% of overrun. This means that beyond a 25% overrun of the estimated budget, the supplier will have to assume all the extra costs until the delivery. A careful and professional supplier will therefore prepare his proposal accordingly and have confidence to assume that risk.

Let us look at the details of this contract. Table 12.1 describes the change in percentages and their effects. To calculate the values in this table, a simple slope formula is used: for example, for the supplier $y = 4.7x - 146.8$, where x is the number of days to complete the project and y is the percentage of cost over budget. This formula is derived by considering the values from two data points in Figure 12.3: $y_2 = 100\%$, $y_1 = 60\%$, $x_2 = 52.5$ days, and $x_1 = 44$ days. For example, if the project exceeds 14% from the target date, the integrator will assume $y = (4.7 \times 48) - 146.8 = 78.8\%$ of cost overrun. Alternatively, the customer will only pay the remaining 21.2%. The interesting feature of this type of contract is that there is a maximum amount of risk (i.e., cost) that can be identified at 12% = \$36,226.

We know that schedule overruns are unfortunately very likely with these types of software projects. Since all cost overruns beyond 25% will be borne by the supplier, it is possible to calculate a maximum guaranteed budget for the customer should things

Table 12.1 Maximum client budget of a risk sharing agreement for a software project.

Price of contract (set by RFQ)			$1174902				
Estimated effort (set by RFQ)			42 days			Slope	4,7
Daily cost			$27974			Intercept	−146,8
% late (days)	Day	% risk of supplier	% risk of customer	Late ($)		Cost to supplier	Cost to client
On time	**42**	**0**	**0**	**$0**		**$0**	**$0**
2%	43	0	100	$27974		$0	$27974
5%	44	60	40	$55948		$33569	$22379
7%	45	65	35	$83922		$54297	$29624
10%	46	69	31	$111895		$77655	$34240
12%	47	74	26	$139869		$103643	**$36226**
14%	48	79	21	$167843		$132260	$35583
17%	49	84	17	$195817		$163507	$32310
19%	50	88	12	$223791		$197384	$26407
21%	51	93	7	$251765		$233889	$17875
24%	52	98	2	$279739		$273025	$6714
25%	52,5	100	0	$293726		$293726	$0
Client budget for worst case = 1174902+$36226 =				**$1211128**			

go very wrong. Looking at the cost to the client, the maximum value is reached at a 12% overrun. With this type of contract, the customer's project manager is fully confident that with a budget of $1,211,128, he incurs no additional risk of overruns. With this assurance, the project manager and SQA team can focus on the quality of the deliverables.

To advise project teams and assist in the establishment of well-designed contracts, the SQA specialist needs to be familiar with software contract types and clauses. After planning a contract strategy and before signing this agreement, contract reviews are required to ensure the alignment of the project plan, quality plan and contract clauses. The next section presents the recommended contract review activities.

12.8 SOFTWARE CONTRACT REVIEWS

There are two major contract reviews mandated to ensure the quality of a software agreement. These two reviews (initial and final) aim to improve the odds of meeting budget, schedule, and target quality. The requirement for a contract review process should originate from the customer. Suppliers should make a substantial contribution to these activities.

Contract review objectives are:

– to identify the factors that influence the magnitude of each review;
– identify the difficulties in conducting the contract review;
– to explain the activities and objectives for the implementation of each contract review;
– to discuss the importance of conducting these reviews.

12.8.1 Two Reviews: Initial and Final

Many situations require the signing of a contract between a customer and a supplier. The most common are:

– participation in a tender process;
– presentation of a proposal in response to a client request;
– receipt of a request or an order from another organization department.

The review process and its usefulness for detecting defects were discussed in a previous chapter. The contract review aims at ensuring the quality of the software contract and supporting documents. If applicable, the study of the project plans and contract will ensure that a suitable type of contract has been selected and that sufficient details have been included in the agreement concerning SQA, for all stakeholders involved. The contract review process is typically conducted in two separate steps:

– first step: initial review of the project proposal. This step looks carefully at the original project proposal right at the time of supplier selection. This review establishes:
  o that the customer requirements list and the accompanying documents offer enough details;
  o that the supplier description concerning cost estimates, schedule, and resources is detailed enough and contains SQA activities;
  o the contract type recommended by the supplier;
  o the supplier and subcontractor's responsibilities.

– second step: final review of each of the proposed contract clauses before signature. This second contract review allows for the detailed review of contract terms, budgets, deadlines, warranty, and quality level including the amendments agreed upon during contract negotiations meetings.

The contract review process can start once the project plans are submitted. Persons conducting the review must have, on hand, a checklist to ensure the completeness of items to be reviewed. After the completion of a contract review, it is necessary to confirm that the modifications, additions and corrections are made by the supplier to the contract. Proper contract configuration management must be applied. Some organizations will also want to seek the participation of their legal department. These delays must be taken into account.

Risks Related to Contracting Software Projects:
– delays in delivery;
– not having the expertise in the required domain;
– delivery within budget;
– in terms of software product quality;
– poorly written and missing documentation;
– project success;
– supplier bankruptcy;
– litigation;
– with regards to the intellectual property of software developed by a supplier;
– in terms of support after delivery, for example, many years of support required for an organization that manages a subway, medical software, and so on.

Adapted from CEGELEC (1990) [CEG 90]

12.8.2 Initial Contract Review

As might be expected, the initial and final contract reviews that are proposed will have different objectives. The objective of an initial contract review is to verify that the following items have been satisfactorily completed:

– the customer requirements have been clarified and documented to a level that provides a good comprehension of the task ahead. Some project planning documents and technical documents can be too general or imprecise. Therefore, additional information should be obtained to refine and clarify expectations and requirements. This document is a paramount component (i.e., attachment) of a software contract;

– alternative options to acquisition have been investigated. Often, these alternatives have not been thoroughly considered at the onset of the project. This initial contract review allows for alternative options to be considered one last time: building a solution from the ground up, reusing or updating the existing software, partnerships with other organizations to use their solution and finally, when acquisition is imminent, contract types that would better fit the particular situation;

– the planned responsibilities of project participants and stakeholders as well as the planned approval process and communications channels are reviewed. The final proposal should identify, as clearly as possible, formalities like: roles and responsibilities, stakeholder activities and communication channels, interface between groups involved, acceptance criteria (e.g., intermediary deliverables and final solution) for users, maintainers, and infrastructure organizations, the approval process stages and steps, as well as the inevitable change control process;

– the risk management approach should be reviewed. We have already seen, in this chapter, that the choice of a contract type already mitigates some risks. Other areas of risks should be examined during this review: missing description of complex requirements, missing comprehension of existing processes/business rules, interdependencies to other ongoing projects, missing expertise, and planned use of new and unproven technologies, techniques, and tools. A previous chapter covered this topic in more detail and proposed approaches;

– a review of the proposed estimates for resources, schedule, and budget. This is key as historically, at RFQ response time, answers from suppliers do not provide much detail and do not always inspire confidence that all of what is needed has been understood correctly and planned for in the project, for example: interfaces to other systems, data conversion, process re-engineering, training, and change management.

In many software acquisition projects in which Dr. April has participated, the supplier deliberately presents the lowest bid to win the contract. Then, as the project progresses, he will raise a number of change requests to compensate. To try to contain this type of behaviour, customers should use solid contract clauses and project management processes that protect the client. Visit our website, in the contract section, to get useful contract clauses that are adaptable to your particular situation. When knowledgeable and professional organizations are involved, the contract is sound, the estimates and contingencies are realistic, there is no need to play games, and it has been demonstrated that fixed price contracts can be successful for both parties when a risk sharing contract is used.

– a review of the ability of the supplier to meet his obligations. This area of the review investigates the financial health, previously demonstrated track record, and current ability of the chosen external provider that will lead the project. At this time, we need to have a good level of confidence that the supplier personnel planned to be assigned to the project have the sustained availability, experience, and competence for the task ahead. You want to focus your attention on the key personnel that are planned to be assigned and their previous expertise on similar projects. You will also want to ensure that these key personnel will not be removed during execution, that personnel that you find inadequate can be replaced, and that there is a good understanding of the organization chart proposed. This includes a review of the understanding of the notion of integrator versus supplier. The term we like to use, in these types of complex and risky software acquisition projects (for an example, refer to the contract example on our website), is integrator. This is critical because an integrator is responsible for the coordination of the work of other suppliers for a turnkey solution. This gives a lot of responsibilities to the external provider playing this role and therefore, in this part of the initial contract review, it is necessary to validate that this is well understood. A typical subcontractor will have limited responsibilities which creates the possibility that an important activity or deliverable is not assigned to anyone;

– a review of the ability of the customer organization to meet their project obligations. It is common to ask questions about the suppliers' ability to deliver but have we done the same exercise with our own organization? In this part of the initial contract review, consider internal expertise and the availability of: a dedicated and committed customer representative, competent and available IT staff from software architecture, business analysis, interface development, data understanding/conversion, maintenance and infrastructure, and the support of the legal department;

– the definition of integrator versus subcontractor. The term used to describe the external provider is important because it impacts his responsibilities. An integrator will be responsible for the totality of the turnkey deliverables. A subcontractor is typically responsible for more limited work. It is necessary to clarify these responsibilities in the contract;

– the definition and protection of the ownership rights. At this time all licensing matters are reviewed to understand how the final software can be used and by what organizations and users. Some organizations also take the time to review the security issues.

As we have seen, the initial contract review is quite intensive and can be summarized as described below.

A checklist you can use to ensure the eight topics of the initial contract review have been addressed:

– the requirements, both functional and technical (e.g., the information technology requirements), have been clarified and are documented in the planned contract attachments;
– alternatives to acquisition have been considered, documented, and discarded;
– the relationship between the customer and the integrator is documented and roles and responsibilities are clear;
– the risk management approach has been clarified;
– estimates have been reviewed and provide good confidence;
– the integrator's abilities have been verified and provide good confidence, including the understanding of the role of an integrator;
– the customer and the information technology abilities have been verified and provide good confidence;
– rights of use and licensing have been reviewed.

12.8.3 Final Contract Review

Once a proper initial review has been completed, confidence that the software acquisition will be a success improves greatly. This second review is concerned with the detailed clauses of the contract:

– areas that have not been clarified in the attachments or in the text of the main contract;
– all contract clauses that describe the agreed upon key processes to be executed by both the customer and the integrator in this project;
– there has not been a last-minute change, addition, or omission in the contract. All changes have been thoroughly discussed and agreed.

Avoid Ending up in Court

Instead of using the contract to constrain the parties, it is important that both parties completely understand their obligations before beginning the relationship. When both parties develop a good partnership, the project will be a success. Litigation then becomes less likely.

How can you know if you are partnering and not simply signing a contracting you may later regret? The best indicator is that you feel that one party is trying to win something while the other party is losing. Adversarial attitudes appear early on with these types of projects and you should look out for their signs. Win–lose situations are precursors of litigation.

Adapted from DeMarco and Lister (2000) [DEM 00]

12.9 SUPPLIER AND ACQUIRER RELATIONSHIP AND THE SQAP

IEEE 730 [IEE 14] has been created for the situation where software development is considered to be carried out by a supplier and delivered to an acquirer. For this reason, the entire standard is pertinent when a supplier and an acquirer enter into a relationship for a software acquisition. This complex software project needs to describe, in its SQAP, how suppliers will be managed to ensure a quality delivery. This standard requires that means to achieve quality be described in a plan and a contract to ensure that both supplier and acquirer clearly understand the requirements as well as their responsibilities.

IEEE 730 explains, in detail, every SQA activity presented in the SQA process of the ISO 12207 standard.

Warning Signs of Acquisition Headaches Ahead

– regularly scheduled status reports that are not delivered on time, lack of expected information, or do not jibe with visible signs of progress, such as completed deliverables;

– uncompleted action items, unresolved issues, failed dependencies, conflicts that are not being resolved effectively or other unfulfilled commitments;

– unqualified supplier or acquirer staff being assigned, or key supplier or acquirer staff being replaced by other individuals;

– acquirer not actively managing and monitoring the relationship with the supplier;

– unrequested requirements being implemented or requested requirements being omitted without negotiation and agreement;

– scheduled reviews that did not take place, or reviews that should have been scheduled but were not;

– decisions not being made in a timely fashion by the right people, or decisions that are not communicated promptly to the affected individuals;

– incomplete deliverables received, or contractually required deliverables that do not appear;

– documents being received, but no working software being delivered;

– processes that are not working well or are being bypassed inappropriately;

– project-tracking trend charts (such as earned-value, defect-detection, defect-closure, and requirements-change charts) that do not show signs of completion being forthcoming;

– actual cost, schedule, or effort results that deviate significantly from estimates without explanation;

– missed early milestones, which do not bode well for completion of future milestones.

Adapted from Wiegers (2003) [WIE 03]

12.10 SUCCESS FACTORS

We summarize the factors that affect quality during the software acquisition process in the next text box.

Factors that Foster Software Quality

1) Software acquisition process maps communicated to suppliers in RFQ.

2) A pre-approved adaptable software acquisition contract template is available.

3) Help and support of a knowledgeable QA specialist for contract reviews.

4) Using the appropriate contract type for the situation.

5) Review is done beforehand and there is follow-up during the project.

Factors that may Adversely Affect Software Quality

1) External providers that do not understand the software acquisition process.

2) Using the external supplier contract without having the opportunity to adapt it.

3) A software acquisition contract that was not reviewed by the project manager and the QA specialist.

4) Choosing an inappropriate type of contract.

5) Not receiving progress reports or receiving reports that contain partial information.

6) Action items that stay open, unresolved questions and conflicts.

7) Unqualified or constantly changing personnel.

8) Missing or delaying software product reviews.

12.11 FURTHER READING

EBERT C. *Software Engineering on a Global Scale: Distributed Development, Rightshoring and Supplier Management.* IEEE Computer Society, Los Alamitos, CA, 2011.

TOLLEN D. *The Tech Contracts Handbook: Software Licences and Technology Services Agreement for Lawyers and Businesspeople.* American Bar Association, Chicago, IL, 2011.

VERVILLE J. and HALINGTEN A. *Acquiring Enterprise Software: Beating the Vendors at Their Own Game.* Prentice Hall, Upper Saddle River, NJ, 2000.

12.12 EXERCISES

12.1   One of the objectives of a contract review is to assess the risks of going forward with the agreement.

   a) List the types of risks that are more likely to be present.

   b) What activities can you suggest to mitigate these risks?

12.2   The complexity of a contract review varies according to the complexity of the software acquisition project:

   a) What are the software acquisition project characteristics that justify using the recommendations made in this chapter?

   b) What adaptations of the concepts presented can be made for small software acquisitions? Explain what can be left out and why.

12.3   It is sometimes difficult to successfully conduct contract reviews:

   a) List the difficulties of conducting contract reviews.

   b) Can you create a checklist that will remind you of the important items to be reviewed?

12.4   Explain the difference between a supplier and an integrator.

12.5   Describe the risks incurred by an organization when it wants to acquire a software product from a supplier that himself uses subcontractors.

12.6 Describe two essential activities to ensure the good management of multiple suppliers in a software acquisition project. Explain why you think these two are the most important.

12.7 Explain the differences between a cost plus percentage of cost contract and a cost plus fixed fee contract.

12.8 Describe the terms of a potential risk sharing agreement for a 40–60% before deadline delivery as well as a 50–50% when an overrun occurs (except for overruns between 1% and 10%, where the customer absorbs 100% of the overrun costs). The estimated budget is a million dollars:

a) Draw the contract risk sharing diagram.

b) Develop the cost sharing table as well as a detailed description of the risk shared between the customer and the integrator.

c) What is the maximum overrun possible?

d) What is the maximum price guaranteed by this contract?

Chapter 13

Software Quality Assurance Plan

After reading this chapter you will be able to:

- use the information provided in each chapter to develop a complete SQA plan for a project;
- understand the SQA requirements presented in the IEEE 730 standard;
- refer to the detailed explanations in the appropriate chapter and section of the book.

13.1 INTRODUCTION

This chapter is devoted to the use of concepts and practices presented in this book to implement a software quality assurance plan (SQAP). Figure 13.1, adapted from Daniel Galin's house of software quality, connects all of the book's concepts as components that must come together to achieve software quality.

Before diving into the development of a SQAP, it is worth reviewing the definition of SQA as described in the following text box.

Software Quality Assurance

A set of activities that define and assess the adequacy of software processes to provide evidence that establishes confidence that the software processes are appropriate for and produce software products of suitable quality for their intended purposes. A key attribute

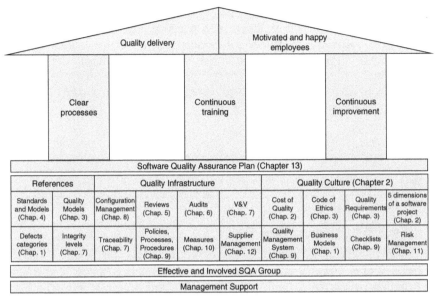

Figure 13.1 The house of quality for software projects.
Source: Adapted from Galin (2017) [GAL 17].

of SQA is the objectivity of the SQA function with respect to the project. The SQA function may also be organizationally independent of the project; that is, free from technical, managerial, and financial pressures from the project.

Function

In a software application, a module that performs a specific action. In an organization, a function is a set of resources and activities that achieve a particular purpose.

IEEE 730 [IEE 14]

IEEE 730 states that the term "SQA function" is not to be interpreted as a particular person, tool, document, job title, or a specific group dedicated to SQA, regardless of how that function is staffed, organized, or executed. The SQA function's responsibility is to produce and collect evidence that forms the basis for giving a justified statement of confidence that the software product conforms to its established requirements [IEE 14].

Two clauses of IEEE 730 are dedicated to the planning and execution of the SQAP: clause 5.3.3 entitled "Document SQA planning," and clause 5.3.4 entitled "Execute the SQA Plan." In the first part of this chapter, we present the SQA planning, and in the second part, we will present the execution of the SQAP.

Thirteen mandatory tasks that are part of clause 5.3.3 describe what has to be done by the SQA function during the software planning stage of a project. IEEE 730 states that SQA activities are planned and executed in a manner that is commensurate with product risk—the higher the product risk, the greater the breadth and depth of SQA activities.

As can be seen in the following text box, the IEEE 730 standard [IEE 14] describes the normative outline of a SQAP.

Software Quality Assurance Plan—Outline

– Purpose and scope
– Definitions and acronyms
– Reference documents
– SQA plan overview
  o Organization and independence
  o Software product risk
  o Tools
  o Standards, practices, and conventions
  o Effort, resources, and schedule
– Activities, outcomes, and tasks
  o Product assurance
  o Evaluate plans for conformance
  o Evaluate product for conformance
  o Evaluate plans for acceptability
  o Evaluate product life cycle support for conformance
  o Measure products
  o Process assurance
    • Evaluate life cycle processes for conformance
    • Evaluate environments for conformance
    • Evaluate subcontractor processes for conformance
    • Measure processes
    • Assess staff skill and knowledge
– Additional considerations
  o Contract review
  o Quality measurement
  o Waivers and deviations
  o Task repetition
  o Risk to performing SQA

- o Communications strategy
- o Non-conformance process
- – SQA records
  - o Analyze, identify, collect, file, maintain, and dispose
  - o Availability of records

IEEE 730 [IEE 14]

One task included in clause 5.3.3, task 12, is quite clear about the content and the development of the SQAP [IEE 14]: address all of the topics listed in the normative SQA plan outline. Every section in the plan is to be included. The topics of the SQAP are normative but the section names and order are informative. If a section in the outline is not applicable to a given project, a placeholder for that section may be included along with a justification for why the topic is not applicable. This clause must be respected if an organization wants to claim conformance to IEEE 730. If an organization does not have to conform to IEEE 730, it may use this chapter as a set of guidelines in the development of a SQAP commensurate with the risks of the product to be developed.

Preparing and obtaining sign off on a SQAP is often the project manager's or the quality assurance manager's responsibility. SQA can help by offering a standardized template, examples, and explanations that can assist with this task. Organizations with an existing quality assurance function should have developed a SQAP template adapted to their internal methodology. A quality plan, which describes the quality activities and tasks to be executed during a software project, should be created from this template, approved by stakeholders, and kept up to date throughout the project. Exemplary SQAPs (i.e., from previous projects) can be made available in the life cycle process library to facilitate the development of a project-specific SQAP and to show newcomers examples of past projects in order to help them understand the typical content required.

"In today's software marketplace, the principal focus is on cost, schedule, and function; quality is lost in the noise. This is unfortunate since poor quality performance is the root cause of most software cost and schedule problems."

Watts S. Humphrey

13.2 SQA PLANNING

The following sections present, in more detail, the contents of each section of the SQAP according to Annex C of the IEEE 730 standard, and refer to specific chapters and sections of this book where detailed information has already been provided on the topic. Annex C also provides "suggested inputs" to help with the preparation and execution of a SQAP.

13.2.1 Purpose and Scope

This first section of a SQAP should describe the purpose, objectives, and scope of quality assurance activities that will be undertaken, including waivers obtained if present. This section should also explicitly identify processes and software life cycle products covered by quality assurance activities. It is a good idea to summarize which business model is associated with the specific project. You can find information about the different IT business models and their influence on quality assurance practices in section 1.6 of this book.

Here are questions that the SQA function can ask during the project planning phase to help refine the purpose and scope of SQA for a new software project (IEEE 730, Table C.3):

- Is the project scope clearly defined and well understood?
- Is the SQA role on this project understood by the acquirer, the organization, the project team and the SQA team?
- Are potential product risks known and well documented?
- Are potential product risks understood so that SQA activities can be planned in a manner commensurate with product risk?

13.2.2 Definitions and Acronyms

This section of the SQAP ensures that abbreviations and acronyms are explicitly stated. Refer to Chapter 1 to review information about the terminology that should be used consistently by software engineers when discussing quality. This is important because using recognized and standardized terminology that is fully aligned with approved software engineering standards will ensure that you can use your body of knowledge when a problem occurs. If a debate occurs, during final testing and delivery, about the meaning of a deliverable or a responsibility, you will always be able to fall back on your standards and body of knowledge to explain the intended meanings and solve these issues rapidly. It also ensures definitions, acronyms, and abbreviations are clarified for the project participants.

13.2.3 Reference Documents

This section of the SQAP identifies all applicable standards, industry-specific regulations and compliance clauses of contracts, other documents referenced by the SQAP, and any relevant supporting documents. Supporting documents may include applicable professional, industry, government, corporate, organizational, and project-specific references. Here are the questions that should be answered during the project-planning phase (IEEE 730, Table C.4):

- What government regulations are applicable to this project?
- What specific standards are applicable to this project?
- What organizational reference documents (such as standard operating procedures, coding standards, and document templates) are applicable to this project?
- What project-specific reference documents are applicable to this project?
- Is SQA expected to assess compliance with applicable regulations, standards, organizational documents, and project reference documents?
- What reference documents are appropriate to include in the SQAP?

The configuration management chapter of this book has presented how to refer, safeguard and manage successive versions of key project documents and life cycle artifacts. Documents referred to may include industry, legislative, or contractual documents pertaining to the project. The SQAP should also identify the origin of the project's requirements such as contracts, specifications documents, or deliverables list. Key project documentation and mandatory deliverables should be described with more detail, such as:

- the mandatory project deliverables that will be monitored, reviewed, and authorized during this project;
- where these documents/deliverables can be accessed;
- a reference to existing templates and examples, when available, that will help participants better understand the expected scope and content to avoid misunderstandings;
- identification of the role or individual who will be responsible for creating the deliverables and authorizing changes once official versions are published and finalized.

The benefit of this section of the SQAP is that by trying to produce Table 13.1 at the beginning of a project, the project manager will be forced to: (1) identify mandatory deliverables and (2) see if all the resources are available for the task ahead.

It becomes a useful checklist during project planning. Table 13.1 shows an example of the description of two mandatory software project deliverables. Organizations

Table 13.1 Example of Key Documents and Mandatory Deliverables Checklist for the Project

Document name	File name	Location	Reference template used	Authors	Approbation
Project plan	Project_143_plan.doc	C :Project_143-\Plan	Yes	M. Smith	A.Anderson
Functional specifications	Project_143_specs.doc	C :Project_143-\Specs	Yes	D. Connor	B.Thomas, P.Rodriguez
...	...	...	...	...	...

must decide on their minimum list of mandatory deliverables to be produced during a software project to ensure that the software meets the internal requirements of their internal software development methodology.

When reviewing the SQAP and during project reviews, the following questions should be asked:

– Is this table complete? Are all the mandatory documents/deliverables listed (i.e., refer to the mandatory deliverables mentioned in the organization's methodology) for the project?

– Are the documents/deliverable templates identified and used by the project?

– Is there a SQA activity in the project schedule to review the quality of documents/deliverables?

– Can I easily find and access the content of the documents/deliverables?

– Who has access to the documents/deliverables? Is it the authorized and most recent version?

– Are the project documents/mandatory deliverables under version control?

13.2.4 SQAP Overview—Organization and Independence

Software project organization and project management, as a whole, are topics that have not been covered by this book. They are important topics that impact software quality.

You can refer to the Software Engineering Management knowledge area of the SWEBOK (www.swebok.org) as well as the PMBOK® Guide of the Project Management Institute (www.pmi.org) for recommendations and more information on these topics.

This section of the SQAP aims at clarifying the parties responsible for performing SQA, for a project, and seeks to present the interrelationship with the project management team. A functional organization chart often represents these relationships. If subcontractors are involved, the relationships and information flow between SQA and each subcontractor should also be shown. By clarifying relationships in this section of the SQAP it will be possible to see if the required roles and responsibilities are present. Here are the questions that SQA can ask during the project-planning phase concerning this topic (IEEE 730, Table C.5):

- Have deficiencies in the organization's SQA policy been identified and documented?
- Has the project manager established a method for monitoring the execution of SQA activities, tasks, and outcomes along with a method for providing feedback to the independent SQA function (when present)?
- Has the project manager established an effective and appropriate policy defining and governing SQA roles and responsibilities during the project?
- Has the organization established an independent SQA function with sufficient influence over software processes, including an effective reporting line independent of the software development project?
- Has the organization established responsibility for supervising a project's SQA function by an individual independent of both the project manager and the software development manager?
- Has the organization established a method to enable projects to learn from the experiences of previous projects, if the SQA function is being established for more than one project?
- Does the independent SQA review exist independently of SQA processes established in individual projects?
- Have adequate resources, including sufficient numbers of suitably skilled and trained people as well as sufficiently capable tools and equipment, been identified for the project?
- Has the degree of independence (i.e., technical, managerial, and financial) been defined? (see Figure 13.2)
- Is the defined degree of independence appropriate, given the potential product risk and the requirements when the project uses external suppliers?

It is therefore suggested that the organizational structure of the software project be described and that the role of each participant is identified in this section of the plan. Figure 13.2 shows individual part of the management committee, project committee, and those supporting the project that will be asked to play a role.

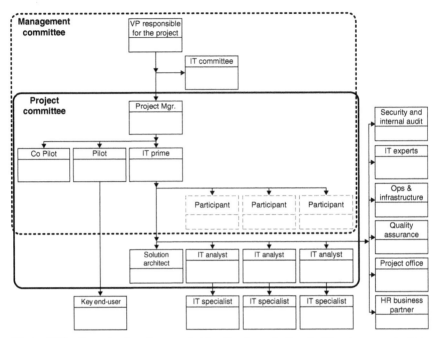

Figure 13.2 Example of a software project functional organization chart for a large organization.

Creating a visual representation of a project organization will quickly highlight if the numbers of individuals that have been assigned to the project are sufficient, if they are located at the right place and if they are qualified to do the work.

RACI Chart or Matrix

Clarifies roles and responsibilities, making sure that nothing falls through the cracks. RACI (Responsible, Accountable, Consulted, Informed) charts also eliminate redundant activities (two or more people working on the exact same thing, without knowing it) and confusion by assigning clear ownership for each task or decision.

Adapted from http://racichart.org

A RACI chart can also be used to describe each individual role. Rarely does everyone on a team have the same understanding of who is responsible for what. A RACI chart, also known as a RACI matrix, clarifies roles and responsibilities, making sure that nothing is forgotten. RACI charts also eliminate redundant activities and confusion by assigning clear ownership for each task or decision. It is important to identify early if a person is playing too many roles or, conversely, if a role is

Table 13.2 Specifying the Names of Individuals Who are Involved in a Project

Role	Assigned to	Responsibilities
Project committee	A. Lopez (Pilot), G. Wright (IT Responsible), M. Thomas (SQA), P. Smith (IT expert)	Follow the progress, address issues, set project directions and priorities, authorize budgets and changes.
Key end-user	P. Clark, H. Johnson	Validate requirements, functional experts, conduct functional acceptance of system.

currently vacant/unassigned. Table 13.2 is another simple way of documenting roles and responsibilities of individuals in a software project.

This section of the SQAP should also specify the degree of independence between the organization performing the SQA function and the project team members. There are three parameters that can be used to define independence: (1) technical independence, (2) managerial independence, and (3) financial independence. Technical independence requires that SQA utilize personnel who are not involved in the development of the system or its elements. SQA forms its own assessments of all project activities. Technical independence is an important method to detect subtle errors overlooked by those too close to the solution. Managerial independence requires that responsibility for SQA be vested in an organization separate from the software development and program management organizations. Managerial independence also means that SQA independently selects segments of software to analyze and test, chooses techniques, defines the schedule of SQA activities, and selects the specific technical issues and problems to act upon. The SQA effort provides its findings in a timely fashion simultaneously to both the software development and program management organizations. Financial independence requires that control of the SQAP budget be vested in an organization independent of the software development organization. This independence prevents situations where SQA cannot complete its activities because funds have been diverted or adverse financial pressures or influences have been exerted.

"Quality is not only right, it is free! And it is not only free, it is the most profitable product we have! The real question is not how much a quality management system will cost, but how much the lack of one will cost."

Harold Geneen
CEO ITT Corporation

13.2.5 SQAP Overview—Software Product Risk

The chapter on risk management has presented how risk relates to SQA. Software product risk refers to the inherent risks associated with the use of the software product (e.g., safety risk, financial risk, security risk). Software product risk is different from project management risk, which is addressed later in this plan. Specific V&V techniques may be required to address software product risk. Table 13.3 lists the questions and suggested inputs, such as a RMP, that SQA can discuss during the project-planning phase (IEEE 730, Table C.7).

To explain the extent of risk management activities, this section of the SQAP should clearly state what level of integrity the software being worked on is at. Criticality analysis of software components, which includes four levels, was presented in the chapter on V&V. The SQAP must ensure that the tasks and activities identified are carried out in a way that is commensurate with the level of criticality of the assigned software. Depending on the software criticality, projects apply more or less strict engineering and SQA requirements. This important information will allow the reader to assess the number of compensating provisions required to mitigate the consequences of software failure.

Table 13.3   Questions and Suggested Inputs Related to Software Product Risk to Consider Asking During Project Planning Phase [IEE 14]

Questions	Suggested inputs
– Are potential product risks known and well documented?	– Acquisition plan
– Are potential product risks understood so that SQAP activities can be planned in a manner commensurate with product risk?	– Contract – Concept of operations – Risk management plan
– Has the scope of product risk management to be performed been determined?	
– Have appropriate product risk management strategies been defined and implemented?	
– Is a criticality analysis planned?	
– Does the project team have adequate training in product risk management techniques?	
– Is the project team planning to adjust their activities and tasks in a manner commensurate with product risk?	
– Are the breadth and depth of planned SQAP activities commensurate with product risk?	

13.2.6 SQAP Overview—Tools

This section of the SQAP describes tools to be used by SQA to perform specific tasks. These may include a variety of software tools to be used as part of the SQA process. Appropriate acquisition, documentation, training, support, validation, and qualification information for each tool is included in the SQAP. Here are the questions that SQA can ask during the project-planning phase:

- Have adequate resources, including sufficiently capable tools and equipment, been identified for the project as well as for other projects if the SQA function is being established for multiple projects?
- Are all tools planned to be used by SQA on this project completely identified, including supplier, version or release number, system platform requirements, tool description, and numbers of concurrent users?
- Based on product risk, do these tools require validation before they can be used on this project?
- Is training in the effective use of SQA tools required and if so, is training planned?

Regarding the tools, do not forget to specify the version of the tools used for a given project. It is also in this section of the plan that you should describe the tools and techniques that will be identified by the team for this project.

13.2.7 SQAP Overview—Standards, Practices, and Conventions

Chapter 4 of the book described software engineering standards and how they relate to SQA. This section of the SQAP identifies standards, practices, and conventions to be used in performing activities and tasks and for creating outcomes:

- Have all laws, regulations, standards, practices, conventions, and rules invoked in the contract been identified?
- Have specific criteria and standards against which all project plans are to be evaluated been identified and shared within the project team?
- Have specific criteria and standards against which software life cycle processes (supply, development, operation, maintenance, and support processes including quality assurance) are to be evaluated been identified and shared with the project team?

Describe the techniques and methods to be used for the project. Chapter 9 discussed how policies, processes, and procedures are used by a project. For software projects involving external suppliers, it is important to clarify which methodology (e.g., Oracle Unified Method, IBM Rational Unified Process, Scrum) will be used

Table 13.4 Reference Table for Standards, Practices, and Conventions

Step of the life cycle	Intermediary deliverable	Standard, practice, or convention
Planning	Project plan	PMI - PMBOK® Guide
Planning	SQAP	IEEE 730
Programming	Source code	Java programming rules revised April, 20 1999
Transition to production	Technical documentation	A local production criteria and checklist

and also to clarify the obligations to produce certain mandatory deliverables (see Table 13.2).

Methodology

A system of practices, techniques, procedures, and rules used by those who work in a discipline.

PMBOK® Guide [PMI 13]

This section also lists which standards, practices, and conventions apply to this project. It will specify the stage of the life cycle, the list of intermediate deliverables as well the standards, guides, or conventions used to ensure that quality will be met during reviews, inspections, and final acceptance steps of the project. We have discussed many standards and models in Chapter 4.

Table 13.4 provides an example of a simple presentation of these concepts.

By specifying these items in advance, we facilitate the task of reviewing the quality plan and project reviews for the project manager and SQA specialist.

13.2.8 SQAP Overview—Effort, Resources, and Schedule

This section includes estimates of the effort required to complete the activities, tasks, and outcomes as defined in the SQAP. This section identifies appropriately qualified SQA personnel and defines their specific responsibilities and authority within the context of the project.

This section also identifies additional SQA resources, including facilities, lab space, and special procedural requirements (e.g., security access rights and documentation control) that are required to perform SQA activities. This section should also include a list of critical SQA project milestones and a schedule of planned SQA activities, tasks, and outcomes. Here are the questions that SQA can ask during the project planning phase:

- – Can estimated effort and schedules be based on past projects?
- – Can resource requirements for this project be determined based on past projects?
- – What resources (lab spaces, servers, software, databases, operating systems, security rights, document control, etc.) are required for this project?
- – Is effort based on factual information rather than "gut feel"?
- – What estimating and scheduling techniques can be used for this project?

Typically, the estimates describe the software activities pretty well (i.e., feasibility, requirements, design, construction, testing, and delivery). But, have all the efforts associated with the cost of quality been considered in these estimates and the schedule (prevention costs, appraisal costs, failure costs)? Refer to section 2.2 on cost of quality for details.

Software Estimation

We could not cover the software estimation process in detail in this book. If you are looking for help with understanding and improving this practice, we recommend that you consult the book published by Dr. Alain Abran: *"Software Project Estimation: The Fundamentals for Providing High Quality Information to Decision Makers"* [ABR 15]. The book introduces concepts and examples to explain the fundamentals of the design and evaluation of software estimation models.

For the assessment of the project schedule, this section of the SQAP should describe the project milestones and the planned SQA in the project schedule. Specifically, it indicates the deliverables that will be produced by the SQA activities required by the project manager and the project committee. A good schedule reflects all project activities and shows the individual assignment of personnel. To verify whether a scheduled assignment is realistic and reviewed, an assignment email could be sent to individuals during planning. A letter of assignment (also called a work assignment email) is to formally contact a project resource asking:

- – if a review of the planned activities represented in the proposed schedule has been done;
- – to confirm the durations;
- – if this planned assignment of work is correct and to confirm with a reply.

This simple approach validates the information described by the schedule. It works especially well with team members that do not report directly to the project manager. Finally, this section of the SQAP is also used to describe the qualifications of staff who will perform SQA activities during the project and to clarify their responsibilities, their degree of independence, and their authority to perform reviews, inspections, and ask for corrections. The plan should also identify additional resources, including facilities, tools, test labs and other requirements (e.g., security permissions and code/document repository access rights).

Assumption

A factor in the planning process that is considered to be true, real, or certain, without proof or demonstration.

Assumptions Analysis

A technique that explores the accuracy of assumptions and identifies risks to the project from inaccuracy, inconsistency, or incompleteness of assumptions.

Basis of Estimates

Supporting documentation outlining the details used in establishing project estimates such as assumptions, constraints, level of detail, ranges, and confidence levels.

PMBOK® Guide [PMI 13]

Every project and its plan is conceived and developed based on a set of hypotheses, scenarios, or assumptions. Assumptions analysis explores the validity of assumptions as they apply to the project. It identifies risks to the project from inaccuracy, instability, inconsistency, or incompleteness of assumptions [PMI 13]. The PMBOK® Guide recommends that assumptions made in developing the activity duration estimate, such as skill levels and availability, as well as a basis of estimates for durations, be documented in project documents. When developing a SQAP, assumptions should be documented and analyzed when estimating effort, resources, and schedules.

13.2.9 Activities, Outcomes, and Tasks—Product Assurance

Product assurance activities provide confidence that software products are developed in conformance to established product requirements, project plans, and contractual requirements. An important aspect of SQA is the establishment of confidence in the quality of the software products produced by the project. These products include not only the software and related documentation but also the plans associated with the development, operation, support, maintenance, and retirement of the software. A product may also be a software service provided to the acquirer. Product assurance activities may include SQA personnel participating in project technical reviews, software development document reviews, and software validation testing.

The outcome of the product assurance activities provides evidence that the software services, products, and any related documentation are identified in and comply with the contract and any non-conformances are identified and addressed. Product assurance is comprised of three activities:

- Evaluate plans for conformance: This section of the SQAP should include activities and tasks for evaluating the degree to which all plans required by contract have been prepared and are consistent with the contract and with each other. Use reviews as described in an earlier chapter to do this evaluation;

- Evaluate product for conformance: This section of the SQAP should identify activities and tasks related to the evaluation of the degree to which the software product and related documentation conform to established requirements, plans, and agreement. Refer to Chapter 7 for V&V and Chapter 5 on reviews for the required outcomes for this activity.

- Evaluate product for acceptability: This section of the SQAP should identify activities and tasks for evaluating the level of confidence that the software products and related documentation will be acceptable to the acquirer prior to delivery. Refer to Chapter 7 for V&V and Chapter 12 on supplier and contract management.

Product assurance requires that some measurement be performed on software products. This section of the SQAP should also identify activities and tasks for evaluating whether the measurements objectively demonstrate the quality of the products in accordance with established standards and processes. Section 3.3 of the book describes a process to formally define software quality requirements in a project so that the measurement can be used during acceptance. Chapter 10 also presents measurement topics.

13.2.10 Activities, Outcomes, and Tasks — Process Assurance

Similar to the product assurance, process assurance aims to ensure that the processes used by the project are appropriate, depending on the level of integrity, and are followed:

- Evaluate life cycle processes for conformance: Ensuring that the correct life cycle has been selected confirms that adaptation, if any, of the life cycle activities and deliverables are appropriate. This section of the SQAP should identify activities and tasks for determining the degree to which project life cycle processes and plans conform to the contract and the degree to which the execution of project activities conforms to project plans (e.g., life cycles selected and adapted, procedures, deliverables templates, gating, project plan, and responsibilities alignment with project/contract requirements). Refer to Chapter 9 on policies, processes and procedures as well as Chapter 12 on contract reviews.

– Evaluate environments for conformance: This section of the SQAP should identify activities and tasks for evaluating whether software development environments and test environments conform to project plans (e.g., suitability of development and test platforms, i.e., planning, sizing, and security), tools validation, configuration management toolset, project documentation server/folders). Refer to Chapter 8 for guidance on configuration management;

– Evaluate supplier's processes for conformance: This section of the SQAP should identify activities and tasks for evaluating whether supplier software processes conform to requirements passed down from the acquirer (e.g., contract reviews, supplier quality audit, supplier responsibilities review, subcontract work traced and managed by the system integrator correctly). Reviews, audits as well as supplier management have already been presented in earlier chapters.

This section also requires the use of reviews and audits in the project. The reviews were presented in Chapter 5 and audits in Chapter 6. Table 13.4 provides an example of a simple presentation of reviews and audits for the project. Configuration management concepts were also discussed in Chapter 8.

Process assurance requires that some measurement be done on software processes. This section of the SQAP should identify activities and tasks for evaluating whether the measurements support effective management of the processes in accordance with established standards and processes (e.g., appropriate set of measures identified and planned, a data collection and analysis process is selected, measurement is consistent with product risk and overall quality goals). Measurement was discussed in Chapter 10.

Finally, this section of the plan should assess staff knowledge and skill. It is necessary that the plan outlines whether the staff assigned to the project has the qualifications and training necessary. If a training plan is to be prepared and executed it typically has two specific audiences: (1) technical specialists who will need specific training and (2) user representatives who will perform the final acceptance as well as future end-users.

In this part of the SQA plan, describe the scope of training needs and the scope covered by the project. Here are the questions that SQA can ask during the project planning phase:

– Have adequate resources, including sufficient numbers of suitably skilled and trained people, as well as sufficiently capable tools and equipment, been identified for the project as well as for other projects if the SQA function is being established for multiple projects?

– Have required project skills been identified?

– Have staff and subcontractor training records been reviewed against required skills?

13.2.11 Additional Considerations

This section of the SQAP identifies any additional considerations, such as SQA processes that support both Project Management and Organizational Quality Management, that are not described elsewhere. The following topics are included in this section of the SQAP:

– Contract review process: This section of the SQAP identifies or references the contract review process and describes SQA roles and responsibilities with regard to contract reviews. In the case where there are suppliers, subcontractors, consultants, or integrators, this section will specify the most important clauses of this relationship. Chapter 5 review techniques can be used here. Supplier and contract management concepts were discussed in Chapter 12 such as:
  o describe the software acquisition process for the project;
  o identify the reviews that will be conducted by the supplier during the project;
  o identify the appropriate contract type;
  o specify which contract template will be used and the planned contract reviews.

– Quality measurements: This section of the SQAP identifies quality measurements that are appropriate for the project, specific data collection requirements associated with intended quality measurements activities as well as responsibilities for data collection, measurement, and reporting. It should identify software quality objectives, measurement processes, and tools that are appropriate for this project. Before quality can be measured, it needs to be specified and accepted as part of the project requirements. This section of the SQAP should refer to the functional and non-functional quality requirements. Recommendations on how to define software quality requirements have also been presented in section 3.3 and show how to:
  o select and describe the quality characteristics and sub-characteristics to be evaluated;
  o specify how this will be measured and clarify the software attributes to be measured;
  o set a quality objective target for the project;
  o describe, with an example, how the calculation is done;
  o explain how and what tools will be used to evaluate this measure at the acceptance stage.

Table 13.5 Overview of Planned Reviews of a Project

Review name	Objective	Life cycle phase	Intermediary product	Type of review
Requirements	Ensure the complete-ness/testability of requirements	Requirements stage	Functional specifications	Document reviews
Architecture and design	Ensure the maintainability and traceability of design to requirements	Architecture and design stage	Design documents	Document reviews
Source code	Ensure the conformity to local programming	Programming stage	Source code & unit test plans and results	Code and documents inspection
Quality audit just before system tests	Progress and readiness of products	Integration tests stage	Overall project	Quality audit

- In addition to setting specific targets for quality measures, the plan should identify the measures that will be used to report:
  ○ project progress for each of the five project dimensions of a software project (e.g., schedule, staff, features, cost and quality) as presented in section 2.4 of Chapter 2 and referred to by the PSM in Chapter 11;
  ○ defects density and reviews/audits of non-conformances correction status.
- Finally, this section of the SQAP should identify responsibilities for collection, validation of this data, and reporting. Table 13.5 outlines the planned reviews for a project.
- Waivers and deviations: This section of the SQAP defines or references the criteria used to review and approve waivers and deviations to the contract and project management controls. The SQAP describes SQA roles and responsibilities with regard to reviewing and approving waivers and deviations. The following text box provides a text that could be added to a SQAP.

A waiver request has been submitted to SQA—A waiver request should be reviewed and approved by a manager of SQA. The waiver request should indicate which part(s) of the SQAP is(are) requested to be waived. Once approved, the waiver request should be stored in the configuration management system and all stakeholders should be informed about the approved changes to the SQAP.

> A non-compliance with the requirements of the SQAP has been recorded by SQA—Typically, a non-compliance report, resulting from an audit, with the SQAP is forwarded to the manager of SQA and to the manager of the project. The project manager should quickly respond to the non-conformity report by following the procedures of the organization.

- Task repetition: This section of the SQAP defines or references the criteria used to determine when and under what conditions, SQA tasks previously completed need to be repeated. This section will describe, if present, the iteration tasks policy recommended after the identification of a defect (e.g., the rework needed to correct a defect). For example, we know that each functional unit of the software is subject to acceptance testing. At this time, the customer is making reasonable efforts to continuously accept and validate this basic functionality for each functional unit. The successful execution of the acceptance test on each functional unit is the means that constitute validation of the basic functionality of software. Should a defect or failure occur, it is necessary to specify in this section of the SQAP, the process that will be initiated to document the defect, assign a severity, assess the effort to correct the defect and retest this functional unit. Refer to rework in section 2.2 of Chapter 2.

- Risks to performing SQA: This section of the SQAP identifies potential projects risks that could prevent SQA from accomplishing its defined purposes, activities, and tasks. Examples include inadequate staffing levels, insufficient resources, and lack of training. Also included in this section are actions taken to mitigate identified project risks. Chapter 11 presents the risk activities as they relate to SQA.

- Communications strategy: This section of the SQAP defines the strategies for communicating SQA activities, tasks, and outcomes to the project team, management, and organizational quality management.

- Document acceptance process: Often, and especially when suppliers are involved, the project manager will have to decide on a process to review/accept intermediary deliverables (i.e., requirements document, design documents, test plans, and many other deliverables that are in the form of documents). At the review meeting, a moderator may be assigned to record the minutes, which detail all corrections and improvements agreed to during the meeting (refer to Chapter 5 on reviews). Based on the severity of the required changes, a unanimous decision must be reached to:
  o accept the document as is (acceptance);
  o accept the document with minor changes (conditional acceptance);
  o repeat the review when the significant changes agreed to in the review meeting have been incorporated.

Table 13.6 Defect Severity Classification Example

Category	Description
1	A defect that prevents the project from testing further as this defect needs to be corrected before going ahead. A solution must be found immediately and corrected before continuing to test this functionality.
2	An important defect that prevents the project from testing parts of a functionality but other parts can continue to be tested. A solution needs to be found in a reasonable delay.
3	A defect that does not cause a major problem for testing to continue. A solution is required but can be placed in a priority list to be treated after category 1 and 2 are done.
4	A minor defect to be placed in the list of things to do.

- The review minutes are then signed by all participants. If the document has been conditionally accepted, an update of the document is prepared and the moderator reviews the minor changes to ensure all items have been addressed. The moderator then declares acceptance for this deliverable.

- Non-conformance process: This section of the SQAP defines activities and tasks related to the process for reporting non-conformances for the project. Non-conformances can be reported by any project member but, at the acceptance stage, will be controlled by a configuration control board (CCB), as presented in section 8.7.2.

- We have learned by experience that it is often necessary to clarify the software Verification and Validation (V&V) terminology, processes, and techniques (including testing details) that are planned for a project, especially if it involves third-party suppliers. Contracts and third-party project plans notoriously omit details about this important quality activity. The technical details about V&V should be presented in the quality and test plans of the project. Use the SQAP to provide a checklist that will identify all of the V&V activities. Chapters 5, 6, and 7 have presented reviews, audit, and V&V topics. What is important for SQA is to ensure that the project team chooses the appropriate techniques depending on the required level of integrity and quality that is expected from this project.

- An initial clarification that needs to be documented in the SQAP is how the project will evaluate the quality of an intermediate deliverable (e.g., a document) during an approval meeting where a review is conducted. It is also important to clarify how a defect will be classified (see example in Table 13.6). Defect severity is often used in contracts with third parties to assess if a milestone has been achieved or not.

- In third-party contracts, it is also necessary to specify the testing levels as it often causes confusion. You will probably need to create a process map of the project acceptance processes to ensure everyone understands how this will be done. For example, Figure 13.3 presents a simple activity sequence diagram that helps with an overall view of how this software will be finally accepted into production and will meet each test level. In this example, three different organizations are involved: the customer for functionality, maintainers for its maintenance and support quality and, finally by the infrastructure team who is concerned with the production readiness and reliability and many other operational aspects (using production criteria).

– Acceptance process: Here is another example of system and integration testing (Figure 13.4). The system and integration test acceptance process consists of the following steps: the supplier provides a system and integration testing plan with an acceptance checklist at least one week in advance of the test. The project team reviews the plan and the checklist within one week (see the document acceptance process). Within 1 week of the supplier notifying the project team of the availability of the deliverable for the system and integration test, the project team attends a demonstration of the deliverable at which the test plan is executed and the acceptance checklist is completed. Should any problems be encountered, they are recorded as incidents (using a category from Table 13.6). The supplier and the project team jointly sign the incident report summary for this test. Based on the incident report summary for this deliverable, acceptance is achieved when no unresolved category 1 and 2 defects (incidents) are open. When any outstanding incidents are repaired, associated tests must be re-run.

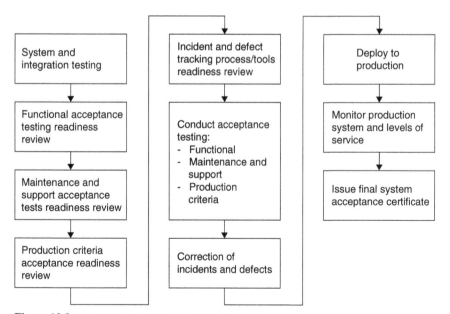

Figure 13.3   Example of a workflow explaining the acceptance steps of a project.

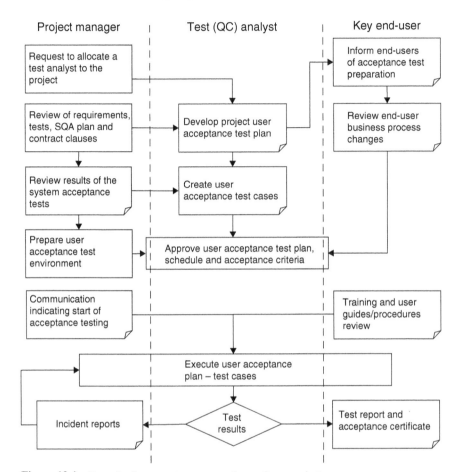

Figure 13.4   Example of an acceptance process for a software project.

- – Specify the acceptance criteria for the software. For example: the project will be accepted by the customer if:
  - o 100% of the functionality approved in the requirements document has been delivered and there is no level 1 or level 2 incident report pending a fix;
  - o The infrastructure and maintenance/support groups have no pending level 1 or 2 incident reports. If there is a level 1 or level 2 incident report, they will have to be addressed before final acceptance of the software.

13.2.12 SQA Records

This section of the SQAP includes activities and tasks for analysis, identification, collection, filing, maintenance, and disposition of quality records. Quality records document that activities were performed in accordance with project plans and the contract. These records enable information sharing and support analysis to identify problems, causes, and eventually result in product and process improvements. The project manager and/or the SQA manager should ask the following questions related to the analysis, identification, collection, filing, maintaining, and disposition of SQA records during the project planning phase:

- What set of records is the project required to produce?
- What set of records is SQA required to produce?
- Is the information required to be included on each record defined?
- What mechanisms will be used to collect, file, maintain, and eventually dispose of quality records?
- Who is responsible for collecting, filing, maintaining, and disposing of records?
- Who is responsible for sharing records?
- What records are to be shared with stakeholders?
- What protections are required to be established in order to share these records, maintain the integrity of the records, and prevent their modification or inadvertent release?
- What records are required from subcontractors?

13.3 EXECUTING THE SQAP

Once a SQAP has been developed and approved, the project has to execute it. The purpose of clause 5.3.4 of IEEE 730 is to "Execute the SQA Plan" in coordination with the project manager, the project team, and organizational quality management. The tasks linked to the execution of the SQAP are quite explicit. To accomplish this activity, the SQA function shall perform the following tasks [IEE 14]:

1) execute the activities and tasks defined in the SQAP, based on project schedules;
2) create the outcomes identified in the SQAP;
3) revise the SQAP in response to project changes;
4) raise non-conformances when actual outcomes do not agree with expectations.

Annex C of IEEE 730 provides, for each section of the SQAP, a list of questions when executing SQA project activities described in the SQAP. Table 13.7 presents an example of such a table for the questions and suggested inputs related to software product risk to consider asking during the project executing phase.

Table 13.7 Questions and Suggested Inputs Related to Software Product Risk to Consider Asking During Project Executing Phase [IEE 14])

Questions	Suggested inputs
– Are risks identified and analyzed as they develop?	– Risk management plan
– Has the priority in which to apply resources to treatment of these risks been determined?	– Improvement plan
– Are risk measures appropriately defined, applied, and assessed to determine changes in the status of risk and the progress of the treatment activities?	– Monitoring and control report – Risk action request
– Has appropriate treatment been taken to correct or avoid the impact of risk based on its priority, probability, and consequence or other defined risk threshold?	
– Has a software integrity level scheme been defined for the project?	
– Has the software integrity level scheme been reviewed and determined to be appropriate?	
– Has a software integrity level been established, if appropriate?	
– Has a set of assurance cases been prepared?	
– Have the assurance cases been reviewed and determined to be appropriate and complete?	
– Has an appropriate risk assessment been performed and documented?	

Task 3 of clause 5.3.4 states that the SQA function shall revise the SQAP in response to project changes. A project is executed in an organization where processes, such as configuration management, have been documented and implemented. The SQAP must follow the organizational configuration management process when revising and updating the SQAP. Usually, a SQAP will have a table, called a revision or history table, listing the approval and revision information. The following text box briefly describes the tasks when producing an updated version of a SQAP.

Revision of the SQAP

A revision of the SQAP may consist of the following tasks:

– identify and record, on the "Revisions table", the modifications made to the sections and paragraphs concerned;

– review and obtain approval of the new version of the SQAP;
– put the new version of SQAP under configuration management;
– distribute the new SQAP to all stakeholders (e.g., project participants, SQA, customer).

13.4 CONCLUSION

The SQAP is the cornerstone of any software project aiming at producing a quality product for external or internal customers. The SQAP pulls together all quality concerns of an organization, its policies, its people, its processes, and tools in producing quality products while building a competitive software capability.

While supporting the development of quality products, a robust SQAP should minimize avoidable rework. In Chapter 2, we presented the concept of cost of quality, and in many chapters we have illustrated how SQA could help in reducing the cost of rework. Hopefully, the following statement made by Dr. Robert Charette will no longer be true for organizations that have implemented the SQA practices presented in this book.

"Studies have shown that software specialists spend about 40 to 50 percent of their time on avoidable rework rather than on what they call value-added work, which is basically work that's done right the first time."

Dr. Robert Charette,
Why Software Fails
IEEE Spectrum, September 2005

13.5 FURTHER READING

GALIN D. *Software Quality: Concepts and Practice*. Wiley-IEEE Computer Society Press, Hoboken, New Jersey, 2017, 726 p.
SCHULMEYER G. G. (Ed.). *Handbook of Software Quality Assurance*. 4th edition. Artec House, Norwood, MA, 2008.

13.6 EXERCISES

13.1 Name five topics to be addressed in a SQAP.

13.2 Describe the review and acceptance of documents approach proposed to ensure quality.

13.3 Describe the process of acceptance of an acceptance test and how to use the categories of defects to ensure quality of delivery.

13.4 How important is the description of the criticality of the software at the beginning of a SQA plan?

13.5 Develop a checklist that will allow you to ensure that the SQA plan conforms to IEEE 730.

13.6 What is the difference between process assurance and product assurance?

13.7 What could be the consequences of not having a SQA plan for your organization?

13.8 What could be the consequences of not requiring a SQA plan from your supplier?

Appendix 1

Software Engineering Code of Ethics and Professional Practice (Version 5.2)

Software Engineering Code of Ethics and Professional Practice (Version 5.2) as recommended by the IEEE-CS/ACM Joint Task Force on Software Engineering Ethics and Professional Practices and jointly approved by the ACM and the IEEE-CS as the standard for teaching and practicing software engineering.

PREAMBLE

Computers have a central and growing role in commerce, industry, government, medicine, education, entertainment, and society at large. Software engineers are those who contribute by direct participation or by teaching, to the analysis, specification, design, development, certification, maintenance, and testing of software systems. Because of their roles in developing software systems, software engineers have significant opportunities to do good or cause harm, to enable others to do good or cause harm, or to influence others to do good or cause harm. To ensure, as much as possible, that their efforts will be used for good, software engineers must commit themselves to making software engineering a beneficial and respected profession. In accordance with that commitment, software engineers shall adhere to the following Code of Ethics and Professional Practice.

The code contains eight principles related to the behavior of and decisions made by professional software engineers, including practitioners, educators, managers, supervisors, and policy makers, as well as trainees and students of the profession. The principles identify the ethically responsible relationships in which individuals, groups, and organizations participate and the primary obligations within these

Software Quality Assurance, First Edition. Claude Y. Laporte and Alain April.
© 2018 the IEEE Computer Society, Inc. Published 2018 by John Wiley & Sons, Inc.

relationships. The clauses of each principle are illustrations of some of the obligations included in these relationships. These obligations are founded in the software engineer's humanity, in special care owed to people affected by the work of software engineers, and in the unique elements of the practice of software engineering. The code prescribes these as obligations of anyone claiming to be or aspiring to be a software engineer.

It is not intended that the individual parts of the code be used in isolation to justify errors of omission or commission. The list of principles and clauses is not exhaustive. The clauses should not be read as separating the acceptable from the unacceptable in professional conduct in all practical situations. The code is not a simple ethical algorithm that generates ethical decisions. In some situations, standards may be in tension with each other or with standards from other sources. These situations require the software engineer to use ethical judgment to act in a manner that is most consistent with the spirit of the Code of Ethics and Professional Practice, given the circumstances.

Ethical tensions can best be addressed by thoughtful consideration of fundamental principles, rather than blind reliance on detailed regulations. These principles should influence software engineers to consider broadly who is affected by their work; to examine if they and their colleagues are treating other human beings with due respect; to consider how the public, if reasonably well informed, would view their decisions; to analyze how the least empowered will be affected by their decisions; and to consider whether their acts would be judged worthy of the ideal professional working as a software engineer. In all these judgments, concern for the health, safety, and welfare of the public is primary; that is, the "Public Interest" is central to this code.

The dynamic and demanding context of software engineering requires a code that is adaptable and relevant to new situations as they occur. However, even in this generality, the code provides support for software engineers and managers of software engineers who need to take positive action in a specific case by documenting the ethical stance of the profession. The code provides an ethical foundation to which individuals within teams and the team as a whole can appeal. The code helps to define those actions that are ethically improper to request of a software engineer or teams of software engineers.

The code is not simply for adjudicating the nature of questionable acts; it also has an important educational function. As this code expresses the consensus of the profession on ethical issues, it is a means to educate both the public and aspiring professionals about the ethical obligations of all software engineers.

PRINCIPLES

Principle 1: *Public*: Software engineers shall act consistently with the public interest. In particular, software engineers shall, as appropriate:

1.01. Accept full responsibility for their own work.

1.02. Moderate the interests of the software engineer, the employer, the client, and the users with the public good.

1.03. Approve software only if they have a well-founded belief that it is safe, meets specifications, passes appropriate tests, and does not diminish quality of life, diminish privacy, or harm the environment. The ultimate effect of the work should be to the public good.

1.04. Disclose to appropriate persons or authorities any actual or potential danger to the user, the public, or the environment, that they reasonably believe to be associated with software or related documents.

1.05. Cooperate in efforts to address matters of grave public concern caused by software, its installation, maintenance, support, or documentation.

1.06. Be fair and avoid deception in all statements, particularly public ones, concerning software or related documents, methods, and tools.

1.07. Consider issues of physical disabilities, allocation of resources, economic disadvantage, and other factors that can diminish access to the benefits of software.

1.08. Be encouraged to volunteer professional skills to good causes and to contribute to public education concerning the discipline.

Principle 2: *Client and Employer*: Software engineers shall act in a manner that is in the best interests of their client and employer, consistent with the public interest. In particular, software engineers shall, as appropriate:

2.01. Provide service in their areas of competence, being honest, and forthright about any limitations of their experience and education.

2.02. Not knowingly use software that is obtained or retained either illegally or unethically.

2.03. Use the property of a client or employer only in ways properly authorized, and with the client's or employer's knowledge and consent.

2.04. Ensure that any document upon which they rely has been approved, when required, by someone authorized to approve it.

2.05. Keep private any confidential information gained in their professional work, where such confidentiality is consistent with the public interest and consistent with the law.

2.06. Identify, document, collect evidence, and report to the client or the employer promptly if, in their opinion, a project is likely to fail, to prove too expensive, to violate intellectual property law, or otherwise to be problematic.

2.07. Identify, document, and report significant issues of social concern, of which they are aware, in software or related documents, to the employer or the client.

2.08. Accept no outside work detrimental to the work they perform for their primary employer.

2.09. Promote no interest adverse to their employer or client, unless a higher ethical concern is being compromised; in that case, inform the employer or another appropriate authority of the ethical concern.

Principle 3: *Product*: Software engineers shall ensure that their products and related modifications meet the highest professional standards possible. In particular, software engineers shall, as appropriate:

3.01. Strive for high quality, acceptable cost, and a reasonable schedule, ensuring significant tradeoffs are clear to and accepted by the employer and the client, and are available for consideration by the user and the public.

3.02. Ensure proper and achievable goals and objectives for any project on which they work or propose.

3.03. Identify, define, and address ethical, economic, cultural, legal, and environmental issues related to work projects.

3.04. Ensure that they are qualified for any project on which they work or propose to work, by an appropriate combination of education, training, and experience.

3.05. Ensure that an appropriate method is used for any project on which they work or propose to work.

3.06. Work to follow professional standards, when available, that are most appropriate for the task at hand, departing from these only when ethically or technically justified.

3.07. Strive to fully understand the specifications for software on which they work.

3.08. Ensure that specifications for software on which they work have been well documented, satisfy the users' requirements, and have the appropriate approvals.

3.09. Ensure realistic quantitative estimates of cost, scheduling, personnel, quality, and outcomes on any project on which they work or propose to work and provide an uncertainty assessment of these estimates.

3.10. Ensure adequate testing, debugging, and review of software and related documents on which they work.

3.11. Ensure adequate documentation, including significant problems discovered and solutions adopted, for any project on which they work.

3.12. Work to develop software and related documents that respect the privacy of those who will be affected by that software.

3.13. Be careful to use only accurate data derived by ethical and lawful means, and use it only in ways properly authorized.

3.14. Maintain the integrity of data, being sensitive to outdated or flawed occurrences.

3.15 Treat all forms of software maintenance with the same professionalism as new development.

Principle 4: *Judgment*: Software engineers shall maintain integrity and independence in their professional judgment. In particular, software engineers shall, as appropriate:

4.01. Temper all technical judgments by the need to support and maintain human values.

4.02 Only endorse documents either prepared under their supervision or within their areas of competence and with which they are in agreement.

4.03. Maintain professional objectivity with respect to any software or related documents they are asked to evaluate.

4.04. Not engage in deceptive financial practices such as bribery, double billing, or other improper financial practices.

4.05. Disclose to all concerned parties those conflicts of interest that cannot reasonably be avoided or escaped.

4.06. Refuse to participate, as members or advisors, in a private, governmental or professional body concerned with software related issues, in which they, their employers or their clients have undisclosed potential conflicts of interest.

Principle 5: *Management*: Software engineering managers and leaders shall subscribe to and promote an ethical approach to the management of software development and maintenance. In particular, those managing or leading software engineers shall, as appropriate:

5.01 Ensure good management for any project on which they work, including effective procedures for promotion of quality and reduction of risk.

5.02. Ensure that software engineers are informed of standards before being held to them.

5.03. Ensure that software engineers know the employer's policies and procedures for protecting passwords, files, and information that is confidential to the employer or confidential to others.

5.04. Assign work only after taking into account appropriate contributions of education and experience tempered with a desire to further that education and experience.

5.05. Ensure realistic quantitative estimates of cost, scheduling, personnel, quality, and outcomes on any project on which they work or propose to work, and provide an uncertainty assessment of these estimates.

5.06. Attract potential software engineers only by full and accurate description of the conditions of employment.

5.07. Offer fair and just remuneration.

5.08. Not unjustly prevent someone from taking a position for which that person is suitably qualified.

5.09. Ensure that there is a fair agreement concerning ownership of any software, processes, research, writing, or other intellectual property to which a software engineer has contributed.

5.10. Provide for due process in hearing charges of violation of an employer's policy or of this code.

5.11. Not ask a software engineer to do anything inconsistent with this code.

5.12. Not punish anyone for expressing ethical concerns about a project.

Principle 6: *Profession*: Software engineers shall advance the integrity and reputation of the profession consistent with the public interest. In particular, software engineers shall, as appropriate:

6.01. Help develop an organizational environment favorable to acting ethically.

6.02. Promote public knowledge of software engineering.

6.03. Extend software engineering knowledge by appropriate participation in professional organizations, meetings, and publications.

6.04. Support, as members of a profession, other software engineers striving to follow this code.

6.05. Not promote their own interest at the expense of the profession, client, or employer.

6.06. Obey all laws governing their work, unless, in exceptional circumstances, such compliance is inconsistent with the public interest.

6.07. Be accurate in stating the characteristics of software on which they work, avoiding not only false claims but also claims that might reasonably be supposed to be speculative, vacuous, deceptive, misleading, or doubtful.

6.08. Take responsibility for detecting, correcting, and reporting errors in software and associated documents on which they work.

6.09. Ensure that clients, employers, and supervisors know of the software engineer's commitment to this code of ethics, and the subsequent ramifications of such commitment.

6.10. Avoid associations with businesses and organizations which are in conflict with this code.

6.11. Recognize that violations of this code are inconsistent with being a professional software engineer.

6.12. Express concerns to the people involved when significant violations of this code are detected unless this is impossible, counter-productive, or dangerous.

6.13. Report significant violations of this code to appropriate authorities when it is clear that consultation with people involved in these significant violations is impossible, counter-productive, or dangerous.

Principle 7: *Colleagues*: Software engineers shall be fair to and supportive of their colleagues. In particular, software engineers shall, as appropriate:

7.01. Encourage colleagues to adhere to this code.

7.02. Assist colleagues in professional development.

7.03. Credit fully the work of others and refrain from taking undue credit.

7.04. Review the work of others in an objective, candid, and properly documented way.

7.05. Give a fair hearing to the opinions, concerns, or complaints of a colleague.

7.06. Assist colleagues in being fully aware of current standard work practices including policies and procedures for protecting passwords, files and other confidential information, and security measures in general.

7.07. Not unfairly intervene in the career of any colleague; however, concern for the employer, the client, or public interest may compel software engineers, in good faith, to question the competence of a colleague.

7.08. In situations outside of their own areas of competence, call upon the opinions of other professionals who have competence in that area.

Principle 8: *Self*: Software engineers shall participate in lifelong learning regarding the practice of their profession and shall promote an ethical approach to the practice of the profession. In particular, software engineers shall continually endeavor to:

8.01. Further their knowledge of developments in the analysis, specification, design, development, maintenance, and testing of software and related documents, together with the management of the development process.

8.02. Improve their ability to create safe, reliable, and useful quality software at reasonable cost and within a reasonable time.

8.03. Improve their ability to produce accurate, informative, and well-written documentation.

8.04. Improve their understanding of the software and related documents on which they work and of the environment in which they will be used.

8.05. Improve their knowledge of relevant standards and the law governing the software and related documents on which they work.

8.06 Improve their knowledge of this code, its interpretation, and its application to their work.

8.07 Not give unfair treatment to anyone because of any irrelevant prejudices.

8.08. Not influence others to undertake any action that involves a breach of this code.

8.09. Recognize that personal violations of this code are inconsistent with being a professional software engineer.

This code was developed by the IEEE-CS/ACM joint task force on Software Engineering Ethics and Professional Practices (SEEPP):

Executive Committee: Donald Gotterbarn (Chair), Keith Miller, and Simon Rogerson;

Members: Steve Barber, Peter Barnes, Ilene Burnstein, Michael Davis, Amr El-Kadi, N. Ben Fairweather, Milton Fulghum, N. Jayaram, Tom Jewett, Mark Kanko, Ernie Kallman, Duncan Langford, Joyce Currie Little, Ed Mechler, Manuel J. Norman, Douglas Phillips, Peter Ron Prinzivalli, Patrick Sullivan, John Weckert, Vivian Weil, S. Weisband, and Laurie Honour Werth.

©1999 by the Institute of Electrical and Electronics Engineers, Inc. and the Association for Computing Machinery, Inc.

This code may be published without permission as long as it is not changed in any way and it carries the copyright notice.

Appendix 2

Incidents and Horror Stories Involving Software

In 1962, the exploration module Mariner 1 of NASA crashes. It was discovered that a period was placed where a parenthesis should have been in a FORTRAN DO-Loop statement.

In 1979, the problems at the Three Mile Island nuclear plant were caused by a misinterpretation of a user interface.

In 1982, a Russian pipeline explodes because of a defect in the software. This explosion is the largest since the nuclear explosions at the end of World War II.

In 1982, during the Falklands War, the HMS Sheffield was sunk by French Exocet missiles launched by the Argentine army. The radar on the Sheffield was programmed to classify the Exocet as "friend" since these missiles were also used by the British army.

In 1983, a false alarm in a nuclear attack detection system came close to causing a nuclear war.

Between 1985 and 1987, the Therac-25 killed six people with an overdose of radiation.

In 1988, a civilian Airbus aircraft full of passengers was shot down due to the pattern recognition software that had not clearly identified the aircraft.

In 1990, the loss of long distance access at AT&T plunged the United States into an unparalleled telephone crisis.

In 1992, London ambulances changed their call-tracking software and lost control of the situation when the number of calls increased.

The maiden flight of the Ariane 5, which took place June 4, 1996 resulted in failure. About 40 seconds after the start of the flight sequence, the rocket that was then at an altitude of some 3700 meters, deviated from its trajectory, crashed, and exploded. It was the total loss of the guidance and altitude information 37 seconds after starting the ignition sequence of the main engine (30 seconds after takeoff),

Software Quality Assurance, First Edition. Claude Y. Laporte and Alain April.
© 2018 the IEEE Computer Society, Inc. Published 2018 by John Wiley & Sons, Inc.

which was responsible for the failure of Ariane 5. This loss of information was due to errors in specification and design of the inertial reference system software.

The objective of the review process, in which all major partners involved with Ariane 5 participated, was to validate design decisions and get the flight certification. During this process, the limitations of the alignment software were not fully analyzed and the consequences of not maintaining this function in flight were not measured. Trajectory data of Ariane 5 were not specifically provided for in the specification for the inertial reference system or in testing at the equipment level. Consequently, the realignment function was not tested under simulated flight conditions for Ariane 5 and the design error was not detected.

Based on its analysis and conclusions, the Commission of Inquiry issued the following recommendations:

- organize a special qualification review for all equipment that includes software;
- redefine critical components taking into account failures that can originate from software;
- treat qualification documents with as much attention as the code;
- improve techniques to ensure consistency between the code and its qualification;
- include external project participants in the specifications review, code review, and document reviews. Review all flight software;
- establish a team that will be responsible for developing the software qualification process to propose strict rules to confirm the qualification and ensure that the specification, verification, and software testing will be at a consistently high level of quality in the Ariane 5 program.

In 1996, the crash of China Airlines flight B1816 was caused by a misinterpretation of an interface by the pilot.

In 1998, a member of the crew of the USS Yorktown entered a zero by mistake, which resulted in a division by zero. This error caused a failure of the navigation system and the propulsion system was stopped. This outage lasted several hours as there was no means to restart the system.

In 1999 the orbital module MARS, worth $125 million, was lost because different development teams used different measuring systems (metric vs. imperial).

In January 2000, a US spy satellite stopped working due to the Year 2000 problem.

In March 2000, a communication satellite for cellular telephones was destroyed 8 minutes after launch. The explosion was caused by a logic error in one line of code.

In May 2001, Apple warned its users not to use its new iTunes software because it would erase all contents of their hard drive.

In 2002, the unofficial results count for Franklin County, USA showed Bush with 4258 votes and Kerry with 260 votes. An audit of the results showed that there were only 638 people who voted in this county.

The FDA's analysis of 3140 medical device recalls conducted between 1992 and 1998 reveals that 242 of them (7.7%) are attributable to software failures. Of those software related recalls, 192 (or 79%) were caused by software defects that were introduced when changes were made to the software after its initial production and distribution. Software validation and other related good software engineering practices discussed in the FDA's guidance are a principal means of avoiding such defects and resultant recalls [FDA 02].

In 2003, computers of T-Mobile were hacked and a large number of passwords were stolen.

In May 2004, a Soyuz TMA-1 spaceship carrying a Russian Cosmonaut and two American Astronauts landed nearly 500 kilometers off course after the craft unexpectedly switched to a ballistic re-entry trajectory. Preliminary indications are that the problem was caused by software in the guidance computer in the new, modified version of the spaceship [PAR 03].

In September 2004, the Los Angeles area Air Traffic Controllers lost voice contact with over 400 airplanes in the airspace around LA. The main system used to communicate with pilots (called Voice Communications Systems Unit—VCSU) failed. A backup system also failed. Pilots were essentially flying blind, not knowing what other planes were in their path. There were at least five documented cases where planes came within the minimum separation distance mandated by the FAA. Inside the control system unit is a countdown timer that ticks off time in milliseconds. The VCSU uses the timer as a pulse to send out periodic queries to the VSCS. It starts out at the highest possible number that the system's server and its software can handle. It's a number just over 4 billion milliseconds. When the counter reaches zero, the system runs out of ticks and can no longer time itself. So it shuts down. In order for this system to work, it requires that a technician manually reboot the system every 30 days [GEP 04].

In 2005, it was reported that 160,000 Toyota Prius were recalled to obtain an update of the software because there were 13 cases where car engines stopped abruptly without the driver turning off the ignition. Toyota asked its dealers to install a new software version.

In 2005, the Russian rocket Cryosat was destroyed when the second stage of the launch failed. The control software had a flaw.

In 2006, there were 70 processors in the BMW745i and this car also had to undergo a recall because the software could create an incorrect synchronization of valves and suddenly stop the engine when at full speed.

In 2006, Nicolas Clarke reported in the Herald tribune that a problem in the industrial design software of the Airbus A380 had led to bad lengths and connections of cables and that production delays were due to numerous software problems. The cable length problem was caused by the use of two different versions of design software from two manufacturers.

On December 21, 2007, three businessmen from California, New York, and Florida and users of QuickBooks Pro, filed a motion to institute a class action suit against the manufacturer Intuit, which specializes in the design of accounting and financial software. The plaintiffs accuse Intuit of sending erroneous code on the weekend of December 15–16, 2007, which caused the loss of files and sales data. The problem occurred when the QuickBooks program launched an automatic update for its software that was sent by Intuit, causing the loss of files containing financial information. This represented hundreds of hours of re-work. The same situation occurred for all users of this software, who were trying to close their books at year-end. The plaintiffs claim was for compensation for the loss of their data, as well as the time and money spent trying to recover the data for themselves and all other victims. This case shows us how the wrong code, sent in an automatic update, could cause significant damage on a national level and even globally. Permanent loss of financial data can be disastrous for the survival of most of these small businesses.

On May 16, 2007, five IT projects, worth a total of $325 million in the state of Colorado, did not provide results and were canceled.

On May 16, 2007, 2900 students from the state of Virginia had to take their standardized test again because of a software problem.

On May 18, 2007, in London, a judge admits to not knowing what a website is. This revelation immediately stopped the trial he was working on (a case judgment concerning terrorists who used the internet).

On May 19, 2007, Alcatel-Lucent lost a hard drive containing the detailed data of 200,000 customers.

In August 2007, the Boeing 777 of Air Malaysia leaving Perth for Kuala Lumpur had problems with the autopilot software and the aircraft had to be flown manually throughout the flight.

In 2009, the Committee on Gaming in Ontario refuses to pay $42.9 million to Mr. Kusznirewicz on a slot machine. The signal for the win was wrong and should not have exceeded $9,025. A software error was the cause of this faulty display during play.

On 13 May 2009, a plaintiff managed to capture and analyze the source code of a breathalyzer 7110 MKIII-C equipment he believed was the source of errors. He proved that the calculations were incorrect and that thousands of people were judged severely because of these erroneous results.

In November 2009, the first Airbus A380 flying for Air France had to turn back to New York following a "minor" computer failure. The Airbus A380, which was flying to Paris from New York with 530 passengers on board, had to turn back an hour and half after takeoff. According to the National Union of Airline Pilots (Syndicat National des Pilotes de Ligne—SNPL) at Air France, the problem was with the autopilot. "It would not come on" said the SNPL spokesman.

On December 1, 2009, the Telegraph newspaper reports the "black screen of death" of Windows 7. This general failure of the new operating system from Microsoft was due to an error in a registry entry.

A software error in the Ford Explorer engine controls limited vehicle speed to 110 miles per hour instead of the specified 99 miles per hour. At 110 miles per hour, the Firestone tires on the Ford Explorers had a rated life of 10 minutes [HUM 02].

Ford warned its dealers that software might disable the continuously variable transmissions in some 30,000 of its new Ford Five Hundred sedans and Freestyle sport wagons. The mechanical parts are fine, but a computer control meant to detect dirty transmission fluid was putting some cars into sluggish "limp home" mode. Ford had to rewrite software to fix the problem, which it says was caught before any vehicles reached customers [MOR 05].

A Boeing 777-200 en route from Perth to Kuala Lumpur presented the pilot with contradictory reports of airspeed: that the aircraft was over-speeding and at the same time was at risk of stalling. The pilot disconnected the autopilot and attempted to descend, but the auto-throttle caused the aircraft to climb 2000 feet. He was eventually able to return to Perth and land the aircraft safely. The incident was attributed to a failed accelerometer. The Air Data Inertial Reference Unit (ADIRU) had recorded the failure of the device in its memory, but because of a software flaw, it failed to recheck the device's status after power cycling [JAC 07].

An Airbus A340-642 en route from Hong Kong to London suffered from a failure in a data bus belonging to a computer that monitors and controls fuel levels and flow. One engine lost power and a second began to fluctuate; the pilot diverted the aircraft and landed safely in Amsterdam. The subsequent investigation noted that a backup slave computer was available that was working correctly but that, due to faulty logic in the software, the failing computer remained selected as the master. A second report recommended an independent low-fuel warning system and noted the risks of a computerized management system that might fail to provide crew with appropriate data, preventing them from taking appropriate actions [JAC 07].

Toyota identified a software flaw that caused Prius hybrid cars to stall or shut down when traveling at high speed; 23,900 vehicles were affected [FRE 05].

"In 1998, researchers were bringing two subcritical chunks of plutonium together in a 'criticality' experiment that measured the rate of change of neutron flux between the two halves. It would be a Real Bad Thing if the two bits actually got quite close, so they were mounted on small controllable cars, rather like a model railway. An operator uses a joystick to cautiously nudge them toward each other. The experiment proceeded normally for a time, the cars moving at a snail's pace. Suddenly both picked up speed, careening towards each other at full speed. No doubt with thoughts of a mushroom cloud in his head, the operator hit the 'shut down' button mounted on the joystick. Nothing happened. The cars kept accelerating. Finally, after he actuated an emergency SCRAM control, the operator's racing heart (happily sans defective embedded pacemaker) slowed when the cars stopped and moved apart. The joystick had failed. A processor reading this device recognized the problem and sent an error message, a question mark, to the main controller. Unhappily, '?' is ASCII 63, the largest number that fits in a 6-bit field. The main CPU interpreted the message as a big number meaning go real fast. Two issues come to mind: the first is to test everything, even exception handlers. The second is that error handling is intrinsically difficult and must be designed carefully." [GAN 04]

"In 1997 Guidant announced that one of its new pacemakers occasionally drives the patient's heartbeat to 190 beats per minute (BPM). Now, I don't know much about cardiovascular diseases, but suspect 190 BPM to be a really bad thing for a person with a sick heart. The company reassured the pacemaker-buying public that there wasn't really a problem; they had fixed the code and were sending disks across the country to doctors. The pacemaker, however, is implanted subcutaneously. There's no Internet connection, no USB port, no PCMCIA slot. Turns out that it's possible to hold an inductive loop over the implanted pacemaker to communicate with it. A small coil in the device normally receives energy to charge the battery. It's possible to modulate the signal and upload new code into flash. The robopatients were reprogrammed and no one was hurt. The company was understandably reluctant to discuss the problem so it's impossible to get much insight into the nature of what went wrong. But there was clearly inadequate testing. Guidant is far from alone. A study in the August 15, 2001 Journal of the American Medical Association ('Recalls and Safety Alerts Involving Pacemakers and Implantable Cardioverter-Defibrillator Generators') showed that more than 500,000 implanted pacemakers and cardioverters were recalled between 1990 and 2000. (This month's puzzler: how do you recall one of these things?) Forty-one percent of those recalls were due to firmware problems. The recall rate increased in the second half of that decade compared with the first. Firmware is getting worse. All five U.S. pacemaker vendors have an increasing recall rate. The study said, 'engineered (hardware) incidents [are] predictable and therefore preventable, while system (firmware) incidents are inevitable due to complex processes combining in unforeseeable ways." [GAN 04].

Glossary – Abbreviations – Acronyms

Acceptance Test (ISO 2382-20) The test of a system or functional unit usually performed by the purchaser on his premises after installation with the participation of the vendor to ensure that the contractual requirements are met.

Acquirer (ISO 12207) Stakeholder that acquires or procures a product or service from a supplier.

Note: The acquirer could be one of the following: buyer, customer, owner, or purchaser.

Agreement (ISO 15288) Mutual acknowledgement of terms and conditions under which a working relationship is conducted.

Alpha testing (ISO 24765) First stage of testing before a product is considered ready for commercial or operational use (cf. beta testing).

Note: Often performed only by users within the organization developing the software.

Assurance case (ISO 15026-1) A reasoned, auditable artifact created that supports the contention that its top-level claim (or set of claims), is satisfied, including systematic argumentation and its underlying evidence and explicit assumptions that support the claim(s).

Note 1 to entry: An assurance case contains the following and their relationships:

– one or more claims about properties;

– arguments that logically link the evidence and any assumptions to the claim(s);

– a body of evidence and possibly assumptions supporting these arguments for the claim(s);

– justification of the choice of top-level claim and the method of reasoning.

Audit

1) An independent examination of a software product, software process, or set of software processes performed by a third party to assess compliance with specifications, standards, contractual agreements, or other criteria (IEEE 1028).

Note: An audit should result in a clear indication of whether the audit criteria have been met.

2) Independent examination of a work product or set of work products to assess compliance with specifications, standards, contractual agreements, or other criteria (ISO 12207).

3) Systematic, independent, and documented process for obtaining audit evidence and evaluating it objectively to determine the extent to which the audit criteria are fulfilled.

Note 1: Internal audits, sometimes called first party audits, are conducted by the organization itself, or on its behalf, for management review and other internal purposes (e.g., to confirm the effectiveness of the management system or to obtain information for the improvement

Software Quality Assurance, First Edition. Claude Y. Laporte and Alain April.
© 2018 the IEEE Computer Society, Inc. Published 2018 by John Wiley & Sons, Inc.

of the management system). Internal audits can form the basis for an organization's self-declaration of conformity. In many cases, particularly in small organizations, independence can be demonstrated by the freedom from responsibility for the activity being audited or freedom from bias and conflict of interest.

Note 2: External audits include second- and third-party audits. Second-party audits are conducted by parties having an interest in the organization, such as customers, or by other persons on their behalf. Third-party audits are conducted by independent auditing organizations, such as regulators or those providing certification.

Note 3: When two or more management systems of different disciplines (e.g., quality, environmental, occupational health, and safety) are audited together, this is termed a combined audit.

Note 4: When two or more auditing organizations cooperate to audit a single auditee, this is termed a joint audit.

    4) An objective examination of a work product or set of work products against specific criteria (e.g., requirements). (See also "objectively evaluate.") This is a term used in several ways in CMMI®, including configuration audits and process compliance audits (CMMI-DEV).

Audit criteria (ISO 19011) Set of policies, procedures, or requirements used as a reference against which audit evidence is compared.

    Note: If the audit criteria are legal (including statutory or regulatory) requirements, the terms "compliant" or "noncompliant" are often used in an audit finding.

Audit evidence (ISO 19011) Records, statements of fact, or other information that are relevant to the audit criteria and verifiable.

    Note: Audit evidence can be qualitative or quantitative.

Audit findings (ISO 19011) Results of the evaluation of the collected audit evidence against audit criteria.

    Note 1: Audit findings indicate conformity or nonconformity.

    Note 2: Audit findings can lead to the identification of opportunities for improvement or recording good practices.

    Note 3: If the audit criteria are selected from legal or other requirements, the audit finding is termed compliance or non-compliance.

Base measure (ISO 15939) Measure defined in terms of an attribute and the method for quantifying it.

    Note: A base measure is functionally independent of other measures.

Baseline (ISO 12207) Formally approved version of a configuration item, regardless of media, formally designated and fixed at a specific time during the configuration item's life cycle.

Beta testing (ISO 24765) Second stage of testing when a product is in limited production use (cf. alpha testing).

    Note: often performed by a client or customer.

Bidirectional traceability (CMMI-DEV) An association among two or more logical entities that is discernable in either direction (i.e., to and from an entity). (See also "requirements traceability" and "traceability.")

Black box (ISO 24765)

1) A system or component whose inputs, outputs, and general function are known but whose contents or implementation are unknown or irrelevant.

2) Pertaining to an approach that treats a system or component whose inputs, outputs, and general function are known but whose contents or implementation are unknown or irrelevant (cf. glass box).

Brainstorming (PMBOK® Guide) A general data gathering and creativity technique that can be used to identify risks, ideas, or solutions to issues by using a group of team members or subject-matter experts.

Branch (ISO 24765)

1) A computer program construct in which one of two or more alternative sets of program statements is selected for execution.

2) A point in a computer program at which one of two or more alternative sets of program statements is selected for execution.

Note: Every branch is identified by a tag. Often, a branch identifies the file versions that have been or will be released as a product release. May denote unbundling of arrow meaning, that is, the separation of object types from an object type set. Also refers to an arrow segment into which a root arrow segment has been divided.

Business model (Wikipedia) A business model describes the rationale of how an organization creates, delivers, and captures value (economic, social, or other forms of value).

The essence of a business model is that it defines the manner by which the business enterprise delivers value to customers, entices customers to pay for value, and converts those payments to profit: it thus reflects management's hypothesis about what customers want, how they want it, and how an enterprise can organize to best meet those needs, get paid for doing so, and make a profit.

Catastrophic (IEEE 1012) Loss of human life, complete mission failure, loss of system security and safety, or extensive financial or social loss.

Change control board (CCB) (PMBOK® Guide) A formally chartered group responsible for reviewing, evaluating, approving, delaying, or rejecting changes to the project, and for recording and communicating such decisions. See also Configuration Control Board.

Change control procedure (ISO 24765) Actions taken to identify, document, review, and authorize changes to a software or documentation product that is being developed.

Note: The procedures ensure that the validity of changes is confirmed, that the effects on other items are examined, and that those people concerned with the development are notified of the changes.

Change management (ISO 24765) Judicious use of means to effect a change, or a proposed change, to a product or service.

Checklist [GIL 93] A specialized set of questions designed to help checkers find more defects, and in particular, more significant defects. Checklists concentrate on major defects. A checklist should be no more than a single page per subject area. Checklist questions interpret specified rules.

Commit (ISO 24765) To integrate the changes made to a developer's private view of the source code into a branch accessible through the version control system's repository.

Commit privileges (ISO 24765) A person's authority to commit changes.

Note: Sometimes privileges are associated with a specific part of the product (e.g., artwork or documentation) or a specific branch.

Commit window (ISO 24765) A period during which commits are allowed for a specific branch.

Note: In some development environments, commit windows for a maintenance branch might only open for short periods a few times a year.

Concept of operations (ConOps) document (IEEE 1362) A user-oriented document that describes a system's operational characteristics from the end-user's viewpoint.

Configuration (ISO 24765) The functional and physical characteristics of hardware or software as set forth in technical documentation or achieved in a product.

Configuration audit (CMMI-DEV) An audit conducted to verify that a configuration item or a collection of configuration items that make up a baseline conforms to a specified standard or requirement. (See also "audit" and "configuration item.")

Configuration baseline (ISO 24765) Configuration information formally designated at a specific time during a product's or product component's life.

Note: Configuration baselines, plus approved changes from those baselines, constitute the current configuration information.

Configuration control (ISO 24765) An element of configuration management, consisting of the evaluation, coordination, approval or disapproval, and implementation of changes to configuration items after formal establishment of their configuration identification. *Synonym*: change control.

Configuration control board (CCB) (ISO 24765) A group of people responsible for evaluating and approving or disapproving proposed changes to configuration items, and for ensuring implementation of approved changes. See also Change Control Board.

Configuration identification (ISO 24765)

1) An element of configuration management, consisting of selecting the configuration items for a system and recording their functional and physical characteristics in technical documentation.

2) The current approved technical documentation for a configuration item as set forth in specifications, drawings, associated lists, and documents referenced therein.

Configuration item (CI) (ISO 12207) Item or aggregation of hardware, software, or both, that is designated for configuration management and treated as a single entity in the configuration management process.

Configuration management (CM)

1) A discipline applying technical and administrative direction and surveillance to: identify and document the functional and physical characteristics of a configuration item, control changes to those characteristics, record and report change processing and implementation status, verify compliance with specified requirements (ISO 24765).

2) A discipline applying technical and administrative direction and surveillance to (1) identify and document the functional and physical characteristics of a configuration item, (2) control changes to those characteristics, (3) record and report change processing and implementation status, and (4) verify compliance with specified requirements (CMMI-DEV).

Configuration status accounting (ISO 24765) An element of configuration management, consisting of the recording and reporting of information needed to manage a configuration effectively.

Note: This information includes a listing of the approved configuration identification, the status of proposed changes to the configuration, and the implementation status of approved changes.

Conflict (ISO 24765) A change in one version of a file that cannot be reconciled with the version of the file to which it is applied.

Note: can occur when versions from different branches are merged or when two committers work concurrently on the same file.

Conformity (ISO 9000) Fulfilment of a requirement.

Contract (ISO 12207) See *Agreement.*

Corrective action (PMBOK® Guide) An intentional activity that realigns the performance of the project work with the project management plan.

Critical (IEEE 1012) Major and permanent injury, partial loss of mission, major system damage, or major financial or social loss.

Critical software (IEEE 610.12) Software whose failure could have an impact on safety, or could cause large financial or social loss.

Criticality (IEEE 1012) The degree of impact that a requirement, module, error, fault, failure, or other item has on the development or operation of a system.

Deactivated code (DO-178) Executable Object Code (or data) that is traceable to a requirement and, by design, is either (a) not intended to be executed (code) or used (data), for example, a part of a previously developed software component such as unused legacy code, unused library functions, or future growth code; or (b) is only executed (code) or used (data) in certain configurations of the target computer environment, for example, code that is enabled by a hardware pin selection or software programmed options. The following examples are often mistakenly categorized as deactivated code but should be identified as required for implementation of the design/requirements: defensive programming structures inserted for robustness, including compiler-inserted object code for range and array index checks, error or exception handling routines, bounds and reasonableness checking, queuing controls, and time stamps.

Dead code (DO-178) Executable Object Code (or data) which exists as a result of a software development error but cannot be executed (code) or used (data) in any operational configuration of the target computer environment. It is not traceable to a system or software requirement. The following exceptions are often mistakenly categorized as dead code but are necessary for implementation of the requirements/design: embedded identifiers, defensive programming structures to improve robustness, and deactivated code such as unused library functions.

Debug (ISO 24765)

1) To detect, locate, and correct faults in a computer program.

2) To detect, locate, and eliminate errors in programs.

Defect

1) A problem which, if not corrected, could cause an application to either fail or to produce incorrect results (ISO 20926).

2) An imperfection or deficiency in a project component where that component does not meet its requirements or specifications and needs to be either repaired or replaced (PMBOK® Guide).

3) A generic term that can refer to either a fault (cause) or a failure (effect) (IEEE 982.1).

Derived measure (ISO 15939) Measure that is defined as a function of two or more values of base measures.

Development testing

1) Formal or informal testing conducted during the development of a system or component, usually in the development environment by the developer (ISO 24765).

2) Testing conducted to establish whether a new software product or software-based system (or components of it) satisfies its criteria. The criteria will vary based on the level of test being performed (IEEE 829).

Effectiveness (ISO 9000) Extent to which planned activities are realized and planned results achieved.

Efficiency

1) The degree to which a system or component performs its designated functions with minimum consumption of resources (ISO 24765).

2) Relationship between the result achieved and the resources used (ISO 9000).

Effort (PMBOK® Guide) The number of labor units required to complete a schedule activity or work breakdown structure component. Usually expressed in hours, days, or weeks.

Error (ISO 24765)

1) A human action that produces an incorrect result, such as software containing a fault.

2) An incorrect step, process, or data definition.

3) An incorrect result.

4) The difference between a computed, observed, or measured value or condition and the true, specified, or theoretically correct value or condition.

Evaluation (ISO 12207) Systematic determination of the extent to which an entity meets its specified criteria.

Exit criteria (CMMI-DEV) States of being that must be present before an effort can end successfully.

Failure

1) Termination of the ability of a product to perform a required function or its inability to perform within previously specified limits (ISO 25000).

2) An event in which a system or system component does not perform a required function within specified limits (ISO 24765).

Financial independence (IEEE 1012) This requires that control of the IV&V budget be vested in an organization independent of the development organization. This independence prevents situations where the IV&V effort cannot complete its analysis or test or deliver timely results because funds have been diverted or adverse financial pressures or influences have been exerted.

Firmware

1) Combination of a hardware device and computer instructions or computer data that reside as read-only software on the hardware device (IEEE 1012).

2) An ordered set of instructions and associated data stored in a way that is functionally independent of main storage, usually in a ROM (ISO 2382-1).

Note 1: The software cannot be readily modified under program control (ISO 24765).
Note 2: This term is sometimes used to refer only to the hardware device or only to the computer instructions or data, but these meanings are deprecated (IEEE 1012).
Note 3: The confusion surrounding this term has led some to suggest that it be avoided altogether (IEEE 1012).

Functional configuration audit (FCA) (ISO 24765) An audit conducted to verify that the development of a configuration item has been completed satisfactorily, that the item has achieved the performance and functional characteristics specified in the functional or allocated configuration identification, and that its operational and support documents are complete and satisfactory.

Functional requirement (IEEE 1220) A statement that identifies what a product or process must accomplish to produce required behavior and/or results.

Glass box (ISO 24765)

1) A system or component whose internal contents or implementation are known.

2) Pertaining to an approach that treats a system or component as in (1).

Synonym: white box.

Hazard (IEEE 1012)

1) An intrinsic property or condition that has the potential to cause harm or damage.

2) A source of potential harm or a situation with a potential for harm in terms of human injury, damage to health, property, or the environment, or some combination of these.

Independent (ISO 24765) Performed by an organization free from control by the supplier, developer, operator, or maintainer.

Independent verification and validation (IV&V) (ISO 24765) Verification and validation performed by an organization that is technically, managerially, and financially independent of the development organization.

Indicator (ISO 15939) Measure that provides an estimate or evaluation of specified attributes derived from a model with respect to defined information needs.

Information Management Process (ISO 15289) The documentation management process shall include these activities:

1) identify the documents to be produced by the organization, service, process, or project;

2) specify the content and purpose of all documents and plan and schedule their production;

3) identify the standards to be applied for development of documents;

4) develop and publish all documents in accordance with identified standards and in accordance with nominated plans;

5) maintain all documents in accordance with specified criteria.

Integration testing (IEEE 1012) Testing in which software components, hardware components, or both are combined and tested to evaluate the interaction between them.

Integrity level (IEEE 1012) A value representing project-unique characteristics (e.g. complexity, criticality, risk, safety level, security level, desired performance, and reliability) that define the importance of the system, software, or hardware to the user.

Integrity level scheme (IEEE 829) A set of system characteristics (such as complexity, risk, safety level, security level, desired performance, reliability, and/or cost) selected as important to stakeholders, and arranged into discrete levels of performance or compliance (integrity levels), to help define the level of quality control to be applied in developing and/or delivering the software.

Internal audit (ISO 9001) The organization shall conduct internal audits at planned intervals to determine whether the quality management system:

1) conforms to the planned arrangements to the requirements of this International Standard and to the quality management system requirements established by the organization;

2) is effectively implemented and maintained.

An audit programme shall be planned, taking into consideration the status and importance of the processes and areas to be audited, as well as the results of previous audits. The audit criteria, scope, frequency, and methods shall be defined. The selection of auditors and conduct of audits shall ensure objectivity and impartiality of the audit process. Auditors shall not audit their own work.

A documented procedure shall be established to define the responsibilities and requirements for planning and conducting audits, establishing records and reporting results.

Records of the audits and their results shall be maintained.

The management responsible for the area being audited shall ensure that any necessary corrections and corrective actions are taken without undue delay to eliminate detected nonconformities and their causes.

Follow-up activities shall include the verification of the actions taken and the reporting of verification results.

Issue (PMBOK® Guide) A point or matter in question or in dispute, or a point or matter that is not settled and is under discussion or over which there are opposing views or disagreements.

Lessons learned (PMBOK® Guide) The knowledge gained during a project that shows how project events were addressed or should be addressed in the future with the purpose of improving future performance.

Life cycle

1) Evolution of a system, product, service, project or other human-made entity from conception through retirement (ISO 12207);

2) The system or product evolution initiated by a perceived stakeholder need through the disposal of the products (IEEE 1220).

Life cycle processes (IEEE 1012) A set of interrelated or interacting activities that result in the development or assessment of system, software, or hardware products. Each activity consists of tasks. The life cycle processes may overlap one another. For verification and validation (V&V) purposes, no life cycle process is concluded until its development products are verified and validated according to the defined tasks in the verification and validation plan (VVP).

Managerial independence (IEEE 1012) This requires that the responsibility for the IV&V effort be vested in an organization separate from the development and program management organizations. Managerial independence also means that the IV&V effort independently selects the segments of the software, hardware, and system to analyze and test, chooses the IV&V techniques, defines the schedule of IV&V activities, and selects the specific technical issues and problems to act on. The IV&V effort provides its findings in a timely fashion simultaneously to both the development and program management organizations. The IV&V effort is allowed to submit to program management the IV&V results, anomalies, and findings without any restrictions (e.g., without requiring prior approval from the development group) or adverse pressures, direct or indirect, from the development group.

Marginal (IEEE 1012) Severe injury or illness, degradation of secondary mission, or some financial or social loss.

Master library (ISO 24765) A software library containing master copies of software and documentation from which working copies can be made for distribution and use. (cf. production library, software development library, software repository, system library.)

Measure (noun) (ISO 15939) Variable to which a value is assigned as the result of measurement.

Note: The plural form "measures" is used to refer collectively to base measures, derived measures and indicators.

Measure (verb) (ISO 15939) Make a measurement.

Measurement function (ISO 15939) Algorithm or calculation performed to combine two or more base measures.

Measurement information model (ISO 15939) A structure linking information needs to the relevant entities and attributes of concern. Entities include processes, products, projects, and resources. The measurement information model describes how the relevant attributes are quantified and converted to indicators that provide a basis for decision making.

Measurement method (ISO 15939) Logical sequence of operations, described generically, used in quantifying an attribute with respect to a specified scale.

Note: The type of measurement method depends on the nature of the operations used to quantify an attribute. Two types can be distinguished:

1) subjective: quantification involving human judgment;

2) objective: quantification based on numerical rules.

Measurement process (ISO 15939) Process for establishing, planning, performing, and evaluating measurement within an overall project, enterprise, or organizational measurement structure.

Measurement process owner (ISO 15939) Individual or organization responsible for the measurement process.

Medical device (ISO 13485) Instrument, apparatus, implement, machine, appliance, implant, reagent for *in vitro* use, software, material, or other similar or related article, intended by the manufacturer to be used, alone or in combination, for human beings for one or more of the specific purpose(s) of: diagnosis, prevention, monitoring, treatment, or alleviation of disease; diagnosis, monitoring, treatment, alleviation of or compensation for an injury; investigation, replacement, modification, or support of the anatomy or of a physiological process; supporting or sustaining life; control of conception; disinfection of medical devices; providing information for medical purposes by means of *in vitro* examination of specimens derived from the human body; and which does not achieve its primary intended action in or on the human body by pharmacological, immunological, or metabolic means, but which may be assisted in its function by such means.

Microcode (IEEE 1012) A collection of microinstructions, comprising part of, all of, or a set of microprograms.

Microprogram (ISO 24765) A sequence of instructions, called microinstructions, specifying the basic operations needed to carry out a machine language instruction.

Negligible (IEEE 1012) Minor injury or illness, minor impact on system performance, or operator inconvenience.

Nonconformity (ISO 9000) Non-fulfilment of a requirement.

Nonfunctional requirement (ISO 24765) A software requirement that describes not what the software will do but how the software will do it. Synonym: design constraints, nonfunctional requirement. (cf. functional requirement.)

Example: software performance requirements, software external interface requirements, software design constraints, and software quality attributes. Non-functional requirements are sometimes difficult to test, so they are usually evaluated subjectively.

Objectively evaluate (CMMI-DEV) To review activities and work products against criteria that minimize subjectivity and bias by the reviewer (See also "audit."). An example of an objective evaluation is an audit against requirements, standards, or procedures by an independent quality assurance function.

Operational testing (IEEE 829) Testing conducted to evaluate a system or component in its operational environment.

Opportunity (PMBOK® Guide) A risk that would have a positive effect on one or more project objectives.

Organizational policy (CMMI-DEV) A guiding principle typically established by senior management that is adopted by an organization to influence and determine decisions.

Organization's process asset library (CMMI-DEV) A library of information used to store and make process assets available that are useful to those who are defining, implementing, and managing processes in the organization. This library contains process assets that include process related documentation such as policies, defined processes, checklists, lessons learned documents, templates, standards, procedures, plans, and training materials.

Path (ISO 24765) In software engineering, a sequence of instructions that may be performed in the execution of a computer program.

Personal Process [SEI 09] A defined set of steps or activities that guide individuals in doing their personal work. It is usually based on personal experience and may be developed entirely from scratch or may be based on another established process and modified according to personal experience. A personal process provides individuals with a framework for improving their work and for consistently doing high-quality work.

Physical configuration audit (PCA) (ISO 24765) An audit conducted to verify that a configuration item, as built, conforms to the technical documentation that defines it.

Policy (ISO 24765)

1) A set of rules related to a particular purpose.

2) Clear and measurable statements of preferred direction and behavior to condition the decisions made within an organization.

Note: A rule can be expressed as an obligation, an authorization, a permission, or a prohibition. Not every policy is a constraint. Some policies represent an empowerment.

Post-mortem [DIN 05] A collective learning activity that can be organized for projects either when they end a phase or are terminated. The main motivation is to reflect on what happened in the project in order to improve future practice for the individuals that have participated in the project and for the organization as a whole. The physical outcome of a meeting is a post-mortem report.

Preventive action (PMBOK® Guide) An intentional activity that ensures the future performance of the project work is aligned with the project management plan.

Procedure

1) Ordered series of steps that specify how to perform a task (ISO 26514).

2) Specified way to carry out an activity or a process (ISO 9000).

Note 1: Procedures can be documented or not (ISO 9000).

Process (ISO 9000) Set of interrelated or interacting activities that use inputs to deliver an intended result.

Process approach (ISO 9000) Any activity, or set of activities, that uses resources to transform inputs to outputs can be considered as a process.

For organizations to function effectively, they have to identify and manage numerous interrelated and interacting processes. Often, the output from one process will directly form the input into the next process. The systematic identification and management of the processes employed within an organization and particularly the interactions between such processes is referred to as the "process approach."

The intent of this International Standard is to encourage the adoption of the process approach to manage an organization.

Process asset (CMMI-DEV) Anything the organization considers useful in attaining the goals of a process area.

Process description (ISO 24765) Documented expression of a set of activities performed to achieve a given purpose.

Note: A process description provides an operational definition of the major components of a process. The description specifies, in a complete, precise, and verifiable manner, the requirements, design, behavior, or other characteristics of a process. It also may include procedures for determining whether these provisions have been satisfied. Process descriptions can be found at the activity, project, or organizational level.

Process owner (ISO 24765) Person (or team) responsible for defining and maintaining a process.

Product

1) Output of an organization that can be produced without any transaction taking place between the organization and the customer (ISO 9000).

Note: Hardware is tangible and its amount is a countable characteristic (e.g., tyres). Processed materials are tangible and their amount is a continuous characteristic (e.g., fuel and soft drinks). Hardware and processed materials are often referred to as goods. Software consists of information regardless of delivery medium (e.g., computer programme, mobile phone app, instruction manual, dictionary content, musical composition copyright, driver's license).

2) An artifact that is produced, is quantifiable, and can be either an end item in itself or a component item (PMBOK® Guide).

Production library (ISO 24765) A software library containing software approved for current operational use.

Program librarian (ISO 24765) The person responsible for establishing, controlling, and maintaining a software development library.

Prototype (ISO 15910)

1) A preliminary type, form, or instance of a system that serves as a model for later stages or for the final, complete version of the system;

2) Model or preliminary implementation of a piece of software suitable for the evaluation of system design, performance or production potential, or for the better understanding of the software requirements.

Prototyping (ISO 24765) A hardware and software development technique in which a preliminary version of part or all of the hardware or software is developed to permit user feedback, determine feasibility, or investigate timing or other issues in support of the development process.

Qualification (ISO 12207) Process of demonstrating whether an entity is capable of fulfilling specified requirements.

Qualification testing (ISO 12207) Testing, conducted by the developer and witnessed by the acquirer (as appropriate), to demonstrate that a software product meets its specifications and is ready for use in its target environment or integration with its containing system.

Quality (ISO 9000) Degree to which a set of inherent characteristics of an object fulfills requirements.

Note 1: The term "quality" can be used with adjectives such as poor, good or excellent.

Note 2: "Inherent." as opposed to "assigned," means existing in the object.

Quality assurance (QA)

1) A planned and systematic pattern of all actions necessary to provide adequate confidence that an item or product conforms to established technical requirements (ISO 24765).

2) A set of activities designed to evaluate the process by which products are developed or manufactured (ISO 24765).

3) Part of quality management focused on providing confidence that quality requirements will be fulfilled (ISO 12207).

Quality audits (PMBOK® Guide) A quality audit is a structured, independent process to determine if project activities comply with organizational and project policies, processes, and procedures.

Quality model (ISO 25000) Defined set of characteristics, and of relationships between them, which provides a framework for specifying quality requirements and evaluating quality.

Quality policy (ISO 9000) Intentions and direction of an organization as formally expressed by top management related to quality.

Note 1: Generally, the quality policy is consistent with the overall policy of the organization, can be aligned with the organization's vision and mission and provides a framework for the setting of quality objectives.

Note 2: Quality management principles presented in this International Standard can form a basis for the establishment of a quality policy.

Release

1) A delivered version of an application which may include all or part of an application (ISO 24765).

2) Particular version of a configuration item that is made available for a specific purpose (e.g., test release) (ISO 12207).

Request for information (RFI) (PMBOK® Guide) A type of procurement document whereby the buyer requests a potential seller to provide various pieces of information related to a product or service or seller capability.

Request for proposal (RFP) or Tender (ISO 12207) Document used by the acquirer as the means to announce its intention to potential bidders to acquire a specified system, software product, or software service.

Request for proposal (RFP) (PMBOK® Guide) A type of procurement document used to request proposals from prospective sellers of products or services. In some application areas, it may have a narrower or more specific meaning.

Request for quotation (RFQ) (PMBOK® Guide) A type of procurement document used to request price quotations from prospective sellers of common or standard products or services. Sometimes used in place of request for proposal and, in some application areas, it may have a narrower or more specific meaning.

Requirements traceability (CMMI-DEV) A discernable association between requirements and related requirements, implementations, and verifications.

Reusable product (IEEE 1012) A system, software, or hardware product developed for one use but having other uses, or one developed specifically to be usable on multiple projects or in multiple roles on one project. Examples include, but are not limited to, commercial off-the-shelf (COTS) software products, acquirer-furnished software products, software products in reuse libraries, and preexisting developer software products. Each use may include all or part of the software product and may involve its modification. This term can be applied to any software product (e.g., requirements, architectures), not just to software itself.

Reuse (ISO 24765) The use of an asset in the solution of different problems.

Review

1) A process or meeting during which a work product, or set of work products, is presented to project personnel, managers, users, customers, or other interested parties for comment or approval (ISO 24765).

2) A process or meeting during which a software product, set of software products, or a software process is presented to project personnel, managers, users, customers, user representatives, auditors or other interested parties for examination, comment, or approval (IEEE 1028).

Risk

1) The combination of the probability of an event and its consequence (ISO 16085).

Note 1: The term "risk" is generally used only when there is at least the possibility of negative consequences.
Note 2: In some situations, risk arises from the possibility of deviation from the expected outcome or event.

2) An uncertain event or condition that, if it occurs, has a positive or negative effect on one or more project objectives (PMBOK® Guide).

Risk action request (ISO 16085) The recommended treatment alternatives and supporting information for one or more risks determined to be above a risk threshold.

Risk management plan (ISO 16085) A description of how the elements and resources of the risk management process will be implemented within an organization or project.

Risk management process (ISO 16085) A continuous process for systematically identifying, analyzing, treating, and monitoring risk throughout the life cycle of a product or service.

Risk mitigation

1) A course of action taken to reduce the probability of and potential loss from a risk factor (ISO 24765).

2) A risk response strategy whereby the project team acts to reduce the probability of occurrence or impact of a risk (PMBOK® Guide).

Risk profile (ISO 16085) A chronological record of a risk's current and historical risk state information.

Risk state (ISO 16085) The current project risk information relating to an individual risk.

Note: The information concerning an individual risk may include the current description, causes, probability, consequences, estimation scales, confidence of the estimates, treatment, threshold, and an estimate of when the risk will reach its threshold.

Risk threshold (ISO 16085) A condition that triggers some stakeholder action.

Note: Different risk thresholds may be defined for each risk, risk category, or combination of risks based upon differing risk criteria.

Risk treatment (ISO 16085) The process of selection and implementation of measures to modify risk.

Note 1: The term "risk treatment" is sometimes used for the measures themselves.
Note 2: Risk treatment measures can include avoiding, optimizing, transferring or retaining risk.

Security branch (ISO 24765) A branch, created at the time of a release, to which only security commits are made.

Service Management System (SMS) (ISO 20000-1) Management system to direct and control the service management activities of the service provider.

Note 1: A management system is a set of interrelated or interacting elements to establish policy and objectives and to achieve those objectives.
Note 2: The SMS includes all service management policies, objectives, plans, processes, documentation, and resources required for the design, transition, delivery, and improvement of services and to fulfill the requirements in this part of ISO/IEC 20000.
Note 3: Adapted from the definition of "quality management system" in ISO 9000:2005.

Software (ISO 24765) All or part of the programs, procedures, rules, and associated documentation of an information processing system. Example: command files, job control language.

Note: includes firmware, documentation, data, and execution control statements.

Software development library (ISO 24765) A software library containing computer readable and human readable information relevant to a software development effort. *Synonym*: project library, program support library.

Software engineering (ISO 24765) The systematic application of scientific and technological knowledge, methods, and experience to the design, implementation, testing, and documentation of software.

Software library (ISO 24765) A controlled collection of software and related documentation designed to aid in software development, use, or maintenance.

Software Package (AQAP 2006) A software package is already developed and usable as is or after adaptation. This type of software can be designated as reusable software, software provided by the state, or commercial software, depending on its origin.

Software quality (ISO 25000) Capability of software product to satisfy stated and implied needs when used under specified conditions.

Software quality assurance (IEEE 730) A set of activities that define and assess the adequacy of software processes to provide evidence that establishes confidence that the software processes are appropriate for and produce software products of suitable quality for their intended purposes. A key attribute of SQA is the objectivity of the SQA function with respect to the project. The SQA function may also be organizationally independent of the project; that is, free from technical, managerial, and financial pressures from the project.

Software repository (ISO 24765) A software library providing permanent, archival storage for software and related documentation.

Examples include inadequate staffing levels, insufficient resources, and lack of training. Also included in this section are actions taken to mitigate identified project risks.

Stable branch (ISO 24765) A branch where stability-disrupting changes are discouraged.

Note: the branch used for releasing the product's stable production version.

Staff-hour (ISO 24765) An hour of effort expended by a member of the staff.

Statement of work (SOW) (ISO 12207) Document used by the acquirer to describe and specify the tasks to be performed under the contract.

Supplier (ISO 12207) Organization or an individual that enters into an agreement with the acquirer for the supply of a product or service.

Note 1: Other terms commonly used for supplier are contractor, producer, seller, or vendor.
Note 2: The acquirer and the supplier sometimes are part of the same organization.

Synchronize (ISO 24765)

1) To pull the changes made in a parent branch into its (evolving) child (e.g., feature) branch.

2) To update a view with the current version of the files in its corresponding branch.

System (ISO 15288) Combination of interacting elements organized to achieve one or more stated purposes.

System library (ISO 24765) A software library containing system-resident software that can be accessed for use or incorporated into other programs by reference.

Systems integration testing (IEEE 829) Testing conducted on multiple complete, integrated systems to evaluate their ability to communicate successfully with each other and to meet the overall integrated systems' specified requirements.

Tag (ISO 24765) A symbolic name assigned to a specific release or a branch.

Note: provides developers and end-users with a unique reference to the code base they are working with.

Technical independence (IEEE 1012) Technical independence requires the V&V effort to use personnel who are not involved in the development of the system or its elements. The IV&V effort should formulate its own understanding of the problem and how the proposed system is solving the problem. Technical independence ("fresh viewpoint") is an important method to detect subtle errors overlooked by those too close to the solution.

For system tools, technical independence means that the IV&V effort uses or develops its own set of test and analysis tools separate from the developer's tools. Sharing of tools is allowable for computer support environments (e.g., compilers, assemblers, and utilities) or for system simulations where an independent version would be too costly. For shared tools, IV&V conducts qualification tests on tools to assure that the common tools do not contain errors that may mask errors in the system being analyzed and tested. Off-the-shelf tools that have extensive history of use do not require qualification testing. The most important aspect for the use of these tools is to verify the input data used.

Template

1) An asset with parameters or slots that can be used to construct an instantiated asset (IEEE 1517).

2) A partially complete document in a predefined format that provides a defined structure for collecting, organizing, and presenting information and data (PMBOK® Guide).

Test (IEEE 829)

1) An activity in which a system or component is executed under specified conditions, the results are observed or recorded, and an evaluation is made of some aspect of the system or component.

2) To conduct an activity as in (1).

3) A set of one or more test cases and procedures.

Threat (PMBOK® Guide) A risk that would have a negative effect on one or more project objectives.

Traceability

1) The degree to which a relationship can be established between two or more products of the development process, especially products having a predecessor-successor or master-subordinate relationship to one another (ISO 24765).

    Example: the degree to which the requirements and design of a given system element match; the degree to which each element in a bubble chart references the requirement that it satisfies.

2) A discernable association among two or more logical entities such as requirements, system elements, verifications, or tasks. (See also "bidirectional traceability" and "requirements traceability.") (CMMI-DEV).

Traceability matrix (ISO 24765) A matrix that records the relationship between two or more products of the development process.

   Example: a matrix that records the relationship between the requirements and the design of a given software component.

Trunk (ISO 24765) The software's main line of development; the main starting point of most branches.

Note: One can often distinguish the trunk from other branches by the version numbers used for identifying its files, which are shorter than those of all other branches.

Unit test

1) Testing of individual routines and modules by the developer or an independent tester.

2) A test of individual programs or modules in order to ensure that there are no analysis or programming errors (ISO/IEC 2382-20).

3) Test of individual hardware or software units or groups of related units (ISO 24765).

Validation

1) Confirmation, through the provision of objective evidence, that the requirements for a specific intended use or application have been fulfilled (ISO 15288).

2) (A) The process of evaluating a system or component during or at the end of the development process to determine whether it satisfies specified requirements. (B) The process of providing evidence that the system, software, or hardware and its associated products satisfy requirements allocated to it at the end of each life cycle activity, solve the right problem (e.g., correctly model physical laws, implement business rules, and use the proper system assumptions), and satisfy intended use and user needs (IEEE 1012).

Vendor branch (ISO 24765) A branch for keeping track of versions of imported software.

Note: Differences between successive versions can then be readily applied to the locally modified import.

Verification

1) Confirmation, through the provision of objective evidence, that specified requirements have been fulfilled (ISO 12207).

2) (A) The process of evaluating a system or component to determine whether the products of a given development phase satisfy the conditions imposed at the start of that phase. (B) The process of providing objective evidence that the system, software, or hardware and its associated products conform to requirements (e.g., for correctness, completeness, consistency, and accuracy) for all life cycle activities during each life cycle process (acquisition, supply, development, operation, and maintenance); satisfy standards, practices, and conventions during life cycle processes; and successfully complete each life cycle activity and satisfy all the criteria for initiating succeeding life cycle activities. Verification of interim work products is essential for proper understanding and assessment of the life cycle phase product(s) (IEEE 1012).

Version (ISO 24765)

1) An initial release or re-release of a computer software configuration item, associated with a complete compilation or recompilation of the computer software configuration item.

2) An initial release or complete re-release of a document, as opposed to a revision resulting from issuing change pages to a previous release.

3) An operational software product that differs from similar products in terms of capability, environmental requirements, and configuration.

4) An identifiable instance of a specific file or release of a complete system.

Note: Modification to a version of a software product, resulting in a new version, requires configuration management action.

Version control (ISO 24765) Establishment and maintenance of baselines and the identification and control of changes to baselines that make it possible to return to the previous baseline.

Very Small Entity (VSE) (ISO 29110) A VSE is an entity (enterprise, organization, department, or project) having up to 25 people.

View (ISO 24765) A developer's copy of a branch.

ABBREVIATIONS – ACRONYMS

AAR	After Action Review
ACM	Association for Computing Machinery
AECL	Atomic Energy Canada Limited
ASQC	American Society for Quality Control
BPMN	Business Process Modeling and Notation
CAR	Causal Analysis and Resolution
CASE	Computer Aided Software/System Engineering
CCB	Configuration Control Board
CEO	Chief Executive Officer
CI	Continuous Improvement
CI	Configuration Item
CM	Configuration Management
CMM®	Capability Maturity Model
CMMI®	Capability Maturity Model Integration (www.sei.cmu.edu/cmmi)
CMMI–DEV	CMMI for Development
CMMI–ACQ	CMMI for Acquisition
CMMI–SVC	CMMI for Services
CobiT	Control Objectives for Information and related Technology
COCOMO	COnstructive COst MOdel (http://sunset.usc.edu/csse/research/COCOMOII/cocomo_main.html)
COTS	Commercial off the shelf
DP	Deployment Package
EIA	Electronic Industries Alliance
ÉTS	École de technologie supérieure (www.etsmtl.ca)
FDA	Food and Drug Administration
FEMA	Failure Mode and Effect Analysis
FSM	Finite State Machine
GE	General Electric
GSEP	Generic Systems Engineering Process
HRMS	Human Resource Management System
IV&VI	Independent Verification and Validation
IBM	International Business Machines
IDEAL	Initiating, Diagnosing, Establishing, Acting, Learning
IEC	International Electrotechnical Commission
IEEE	Institute of Electrical and Electronics Engineers (www.ieee.org)

INCOSE	International Council on Systems Engineering (www.incose.org)
IS	Information System
ISACA	Information Systems Audit and Control Association
ISBG	International Software Benchmarking Standards Group (www.isbsg.org)
ISM	Integrated Software Management
ISO	International Organization for Standardization (www.iso.org). *Note*: ISO is not an abbreviation.
ISO/IEC	International Organization for Standardization/ International Electrotechnical Commission
ISO/IEC JTC 1 SC 7	Sub committee 7 Software and systems engineering
ISSEP	Integrated Systems and Software Engineering Process
IT	Information Technology
JTC 1	Joint Technical Committee 1 (of ISO/IEC)
KLOC	Thousand lines of source code
MLOC	Million lines of source code
MR	Modification Request also Change Request (CR) or Problem Report (PR)
NASA	National Aeronautics and Space Administration (www.nasa.gov)
OO	Object Oriented
OPD	Organization Process Definition
OPF	Organization Process Focus
PAL	Process asset library
PDCA	Plan–Do–Check–Act
PM	Project Management
PMI	Project Management Institute (www.pmi.org)
PMBOK®	Project Management Body of Knowledge (www.pmi.org)
PODCAST	Portable media player digital audio file made available on the Internet
PPQA	Process and Product Quality Assurance
PR	Peer Review
PR	Problem Report
QE	Quality Engineering
QMS	Quality Management System
RAMS	Reliability, Availability, Maintainability, and Safety
RFI	Request For Information
RFP	Request For Proposal
RTCA	Radio Technical Commission for Aeronautics

S^{3m}	Software Maintenance Maturity Model (www.s3m.org)
SAP	System, Anwendunsgen, Produkte/Systems Applications and Products
SCAMPI	Standard CMMI Appraisal Method for Process Improvement
SCE	Software Capability Evaluation
SCM	Software Configuration Management
SCR	Software Change Request
SEI	Software Engineering Institute (www.sei.cmu.edu)
SOX	Sabarnes-Oxley
SOW	Statement of work
SPA	Software Process Assessment
SPICE	Software Process Improvement and Capability dEtermination
SQA	Software Quality Assurance
SQM	Software Quality Management
SRA	Society for Risk Analysis (www.sra.org)
S/W	Software
SW– CMM	Software Capability Maturity Model
SWEBOK®	Software Engineering Body of Knowledge (www.swebok.org)
TOMS	Total Ozone Mapping Spectrometer
TDD	Test Driven Development
TQM	Total Quality Management
VSE	Very Small Entity
V&V	Verification and Validation
WBS	Work Breakdown Structure

References

[ABR 15] ABRAN A. *Software Project Estimation: The Fundamentals for Providing High Quality Information to Decision Makers.* Wiley-IEEE Computer Society Press, Los Alamitos, CA, 2015, 288 p.

[ABR 93] ABRAN A. and NGUYENKIM H. Measurement of the maintenance process from a demand-based perspective. *Journal of Software Maintenance: Research and Practice*, vol. 5, issue 2, 1993, pp. 63–90.

[AGC 06] Auditor General of Canada. *Large IT projects.* Office of the Auditor General, Ottawa, Canada, November 2006, Chapter 3. Available at: http://www.oag-bvg.gc.ca/internet/English/parl_oag_200611_03_e_14971.html

[ALE 91] ALEXANDER M. In: *The Encyclopedia of Team-Development Activities,* edited by J. William Pfeiffer. University Associates, San Diego, CA, 1991.

[AMB 04] AMBLER S. W. *Examining the Big Requirements Up Front Approach.* Ambysoft Inc. Available at: www.agilemodeling.com/essays/examiningBRUF.htm

[ANA 04] ANACLETO A., VON WANGENHEIM C. G., SALVIANO C. F., and SAVI R. Experiences gained from applying ISO/IEC 15504 to small software companies in Brazil. In: 4th International SPICE Conference on Process Assessment and Improvement, Lisbon, Portugal, April 2004, pp. 33–37.

[APR 97] APRIL A., ABRAN A., and MERLO E. Process assurance audits: Lessons learned. In: Proceedings of ICSE 98, Kyoto, Japan, April 19–25, 1997.

[APR 00] APRIL A. and AL-SHUROUGI D. *Software product measurement for supplier evaluation.* In: FESMA 2000, Madrid, Spain, October 18–20, 2000.

[APR 08] APRIL A. and ABRAN A. *Software Maintenance Management: Evaluation and Continuous Improvement.* John Wiley & Sons, Inc., Hoboken, NJ, 2008, 314 p.

[AUS 96] AUSTIN R. *Measuring and Managing Performance in Organizations.* Dorset House Publishing, New York, 1996.

[BAB 15] BABOK, Business Analyst Body of knowledge, v 3.0, International Institute of Business Analysis. Available at: http://www.theiiba.org

[BAS 96] BASILI V. R., CALDIERA G., and ROMBACH H. D. The experience factory. In: *Encyclopedia of Software Engineering*, edited by J. J. Marciniak., John Wiley & Sons, New York, 1996, pp. 469–476.

[BAS 10] BASILI V. R., LINDVALL, M., REGARDIE M., and SEAMAN C. Linking software development and business strategy through measurement. *IEEE Computer*, vol. 43, issue 4, April 2010, pp. 57–65.

Software Quality Assurance, First Edition. Claude Y. Laporte and Alain April.
© 2018 the IEEE Computer Society, Inc. Published 2018 by John Wiley & Sons, Inc.

[BEI 90] BEIZER B. *Software Testing Techniques*, 2nd edition. Van Nostrand Reinhold Co., International Thomson Press, New York, NY, 1990.

[BLO 11] BLOCK P. *Flawless Consulting: A Guide to Getting Your Expertise Used*, 3rd edition. Jossey-Bass/Pfeiffer, San Francisco, CA, 2011.

[BOE 89] BOEHM B. W. *Tutorial: Software Risk Management*. IEEE Computer Society, Los Alamitos, CA, 1989.

[BOE 91] BOEHM B. W. Software risk management: Principles and practices. *IEEE Software*, vol. 8, issue 1, 1991, pp. 32–41.

[BOE 00] BOEHM B. W., ABST C., BROWN A., CHULANI, S., CLARK, B. C., HOROWITZ, E., MADACHY, R., REIFER, D. J., STREECE, B. *Software Cost Estimation with COCOMO II*. Prentice-Hall, Englewood Cliffs, NJ, 2000.

[BOE 01] BOEHM B. W., BASILI V. Software Defect Reduction Top 10 List, *IEEE Computer*, vol. 34, January 2001, pp. 135–137.

[BOL 95] BOLOIX G. and ROBILLARD P. (1995) Software system evaluation framework. *Computer Magazine*, December 1995, pp. 17–26.

[BOO 94] BOOCH G. and BRYAN D. *Software Engineering with Ada*, 3rd edition. Benjamin/Cummings, Redwood City, CA, 1994.

[BOU 05] BOUSETTA A. and LABRECHE P. *Introduction to 6-sygma applied to software, Presentation at Montreal SPIN*, Ecole de Technologie Supérieure (ETS), Montreal, Canada, March 2005.

[BRO 02] BROY M. and DENERT E. (eds.) A history of software inspections. In: *Software Pioneers*. Springer-Verlag, Berlin, Heidelberg, 2002.

[BRI 16] BRIDGES W. *Managing Transitions—Making the Most of Change*, 4th edition. Da Capo Press, Cambridge, MA, 2016.

[BUC 96] BUCKLEY FLETCHER J. *Implementing Configuration Management: Hardware, Software, and Firmware*. IEEE Computer Society Press, Los Alamitos, CA, 1996.

[BYR 96] BYRNES P. and PHILLIPS M. Software Capability Evaluation, Version 3.0, Method Description (CMU/SEI-96-TR-002, ADA309160). Software Engineering Institute, Carnegie Mellon University, Pittsburgh, PA, 1996.

[CAR 92] CARLETON A. D., PARK R. E., GOETHERT W. B., FLORAC W. A., BAILEY E. K., PFLEEGER S. L. Software Management for DoD Systems: Recommendations for Initial Implementation, SEI Technical Report, USA, September 1992.

[CEG 90] CEGELEC, Software Validation Phase Procedure, CEGELEC Methodology, 1990.

[CEG 90a] Software Configuration Management Procedure, CEGELEC Methodology, 1990.

[CEN 01] EN 50128, Railway applications—Communications, signaling, and processing systems—Software for railway control and protection systems, European Standard, 2001.

[CHA 99] CHARRETTE R. R. Building bridges over intelligent rivers. *American Programmer*, September1992, pp. 2–9.

[CHA 06] CHARETTE R. *Focus on Dr. Robert Charette, Master Risk Management Practitioner—A CAI State of Practice Interview*, Computer Aid Inc., Allentown, PA, March 2006.

[CHI 02] CHILLAGERE R., BHANDARI I., CHAAR J., and HALLIDAY, D. Orthogonal defect classification. *IEEE Transactions on Software Engineering*, vol. 18, issue 11, November 2002, pp. 943–956.

[CHR 08] CHRISSIS M. B., KONRAD M., and SHRUM S. *CMMI*, 2nd edition. Pearson Education, Paris, France, 2008.

[COA 03] COALLIER F. International standardization in software and systems engineering. *CrossTalk, The Journal of Defense Soltware Engineering*, February 2003, pp. 18–22.

[COB 12] IT Governance Institute, CobiT, Governance, Control and Audit for Information and Related Technology, version 5, April 2012. Available at: http://www.isaca.org

[COL 10] Commit-monitor for Subversion Repositories, Version 1.7.0, October 24, 2010. Available at: https://sourceforge.net/projects/commitmonitor/

[CON 93] Conseil du Trésor. Politique du Conseil du Trésor, numéro NCTTI-26: Évaluation de logiciels- Caractéristiques d'utilisation-Critères d'applicabilité, version 11, février 1993.

[CRO 79] CROSBY P. B. *Quality Is Free*. McGraw-Hill, New York, 1979.

[CUR 79] CURTIS B. In search of software complexity. In: Proceedings of the IEEE/PINY Workshop on quantitative software models, IEEE catalog no TH00067-9, October 1979, pp. 95–105.

[DAV 93] DAVENPORT T. *Process Innovation*. Harvard Business School Press, Boston, MA, 1993.

[DAV 06] DAVIES I., GREEN P., ROSEMANN M., INDULSKA, M., GALLO S. How do Practitioners Use Conceptual Modeling in Practice? *Data & Knowledge Engineering*, vol. 58, 2006, pp. 358–380.

[DEC 08] DECKER G. and SCHREITER T. OMG releases BPMN 1.1—What's changed?, Inubit AG, Berlin, Germany M., 2008, pp. 1–9, Available at: http://docplayer.net/15316886-Omg-releases-bpmn-1-1-what-s-changed.html

[DEM 00] DEMARCO T. and LISTER T. Both sides always lose: Litigation of software-intensive contracts. *CrossTalk, The Journal of Defense Soltware Engineering*, February 2000, pp. 4–6.

[DES 95] DESHARNAIS J-M. and ABRAN A. How to successfully implement a measurement program: From theory to practice., In: *Metrics in Software Evolution*, edited by M. Müllenberg and A. Abran. R. Oldenbourg Verlag, Oldenburg, 1995, pp. 11–38.

[DIA 02] DIAZ, M. and KING J. How CMM impacts quality, productivity, rework, and the bottom line. *CrossTalk, The Journal of Defense Soltware Engineering*, March 2002, pp. 9–14.

[DIN 05] DINGSØYR T. Postmortem reviews: Purpose and approaches in software engineering. *Information and Software Technology*, vol. 47, issue 5, March 31, 2005, pp. 293–303.

[DIO 92] DION R. Elements of a process improvement program, Raytheon. *IEEE Software*, vol. 9, issue 4, July 1992, pp. 83–85.

[DOD 83] DoD-STD-1679A, *Military Standard—Weapon Systems Software Development*, Department of Defense, Washington D.C., 1983.

[DOD 09] Technology Readiness Assessment (TRA) Deskbook, Department of Defense, United States, July 2009.

[DOR 96] DOROFEE A. J., WALKER, J.A., ALBERTS, C.J., HIGUERA, R.P., MURPHY, R.L., WILLIAMS, R.C. Continuous Risk Management Guidebook, Carnegie Mellon University, Software Engineering Institute, Pittsburgh, PA, 1996.

[DOW 94] DOWN A., COLEMAN M. and ABSOLON P. *Risk Management for Software Projects*. McGraw-Hill Book Company, London, 1994.

[EAS 96] EASTERBROOK S. The role of independent V&V in upstream software development processes. In: Proceedings of the 2nd World Conference on Integrated Design and Process Technology (IDPT), Austin, Texas, USA, December 4, 1996.

[EGY 04] EGYED A. Identifying requirements conflicts and cooperation. *IEEE Software*, November/December, 2004.

[EIA 98] EIA 1998 Electronic Industries Alliance, Systems Engineering Capability Model (EIA/IS-731), Washington, DC, 1998.

[EUR 11] EUROCAE ED-12C - Software Considerations in Airborne Systems and Equipment Certification, EUROCAE, 17 rue Hamelin, 75783, Paris Cedex, France, 2011.

[FAG 76] FAGAN M. E. Design and code inspections to reduce errors in program development. *IBM System Journal*, vol. 15, issue 3, 1976, pp. 182–211.

[FDA 02] General Principles of Software Validation; Final Guidance for Industry and FDA Staff, U.S. Food and Drug Administration, 2002.

[FEN 07] FENTON N. and NEIL M. Software Metrics: Roadmap. Queen Mary University, Department of Computer Science, Harlow, UK, 2007.

[FOR 92] FORNELL G. E. Process for acquiring software architecture, cover letter to draft report, July 10, 1992.

[FOR 05] FORSBERG K., MOOZ H. and COTTERMAN H. *Visualizing Project Management*, 3rd edition. John Wiley & Sons, Inc., New York, NY, 2005.

[FRE 05] FREEMAN S. Toyota attributes Prius shutdowns to software glitch. *Wall Street Journal*, May 16, 2005.

[GAL 17] GALIN D. *Software Quality: Concepts and Practice*. Wiley-IEEE Computer Society Press, Hoboken, New Jersey, 2017, 726 p.

[GAN 04] GANSSLE J. Disaster redux!, Available at: http://www.embedded.com/ electronics-blogs/break-points/4025051/Disaster-redux-

[GAR 84] GARVIN D. What does product quality really mean? *MIT Sloan Management* Review, Fall 1984, pp. 25–45.

[GAR 15] GARCIA L., LAPORTE C. Y., ARTEAGA J., and BRUGGMANN M. Implementation and certification of ISO/IEC 29110 in an IT startup in Peru. *Software Quality Professional Journal*, ASQ, vol. 17, issue 2, 2015, pp. 16–29.

[GAU 04] GAUTHIER R. *Une Force en Mouvement*, La Boule de Cristal, Centre de recherche informatique de Montreal, 22 janvier 2004.

[GEC 98] GECK B., GLOGER M., JOCKUSCH, S., LEBSANFT, K., MEHNER, T., PAUL, P., PAULISCH, F., RHEINDT, M., VOLKER, A., WEBER, N. Software@Siemens: Best practices for the measurement and management of processes and architectures. In: Software Process Improvement Conference, Monte Carlo, Monaco, December 1998.

[GEP 04] GEPPERT L. Lost radio contact leaves pilots on their own. *IEEE Spectrum*, November 2004.

[GHE 09] GHEYSENS P. Drill through merges in TFS2010, Into Visual Studio team System, blogging about the current and upcoming release(s), January 5, 2009. Available at: http://intovsts.net/category/version-control/

[GIL 88] GILB T. *Principles of Software Engineering Management*. Addison-Wesley, Wokingham, UK, 1988.

[GIL 93] GILB T. and GRAHAM D. *Software Inspection*. Addison-Wesley, Wokingham, UK, 1993. ISBN: 0-201-63181-4.

[GIL 08] GILB T. *Review Process Design: Some Guidelines for Tailoring Your Engineering Review Processes for Maximum Efficiency*. International Council on Systems Engineering (INCOSE), The Netherlands, June 2008.

[GOT 99a] GOTTERBARN F. How the new software engineering code of ethics affects you. *IEEE Software*, vol. 16, issue 6, 1999, pp. 58–64.

[GOT 99b] GOTTERBARN D., MILLER K., and ROGERSON S. Computer society and ACM approve software engineering code of ethics. *IEEE Computer*, vol. 32, issue 10, 1999, pp. 84–88.

[GRA 92] GRADY R. *Practical Software Metrics for Project Management and Process Improvement*. Prentice-Hall Inc., Englewood Cliffs, NJ, 1992.

[HAI 02] HAILPERN B. and SANTHANAM P. Software debugging, testing, and verification. *IBM Systems Journal*, vol. 41, issue 1. Humphrey, W. S. *A Discipline for Software Engineering*. Addison-Wesley, Reading, MA, 2002.

[HAL 96] HALEY T. J. Software process improvement at Raytheon. *IEEE Software*, vol. 13, issue 6, 1996, pp. 33–41, Figure abstracted from IEEE Software.

[HAL 78] HALSTEAD M. H. Software science: A progress report. In: Proceedings of the U.S. Army/IEEE Second Life-Cycle Management Conference, Atlanta, August 1978, pp. 174–179.

[HEF 01] HEFNER R. and TAUSER J. Things they never taught you in CMM school. 26th Annual NASA Goddard Software Engineering Workshop, Greenbelt, MD, November 27–29, 2001.

[HEI 14] HEIMANN, D.I. An Introduction to the New IEEE 730 Standard on Software Quality Assurance. *Software Quality Professional (SQP)*, vol. 16, issue 3, 2014, pp. 26–38.

[HOL 98] HOLLAND D. Document inspection as an agent of change. In: *Dare to be Excellent*, edited by A. Jarvis and l. Hayes, Prentice Hall, Upper Saddle River, NJ, 1998.

[HUM 89] HUMPHREY W. *Managing the Software Process*. Addison-Wesley, Boston, MA, 1989.

[HUM 00] HUMPHREY W. S. The Personal Software Process (PSP), CMU/SEI-2000-TR-022, Software Engineering Institute, Carnegie Mellon University, Pittsburgh, PA, 2000.

[HUM 02] HUMPHREY W. S. *Winning with Software, An Executive Strategy*. Addison-Wesley, Reading, MA, 2002.

[HUM 04] HUMPHREY W. S. The Quality Attitude, news@sei newsletter, Number 3, 2004.

[HUM 05] HUMPHREY W. S. *PSP: A Self-Improvement Process for Software Engineers*. Addison-Wesley, Reading, MA, 2005.

[HUM 07] HUMPHREY W. S., KONRAD M., and OVER J. PETERSON, W. Future directions in process improvement. *CrossTalk, The Journal of Defense Soltware Engineering*, February 2007.

[HUM 08] HUMPHREY W. S. The software quality challenge. *CrossTalk, The Journal of Defense Soltware Engineering*, June 2008, pp. 4–9.

[IBE 02] IBERLE K. But will it work for me. In: Proceedings of the Pacific Northwest Software Quality Conference, Portland, United States, 2002, pp. 377–398. Available at: http://www.kiberle.com/publications/

[IBE 03] IBERLE K. They don't care about quality. In: Proceedings of STAR East, Orlando, United States, 2003. Available at: http://www.kiberle.com/publications/

[IEE 98] IEEE 1998, Std. 1320.1-1998. IEEE Standard for Functional Modeling Language - Syntax and Semantics for IDEF0, The Institute of Electrical and Electronics Engineers, New York, NY, 1998.

[IEE 98a] IEEE 1998, Std. 830-1998. IEEE Recommended practice for software requirements, The Institute of Electrical and Electronics Engineers, New York, NY, 1998.

[IEE 98b] IEEE 1998. Std. 1061-1998. IEEE Standard for a Software Quality Metrics Methodology, New York, NY, 1998.

[IEE 99] IEEE-CS, IEEE-CS-1999. Software Engineering Code of Ethics and Professional Practice, IEEE-CS/ACM, 1999, https://www.computer.org/web/education/code-of-ethics

[IEE 07] IEEE Std 1362-2007. IEEE Guide for Information Technology—System Definition—Concept of operations (ConOps) Document, 2007.

[IEE 08a] IEEE 829-2008. IEEE Standard for Software and System Test Documentation, IEEE, The Institute of Electrical and Electronics Engineers, New York, NY, 2008.

[IEE 08b] IEEE 1028, IEEE Standard 1028-2008. IEEE Standard for Software Reviews and Audits, IEEE, The Institute of Electrical and Electronics Engineers, New York, NY, 2008.

[IEE 12] IEEE Std 1012-2012. IEEE Standard for System and Software Verification and Validation, IEEE, The Institute of Electrical and Electronics Engineers, New York, NY, 2012.

[IEE 12b] IEEE Std 828-2012. IEEE Standard for Configuration Management in Systems and Software Engineering, IEEE, The Institute of Electrical and Electronics Engineers, New York, NY, 2012.

[IEE 14] IEEE 730. IEEE Standard for Software Quality Assurance Processes: IEEE, The Institute of Electrical and Electronics Engineers, New York, NY, 2014.

[INC 15] INCOSE Systems Engineering Handbook: A Guide for System Life Cycle Processes and Activities, 4th Edition, Hoboken, NJ, USA: John Wiley and Sons, Inc, ISBN: 978-1-118-99940-0, 304 pages.

[ISO Guide 73] ISO Guide73:2009. Risk management—Vocabulary, International Organization for Standardization (ISO), Geneva, Switzerland, 2009.

[ISO 01] ISO/IEC 9126-1:2001. Software Engineering—Product quality—Part 1: Quality model: 2001, International Organization for Standardization (ISO), Geneva, Switzerland, 2001, 25 p.

[ISO 04a] ISO 17050-1:2004. Conformity assessment–Supplier's declaration of conformity- Part 1: General requirements, International Organization for Standardization (ISO), Geneva, Switzerland, 2004.

[ISO 04b] ISO 17050-2:2004. Conformity assessment—Supplier's declaration of conformity - Part 2: Supporting documentation, International Organization for Standardization (ISO), Geneva, Switzerland, 2004.

[ISO 05c] ISO/IEC 27002 :2005. Information technology—Security techniques—Code of practice for information security management, International Organization for Standardization (ISO), Geneva, Switzerland. 2005.

[ISO 05d] ISO/IEC 17799:2005. Information technology—Security techniques—Code of practice for information security management, International Organization for Standardization (ISO), Geneva, Switzerland, 2005.

[ISO 06a] ISO/IEC/IEEE 16085:2006. Systems and software engineering—Life cycle processes—Risk management, International Organization for Standardization (ISO), Geneva, Switzerland, 2006.

[ISO 08] ISO/IEC 26514:2008. Systems and Software Engineering—Requirements for Designers and Developers of User Documentation, International Organization for Standardization (ISO), Geneva, Switzeerland, 2008.

[ISO 09] ISO/IEC/IEEE 16326. Systems and software engineering—Life cycle processes—Project management, International Organization for Standardization (ISO), Geneva, Switzerland, 2009.

[ISO 09a] ISO 9004:2009. Managing for the sustained success of an organization—A quality management approach, International Organization for Standardization (ISO), Geneva, Switzerland, 2009.

[ISO 10] ISO/IEC TR 24774:2010. Software and systems engineering—Life cycle management- Guidelines for process description, International Organization for Standardization (ISO), Geneva, Switzerland, 2010.

[ISO 11e] ISO/IEC TR 29110-5-1-2:2011. Software Engineering—Lifecycle Profiles for Very Small Entities (VSEs)—Part 5-1-2: Management and Engineering Guide—Basic Profile, International Organization for Standardization (ISO), Geneva, Switzerland. Available for free at: http://standards.iso.org/ittf/PubliclyAvailableStandards/index.html

[ISO 11f] ISO/IEC/IEEE 29148:2011. Systems and software engineering—Life cycle processes—Requirements Engineering, International Organization for Standardization (ISO), Geneva, Switzerland, 2011, 54 p.

[ISO 11g] ISO 19011: 2011. Guidelines for auditing systems, International Organization for Standardization (ISO), Geneva, Switzerland, 2011, 44 p.

[ISO 11h] ISO/IEC 20000-1:2011. Information technology—Service management—Part 1: Service management system requirements, International Organization for Standardization (ISO), Geneva, Switzerland, 2011.

[ISO 11i] ISO/IEC 25010:2011. Systems and software engineering–Systems and Software Quality Requirements and Evaluation (SQuaRE)–System and software quality models, International Organization for Standardization (ISO), Geneva, Switzerland, 2011, 34 p.

[ISO 14] ISO/IEC 90003:2014. Software Engineering—Guidelines for the application of ISO9001:2008 to computer software, International Organization for Standardization (ISO), Geneva, Switzerland, 2014, 54 p.

[ISO 14a] ISO/IEC 25000:2014. System and software engineering—System and Software Quality Requirements and Evaluation (SQuaRE)—Guide to SQuaRE—Guide de SQuaRE, International Organization for Standardization (ISO), Geneva, Switzerland, 2014, 27 p.

[ISO 15] ISO 9001. Quality systems requirement—Requirements, International Organization for Standardization (ISO), Geneva, Switzerland, 2015.

[ISO 15a] ISO 17021-1:2015. Conformity assessment—Requirements for bodies providing audit and certification of management systems—Part 1: Requirements, International Organization for Standardization (ISO), Geneva, Switzerland, 2014, 48 p.

[ISO 15b] ISO 9000, Quality management system-Fundamentals and vocabulary, International Organization for Standardization (ISO), Geneva, Switzerland, 2015.

[ISO 15c] ISO/IEC/IEEE 15288: 2015. Systems and software engineering—System life cycle processes, International Organization for Standardization (ISO), Geneva, Switzerland, 2015.

[ISO 16d] ISO 13485: 2016. Medical devices—Quality management systems—Requirements for regulatory purposes, International Organization for Standardization (ISO), Geneva, Switzerland, 2016.

[ISO 16f] ISO/IEC TR 29110-1:2016. Systems and software Engineering—Lifecycle Profiles for Very Small Entities (VSEs)—Part 1: Overview, International Organization for Standardization (ISO), Geneva, Switzerland, 2016. Available for free at: http://standards.iso.org/ittf/Publicly AvailableStandards/index.html

[ISO 17] ISO/IEC/IEEE 12207:2017. Systems and software engineering—Software life cycle processes, International Organization for Standardization (ISO), Geneva, Switzerland, 2017.

[ISO 17a] ISO/IEC/IEEE 24765:2017. Systems and Software Engineering Vocabulary, International Organization for Standardization (ISO), Geneva, 2017, Available at: www.computer.org/sevocab

[ISO 17b] ISO/IEC/IEEE 15289: 2017. Systems and software engineering—Content of life cycle information items (documentation), International Organization for Standardization (ISO), Geneva, Switzerland, 2017, 88 p.

[ISO 17c] ISO/IEC/IEEE 15939:2017. Systems and software engineering—Measurement process, International Organization for Standardization (ISO), Geneva, Switzerland, 2017.

[ISO 17d] ISO/IEC 20246. Software and Systems Engineering—Work Product Reviews, International Organization for Standardization (ISO), Geneva, Switzerland, 2017, 42 p.

[IST 11] Standard Glossary of terms Used in Software Testing, version 1.3, International Software Testing Qualifications Board, Brussels, Belgium, 2011. Available at: http://www.glossary.istqb.org

[JAC 07] JACKSON D., THOMAS M., and MILLET, L. Software for Dependable Systems: Sufficient Evidence?, Committee on Certifiably Dependable Software Systems, National Research Council, ISBN: 0-309-10857-8, 2007, 120 p.

[JON 00] JONES C. *Software Assessments, Benchmarks, and Best Practices.* Addison-Wesley, Reading, MA, 2000

[JON 03] JONES C. Making measurement work. *CrossTalk, The Journal of Defense Soltware Engineering*, January 2003.

[JPL 00] Report on the Loss of the Mars Polar Lander and Deep Space 2 Missions, JPL Special Review Board, Jet Propulsion Laboratory, March 2000.

[KAS 00] KASSE T. and MCQUAID P. Software Configuration Management for Project Leaders, SQP, *Software Quality Professional*, vol. 2, issue 4, September 2000, pp. 8–19.

[KAS 05] KASUNIC M. *Designing an Effective Survey, Handbook, CMU/SEI-2005-HB-004.* Software Engineering Institute, Pittsburg, PA, 2005.

[KAS 08] KASUNIC M., MCCURLEY J., and ZUBROW D., *Can You Trust Your Data? Establishing the Need for a Measurement and Analysis Infrastructure Diagnostic, Technical Note CMU/SEI-2008-TN-028.* Software Engineering Institute, Pittsburgh, PA, November 2008.

[KER 01] KERTH N. *Project Retrospective: A Handbook for Team Reviews.* Dorset House Publishing, New York, 2001.

[KID 98] KIDWELL P. A. Stalking the elusive computer bug. *IEEE Annals of the History of Computing*, vol. 20, issue 4, 1998, pp 5–9.

[KON 00] KONRAD M. Overview of CMMI Model, Presentation to Montreal SPIN, Montreal, Canada, November 21, 2000.

[KRA 98] KRASNER H. Using the cost of quality approach for software. *CrossTalk, The Journal of Defense Soltware Engineering*, vol. 11, issue 11, November 1998.

[LAG 96] LAGUË B. and APRIL A. 1996. Mapping of the ISO 9125 maintainability internal metrics to an industrial research tool. In: Proceedings of SES 1996, Montreal, Canada, October 21–25, 1996.

[LAN 08] LAND K., HOBART W., and WALZ J. A Practical Metrics and Measurements Guide For Today's Software Project Manager. *IEEE Computer Society ReadyNotes*, 2008.

[LAP 97] LAPORTE C. Y. and PAPICCIO N. L'ingénierie et l'intégration des processus de génie logiciel, de génie systèmes et de gestion de projets. *Revue Génie Logiciel*, vol. 46, 1997.

[LAP 98] LAPORTE C. Y. and TRUDEL S. Addressing the people issues of process improvement activities at Oerlikon Aerospace. *Software Process-Improvement and Practice*, vol. 4, issue 1, 1998, pp. 187–198.

[LAP 03] BOUCHER G. Risk management applied to the re-engineering of a weapon system. *CrossTalk – The Journal of Defense Software Engineering*, January 2003.

[LAP 07a] LAPORTE C. Y., DOUCET M., BOURQUE P., and BELKÉBIR Y. Utilization of a Set of Software Engineering Roles for a Multinational Organization, 8th International Conference on Product Focused Software Development and

Process Improvement, PROFES 2007, Riga (Latvia), July 2–4, 2007, pp. 35–50.

[LAP 07b] LAPORTE C. Y., DOUCET M., ROY D., and DROLET M. Improvement of Software Engineering Performances An Experience Report at Bombardier Transportation—Total Transit Systems Signalling Group, International Council on Systems Engineering (INCOSE) Seventeenth International Symposium, San Diego (CA), USA, June 24–28, 2007.

[LAP 08] LAPORTE C. Y., ALEXANDRE S., and RENAULT A. Developing international standards for very small enterprises, *IEEE Computer*, vol. 41, issue 3, March 2008, pp. 98–101.

[LAP 08a] LAPORTE C. Y., ROY R., and NOVIELI R. La gestion des risques d'un projet de développement et d'implantation d'un système informatisé au Ministère de la Justice du Québec. *Revue Génie Logiciel*, mars 2008, numéro 84, pp. 2–12.

[LAP 12] LAPORTE C. Y., BERRHOUMA N., DOUCET M., and PALZA-VARGAS, E. Measuring the cost of software quality of a large software project at Bombardier Transportation. *Software Quality Professional Journal*, ASQ, vol. 14, issue 3, June 2012, pp 14–31.

[LAP 14] LAPORTE C.Y., O'CONNOR R. Systems and Software Engineering Standards for Very Small Entities Implementation and Initial Results, QUATIC'2014, 9th International Conference on the Quality of Information and Communications Technology, Guimarães, Portugal, September 23–26, 2014, pp. 38–47.

[LAP 16a] LAPORTE C. Y. and O'CONNOR R. V. QUATIC'2016, Implementing process improvement in very small enterprises with ISO/IEC 29110—A multiple case study analysis. In: 10th International Conference on the Quality of Information and Communications Technology (QUATIC 2016), Caparica/Lisbon, Portugal, September 6–9, 2016.

[LEV 00] LEVESON N. G. System safety in computer-controlled automotive systems. Society of Automotive Engineers (SAE) Congress, Detroit, United States, March 2000.

[LEV 93] LEVESON N. and TURNER C. An investigation of the Therac-25 accidents. *IEEE Computer*, vol. 26, issue 7, 1993, pp. 18–41.

[MAY 02] MAY W. A global applying ISO9001:2000 to software products. *Quality Systems Update*, vol. 12, issue 8, August 2002.

[MCC 76] McCABE T. J. A complexity measure. *IEEE Transactions on Software Engineering*, vol. SE-2, issue 4, November 1976, pp. 308–320.

[MCC 77] McCALL J. A., RICHARDS P. K., and WALTERS G. F. *Factors in software quality*. Griffiths Air Force Base, NY: Rome Air Development Center Air Force Systems Command, Springfield, NY, United States, 1977.

[MCC 04] McCONNELL S. *Code Complete: A Practical Handbook of Software Construction*, 2nd edition. Microsoft Press, 2004, 960 p.

[MCF 03] McFALL D., WILKIE F. G., McCAFFERY F., LESTER N. G., and STERRITT R. Software processes and process improvement in Northern Ireland. In: Proceedings of the 16th International Conference on Software & Systems Engineering and their Applications, ICSSEA 2003, Paris, France, December 1–10, 2003, ISSN: 1637-503.

[MCG 02] McGarry J., Card D., Jones C., Layman B., Clark E., Dean J., and Hall F. *Practical Software Measurement: Objective Information for Decision Makers.* Addison-Wesley, Washinghton, DC, USA, 2002.

[MCQ 04] McQuaid P. and Dekkers C. Steer clear of hazards on the road to software measurement success. *Software Quality Professional Journal*, vol. 6, issue 2, 2004, pp. 27–33.

[MIN 92] Mintzberg H. *Structure in Fives: Designing Effective Organizations*, Pearson Education, Boston, MA, 1992, 312 p.

[MOL 13] Moll R. Being prepared – A bird's eye view of SMEs and risk management. *ISO Focus*: International Organization for Standardization, Geneva, Switzerland, February 2013.

[MOR 05] Moran T. What's Bugging the High-Tech Car? *New York Times*, February 6, 2005.

[NAS 04] NASA, Goddard Space Flight Center, Process Asset Library, ETVX Diagram Template, DSTL 580-TM-011-01, v1.0, Greenbelt, Maryland, 2004.

[NOL 15] Nolan A. J., Pickard A. C., Russell J. L., and Schindel W. D. When two is good company, but more is not a crowd. In: 25th Annual INCOSE International Symposium, Seattle, July 13–16, 2015.

[OBR 09] O'Brien, J. Preparing for an Internal Assessment Interview. *CrossTalk Journal of Defense Software Engineering*, vol. 22, issue 7, 2009, pp. 26–27.

[OLS 94] Olson T. G., Reizer N. R., Over J. W. A Software Process Framework for the SEI Capability Maturity Model, CMU/SEI-94-HB-01, 1994

[OLS 06] Olson T. G. Defining short and usable processes. *CrossTalk Journal of Defense Software Engineering*, vol. 19, issue 6, 2006, pp. 24–28.

[OMG 11] Object Management Group, Process Model and Notation (BPMN), version 2.0. Download available at: www.omg.org/spec/BPMN/2.0/PDF//

[OUA 07] Ouanouki R. and April A. IT process conformance measurement: A Sarbanes-Oxley requirement. In: Proceeding of the IWSM Mensura, Palma de Mallorca, Spain, November 4–8, 2007. Available at: http:// publicationslist.org.s3.amazonaws.com/data/a.april/ref-197/1111.pdf

[OZ 94] Oz E. When professional standards are lax: The confirm failure and its lessons. *Communications of the ACM*, vol. 37, issue 10, 1994, pp. 29–36.

[PAR 92] Park R. E. *Software Size Measurement: A Framework for Counting Source Statements (CMU/SEI-92-TR-20)*, Software Engineering Institute, Carnegie Mellon University, Pittburgh, PA, September1992.

[PAR 03] Parnas D., and Lawford M. The role of inspection in software quality assurance. *IEEE Transactions On Software Engineering*, vol. 29, issue 8, August 2003.

[PAU 95] Paulk M., Curtis, B., Chrissis. M.B., Weber, C. V. *The Capability Maturity Model: Guidelines for Improving the Software Process.* Addison Wesley, Reading, MA, 1995.

[PMI 13] *A Guide to the Project Management Body of Knowledge (PMBOK® Guide)*, 5th edition. Project Management Institute, Newtown Square, PA, 2013.

[POM 09] Pomeroy-Huff M., Mullaney Julia L., Cannon R., and Sebern M. *The Personal Software Process Body of Knowledge.* Software Engineering

Institute, Pittsburgh, PA, Carnegie Mellon University. Version 1–2, CMU/SEI-2009- SR-018, Pittsburgh, PA, 2009.

[PRE 14] PRESSMAN R. S. *Software Engineering – A Practitioner's Approach*, 8th edition. McGraw-Hill, 2014, 976 p.

[PSM 00] *Practical Software and Systems Measurement*. Department of Defense and US Army, version 4.0b, October 2000.

[RAD 85] RADICE R. A., ROTH N. K. O'HARA A. C., CIARFELLA W. A. A programming process architecture. *IBM Systems Journal*, vol. 24, issue 2, 1985, pp. 79–90.

[RAD 02] RADICE R. *High Quality Low Cost Software Inspections*. Paradoxicon, Andover, MA, 2002.

[REI 02] REIFER D. Let the numbers do the talking. *CrossTalk, The Journal of Defense Software Engineering*, March 2002.

[REI 04] REICHART G. System architecture in vehicles – The key for innovation, system integration and quality (original in German). In: Proceedings of the 8th Euroforum Jahrestagung, Munich, Germany, February 10–11, 2004.

[RTC 11] RTCA inc., DO-178C. *Software Considerations in Airborne Systems and Equipment Certification*. RTCA, Washington, DC, 2011. Available at: www.rtca.org

[SAR 02] Sarbanes-Oxley act of 2002, public law 107 — July 30, 2002, 107th Congress. Available at: https://www.sec.gov/about/laws/soa2002.pdf

[SCH 00] SCHULMEYER G. and MACKENZI, G. R. *Verification & Validation of Modern Software, Intensive Systems*. Prentice Hall, Upper Saddle River, NJ, 2000.

[SCH 11] SCHAMEL J. How the Pilot's Checklist Came About. Available at: http://www.atchistory.org/History/checklst.htm

[SEI 00] Software Engineering Institute. Overview of CMMI Model, Process Areas, Tutorial Module 7, Software Engineering Institute, Carnegie Mellon University, Pittsburgh, 2000.

[SEI 06] *Standard CMMI® Appraisal Method for Process Improvement* (SCAMPI) A, Version 1.2: Method Definition Document, CMU/SEI-2006-HB-002, Software Engineering Institute, Pittsburgh, PA, 2006.

[SEI 09] Software Engineering Institute. *The Personal Software Process Body of Knowledge*. Carnegie Mellon University, Pittsburgh, PA. Version 1.2, CMU/SEI-2009-SR-018, 2009.

[SEI 10a] Software Engineering Institute. *CMMI® for Development*, Version 1.3. CMMI-DEV, V1.3. Carnegie Mellon University, Pittsburgh, PA. Version 1.3, CMU/SEI-2010-TR-033, Pittsburgh, PA, November 2010.

[SEI 10b] Software Engineering Institute. CMMI for Services, Version 1.3. Carnegie Mellon University, Pittsburgh, PA, 2010. Version 1.3, CMU/SEI-2010-TR-034. Available at: http://www.sei.cmu.edu/reports/10tr034.pdf

[SEI 10c] Software Engineering Institute CMMI for Acquisition, Version 1.3. Carnegie Mellon University, Pittsburgh, PA. Version 1.3, CMU/SEI-2010-TR-032, Carnegie Mellon University, Pittsburgh, 2010. Available at: www.sei.cmu.edu/reports/10tr032.pdf

[SEL 07] SELBY P. and SELBY R. W. Measurement-driven systems engineering using six sigma techniques to improve software defect detection. In: Proceedings of 17th International Symposium, INCOSE, San Diego, United States, June 2007.

[SHE 01] SHEARD S. Evolution of the frameworks quagmire. *IEEE Computer*, vol. 34, issue 7, July 2001, pp. 96–98.

[SHE 97] SHEPEHRD K. Managing risk. In: Proceedings 28th Annual Seminars & Symposium, Project Management Institute (PMI), Chicago, United States, September 1997, pp. 19–27.

[SHI 06] SHINTANI K. Empowered engineers are key players in process improvements. In: Proceedings of the First International Research Workshop for Process Improvement in Small Settings *Software Engineering Institute, Carnegie Mellon University, CMU/SEI-2006-Special Report-001*, Pittsburgh, PA, January 2006, SEI, 2005, pp. 115–116.

[SIL 09] SILVER B. *BPMN Method and Style: A Levels-Based Methodology for BPM Process Modeling and Improvement Using BPMN 2.0*. Cody Cassidy Press, Altadena, CA, US, 2009.

[SPM 10] Software Project Manager Network (SPMN). American Systems, 2010. Available at: http://www.spmn.com

[STS 05] Software Technology Support Center. Configuration management fundamental. *CrossTalk, The Journal of Defense Software Engineering*, July 2005, pp. 10–15.

[SUR 17] SURYN W. ISO/IEC JTC1 SC7 Secretariat Report, Kuantan, Malaysia, May 2017.

[SWE 14] *Guide to the Software Engineering Body of Knowledge*, version 3.0. edited by P. Bourque and R. E. Fairley. IEEE Computer Society, 2014, 335 p.

[TIK 07] *The TickIT Guide*, version 5.5. British Standards Institute, London, UK, November 2007.

[TYL 10] TYLOR E. B. Primitive Culture, Researches into the development of mythology, philosophy, religion, language, art and custom, John Murray, Cambridge University Press, Cambridge, UK, 2010, 440 p.

[USC 16] Statistics About Business Size (including Small Business). US Census Bureau.

[VAN 92] VANSCOY R. L. *Software Development Risk: Opportunity, Not Problem*. SEI, CMU/SEI-92-TR-30, Pittsburgh, PA, September 1992.

[WAL 96] WALLACE D., IPPOLITO L. M., CUTHILL B. B. *Reference Information for the Software Verification and Validation Process*. National Institute of Standards and Technology (NIST), U.D. Department of Commerce, Special Publication 500-234, 1996. Available at: https://www.nist.gov/publications/reference-information-software-verification-and-validation-process

[WES 02] WESTFALL L. Software customer satisfaction. In: Proceedings of the Applications in Software Measurement (ASM) Conference, Anaheim, CA, USA, 2002.

[WES 03] WESTFALL L. Are we doing well, or are we doing poorly? In: Proceedings of the Applications in Software Measurement (ASM) Conference, San Jose, CA, Unites States, June 2–6, 2003. Available at: www.westfallteam.com/Papers/Are_We_Doing_Well.pdf

[WES 05] WESTFALL L. *12 Steps to Useful Software Metrics*. The Westfall Team, Whitepaper, USA, 2005. Available at: http://www.westfallteam.com/Papers/12_steps_paper.pdf

[WES 10] WESTFALL L. *The Certified Software Quality Engineer Handbook*. American Society for Quality, Quality Press, Milwaukee, WI, 2010.

[WIE 96] WIEGERS K. E. *Creating A Software Engineering Culture*. Dorset House, New York, 1996, 358p.

[WIE 98] WIEGERS K. E. Know your enemy: Introduction to risk management. *Software Development*, vol. 6, issue 10, 1998, pp. 38–42.

[WIE 02] WIEGERS K. *Peer Reviews in Software*. Pearson Education, Boston, MA, 2002.

[WIE 03] WIEGERS K. *Software Requirements*, 2nd edition. Microsoft Press, Redmond, WA, 2003, 516 p.

[WIE 13] WIEGERS K. and BEATTY J. *Software Requirements*, 3rd edition. Microsoft Press, Redmond, WA, 2013, 637 p.

Index

acquisition 492
acquisition strategy 494
agreement 492
attribute 414
audit
  according to CMMI 230
  according to IEEE 1028 218–224
  according to ISO 9001 117,
    225–229
  case study 241
  characteristics 185
  CobiT 143
  costs 42, 213–214
  definition 211
  guidance on 109
  internal 212
  process 26, 29
  project 30
  quality 74, 212
  requirements in SQA plans 239
  security 144
  steps 226
  types of 215

base measure 400, 413–418, 430–431
baseline 180, 191, 225, 231, 241, 298, 301,
  305
benchmarking 226, 399, 469, 574
Boehm, Barry 45, 448, 466, 468
BPMN
  artifact 373
  modeling levels 373
  notation 370
  process example 375
brainstorming 11, 87, 193, 195, 460, 468,
  470–471
branching 311
  according to Microsoft 313

commit 314
conflict 313
  with Git and GitHub 314, 317
  simple strategy 315
  strategy 312
  synchronize 314
  tag 313
  trunk 313
  typical strategy 316
business models 23, 108, 205, 214, 257,
  341, 401, 451
  for commercial software 31
  for custom systems written on contract
    27
  for in-house development 30
  for mass market software 31

certification 157
change management
  books recommended 386
  case study 244
  definition 321
  office 321
  policy 322
  process 138, 141, 242, 298
change request 152, 300–301, 319–322,
  330–331
Charrette, Robert 445
checklist 175–176, 282–286, 391
  developing a 283
  improvement 286
  use 285
CMM® for Systems Engineering 130
CMMI®
  history 104
  maturity levels 131
  for Services 130
  validation methods 267

Software Quality Assurance, First Edition. Claude Y. Laporte and Alain April.
© 2018 the IEEE Computer Society, Inc. Published 2018 by John Wiley & Sons, Inc.